On 13 January 1943, when Schrödinger was in Dublin, he wrote in English to Max Born on how he heard about his Nobel Prize:

> I have an amusing recollection of which I must tell you. On the 9th of November 1933 dear George Gordon, President of Magdalen, called me to his office to tell me that the *Times* had said I would be among that year's prize winners. And in his chevaleresque and witty manner he added: "I think you may believe it as the *Times* would not say such a thing unless they really know. As for me I was truly astonished as I thought you had [already] won the prize."

***

Erwin Schrödinger was one of the greatest scientists of all time but it is not widely known that he was a Fellow at Magdalen College, Oxford in the 1930s. This book is an authoritative account of Schrödinger's time in Oxford by Sir David Clary, an expert on quantum chemistry and a former President of Magdalen College, who describes Schrödinger's remarkable life and scientific contributions in a language that can be understood by all.

Through access to many unpublished manuscripts, the author reveals in unprecedented detail the events leading up to Schrödinger's sudden departure from Berlin in 1933, his arrival in Oxford and award of the Nobel Prize, his dramatic escape from the Nazis in Austria to return to Oxford, and his urgent flight from Belgium to Dublin at the start of the Second World War.

Fascinating anecdotes of how this flamboyant Austrian scientist interacted with the President and Fellows of a highly traditional Oxford College in the 1930s are a novel feature of the book. A gripping and intimate narrative of one of the most colourful scientists in history, *Schrödinger in Oxford* also explains how his revolutionary breakthrough in quantum mechanics has become such a central feature in 21st century science.

Schrödinger in Oxford

Schrödinger in Oxford

David C Clary

University of Oxford, UK

NEW JERSEY • LONDON • SINGAPORE • BEIJING • SHANGHAI • HONG KONG • TAIPEI • CHENNAI • TOKYO

Published by

World Scientific Publishing Co. Pte. Ltd.
5 Toh Tuck Link, Singapore 596224
USA office: 27 Warren Street, Suite 401-402, Hackensack, NJ 07601
UK office: 57 Shelton Street, Covent Garden, London WC2H 9HE

Library of Congress Cataloging-in-Publication Data
Names: Clary, David C., author.
Title: Schrödinger in Oxford / David C Clary.
Description: New Jersey : World Scientific, [2022] | Includes bibliographical references and index.
Identifiers: LCCN 2021046171 | ISBN 9789811250002 (hardcover) |
ISBN 9789811249969 (ebook for institutions) | ISBN 9789811249976 (ebook for individuals)
Subjects: LCSH: Schrödinger, Erwin, 1887-1961. | Magdalen College (University of Oxford)--
Biography. | Physicists--Austria--Biography. | Physicists--England--Biography. |
Oxford (England)--Biography. | Europe--Intellectual life--20th century.
Classification: LCC QC16.S265 C53 2022 | DDC 530.092 [B]--dc23/eng/20211214
LC record available at https://lccn.loc.gov/2021046171

British Library Cataloguing-in-Publication Data
A catalogue record for this book is available from the British Library.

For any available supplementary material, please visit
https://www.worldscientific.com/worldscibooks/10.1142/12661#t=suppl

Desk Editor: Shaun Tan Yi Jie

Typeset by Stallion Press
Email: enquiries@stallionpress.com

Printed in Singapore

To Heather

About the Author

Sir David Clary FRS is Professor of Chemistry at the University of Oxford, UK where he was also President of Magdalen College from 2005–2020, Schrödinger's college in Oxford. He is an elected Fellow of many academies including the Royal Society, the Royal Society of Chemistry, the Institute of Physics, the Royal Society of Arts, the American Physical Society, the American Association for the Advancement of Science, the American Academy of Arts and Sciences, and the International Academy of Quantum Molecular Science. He was President of the Faraday Division of the Royal Society of Chemistry from 2006–2008. From 2009–2013, he was the first Chief Scientific Adviser to the UK Foreign and Commonwealth Office. In 2016, he was knighted in the Queen's Birthday Honours for services to international science.

Sir David is a theoretical chemist recognised for his pioneering work on the quantum dynamics of chemical reactions. He has published over 350 papers on developing new theories and computational methods for solving Schrödinger's equation for molecular problems. He was Editor of *Chemical Physics Letters* from 2000–2020 and a Reviewing Editor of *Science* from 2003–2016. He has won many prizes for his research including the Royal Society of Chemistry Meldola, Marlow, Corday-Morgan, Tilden, Polanyi, Chemical Dynamics, Liversidge and Spiers awards, and the Medal of the International Academy of Quantum Molecular Science.

Preface and Personal Acknowledgements

Erwin Schrödinger was one of the greatest scientists of the 20th century. In a paper published in 1926 he introduced the Schrödinger equation and his theory of wave mechanics. This is the central scientific theory for describing the properties of atoms, molecules and condensed phases.

Schrödinger was born in Austria in 1887 and was educated in Vienna. He had many academic positions and was at the University of Zurich in 1926 when he published his great breakthrough with his wave mechanics. He then went to Berlin to take up the Chair previously held by Max Planck. With the rise of the Nazis he left Germany and took up a Fellowship at Magdalen College, Oxford in 1933. He returned to Austria in 1936 but soon had to leave after the Anschluss in 1938. After short periods back in Oxford and then Belgium, he eventually was appointed to a position at the new Institute for Advanced Studies in Dublin. He stayed there until 1956 when he moved back to Vienna. He died in 1961.

There have been several books on Schrödinger's life and work. The most detailed and comprehensive biography is by Walter Moore entitled *Schrödinger: Life and Thought*. There are other books that emphasise a particular topic connected to Schrödinger such as *Erwin Schrödinger and the Quantum Revolution* by John Gribbin and *Einstein's Dice and Schrödinger's Cat* by Paul Halpern. However, none of the books on Schrödinger describe his time in Oxford between 1933 and 1936, and his brief return in 1938, in much detail. The current book *Schrödinger in Oxford* gives emphasis to these periods and his links to Oxford and the

UK. The previous and subsequent aspects of Schrödinger's life are also described to put his career in context. The book does not have mathematical equations and is written with the aim of being understandable to readers who do not have a scientific background but are interested in the lives of great scientists.

On the day that Schrödinger was admitted as a Fellow of Magdalen College in Oxford on 9 November 1933 he heard from the President of the College that it had just been announced that he had won the Nobel Prize for Physics. This book, therefore, has a special emphasis on Nobel Laureates whose works demonstrate significant progress in the physical sciences. Furthermore, like many scientists who left Germany or Austria in the 1930s, Schrödinger was a refugee and his life was directly influenced by the chaotic and dangerous political scene of the 1930s and 40s. Accordingly, the careers and challenges of the scientific refugees during this period are also described in some detail in this book.

The book includes the texts of several letters between Schrödinger and his close friends and colleagues including Albert Einstein, Max Born, Max Planck and Paul Dirac. These letters illustrate the thoughts, hopes, humour and concerns of these remarkable scientists. Many of these letters were written in German and the author bears full responsibility for their translation into English. The book also brings together numerous other writings and comments on Schrödinger. In particular, the acute observations of his wife Anny and his daughter Ruth feature prominently.

I would like to acknowledge the late Ruth Braunizer, Schrödinger's daughter, for her assistance in obtaining material for this book from her archive in Alpbach and for several very helpful discussions in Alpbach and in Oxford, the place of her birth. In addition, the Braunizer family have very kindly granted permissions for publishing the text of Schrödinger's letters and photographs. The late Gustav Born, the son of Max Born, provided helpful information as well. I would like to thank Nancy Greenspan, the biographer of Max Born, and Dr. Dieter Hoffmann at the Max Planck Institute for the History of Science in Berlin for providing a copy of a letter from Schrödinger to Born describing how George Gordon, the President of Magdalen College, informed Schrödinger he had won the Nobel Prize. Furthermore, the Nobel Laureate Walter Kohn, who was also a refugee from Austria, is acknowledged for an informative discussion at Magdalen College on Schrödinger.

It was very helpful to have an email interchange with Ernst Peter Fischer about the origin of the story of "the dark lady" of Arosa where Schrödinger first discovered his equation in 1925. My thanks are also due to Kathrin Peters for translating a handwritten letter from Max Planck to Schrödinger congratulating him on his Nobel Prize. I have been very helpfully assisted by several copyright holders, other individuals and archivists in obtaining, or being granted permissions to publish, texts of letters or photographs, and a summary is made at the end of the book.

New material for this book was obtained from the archives at Magdalen College, Oxford (Archivist Robin Darwell-Smith and Graduate Assistant Tilly Guthrie); the Lindemann Collection at Nuffield College, Oxford; Lincoln College, Oxford; the Churchill Archives Centre, Churchill College, Cambridge (Jessica Collins and Chris Knowles); the Cambridge University Library; the Royal Society; and the Archive of the Society for the Protection of Science and Learning in the Oxford University Library Collection. Material was also obtained from the online archives at the Österreichische Zentralbibliothek für Physik Wien (Christof Capellaro), the Institute for Advanced Study, Princeton (Erica Mosner), the Niels Bohr Library and Archives (Sarah Weirich), the Paul A.M. Dirac Papers at Florida State University (Keila Zayas-Ruiz), the Nobel Prize Archives, the Wellcome Collection, and the Irish Newspaper Archives.

This book was written at home in the lockdown during the pandemic of 2020–21. I give my heartfelt thanks to my wife Heather for her patience and support during this period.

Contents

CHAPTER ONE

Schrödinger's Breakthrough

Equation

On the 9th of November 1933, the great Austrian theoretical physicist Erwin Schrödinger came to Magdalen College in Oxford. He was admitted as a Fellow of the College by the President George Gordon using Latin phrases requesting Schrödinger to obey the Statutes and Bylaws of the College. Schrödinger replied "Do Fidem," which is translated as "I swear." He then shook the right hands of all the Fellows present and each one said to him, "I wish you joy." This ceremony for the admission of Fellows is unchanged today some 88 years later. The ancient bells then rang from the Magdalen Great Tower and Schrödinger and his new colleagues processed to the High Table in the Hall to celebrate his admission over a fine dinner.[1]

At the end of the dinner President Gordon was called to his office in the President's Lodgings to receive a telephone call. It was from the *Times* newspaper saying that the Royal Swedish Academy of Sciences had just announced that Schrödinger had won the Nobel Prize in Physics for 1933 jointly with Paul Dirac from the University of Cambridge. President Gordon then called Schrödinger to his office to inform him of the happy news. In this way, Schrödinger won the Nobel Prize as a Fellow of Magdalen for just a few hours.

Schrödinger was awarded the Nobel Prize for the equation under his own name which he published early in 1926 when he was working at the University of Zurich.[2] In this paper, he wrote down a new equation for the description of the electron in the hydrogen atom and it had the deceptively simple form $H\Psi = E\Psi$. Here H contains mathematical terms representing both the kinetic energy and the potential energy of the electron. At particular quantised energies, contributions from the kinetic and potential energies cancel to leave just the constant energy E of the electron.

Schrödinger's equation gave a very simple form for the possible energies of the electron of the hydrogen atom in terms of integers describing the different quantum states and fundamental constants associated with the mass and charge of the electron together with Planck's constant. This formula for the energy states had been given first by the Danish physicist Niels Bohr who found it fitted perfectly results derived by Balmer and Rydberg from the lines they observed experimentally with different colours in the emission spectrum of hydrogen gas.[3]

It was not this energy formula, however, that seemed so revolutionary to the great scientists of the day but the wave function Ψ, also denoted by Schrödinger as an "orbital". This new mathematical invention gave very accurately all the other observed properties of the electron such as the spectrum for the electron in the presence of an electric field — an observation known as the Stark effect.

At that time there were several researchers, including Werner Heisenberg and his research supervisor Max Born in Göttingen, and Arnold Sommerfeld in Munich, who were all attempting to derive formulae for energies of electrons from their own theories. None, however, realised that a theory could be developed to calculate very accurately all the other properties of the electron which can be observed. Indeed, Max Born stated: "We were taken by surprise by Schrödinger's papers on wave mechanics. This was a new approach to quantum mechanics. It was of fascinating power and elegance. It quickly became the standard theory."[4] Sommerfeld, who within three months of Schrödinger's paper was giving lectures on the breakthrough in London, Oxford, Edinburgh and Manchester, said: "Wave mechanics was the most amazing among all the amazing discoveries of the 20th century."[1]

The term "wave mechanics" was given by Schrödinger to his theory because it relates to the work of the French physicist Louis de Broglie who had proposed that the electron has some properties similar to those of a wave.[5] Accordingly, shortly before his great discovery, it was suggested to Schrödinger by his Zurich colleague Peter Debye that there ought to be a mathematical equation for the electron describing it as a wave.[6] That is exactly what Schrödinger discovered.

In 1926, Schrödinger quickly followed up his great paper on the energy levels of the hydrogen atom with four other papers which were published in the *Annalen der Physik*. He demonstrated that his equation described accurately the harmonic oscillator and rotational motion which can be identified with the vibrating and rotating states of a diatomic molecule.[7] He also showed that his new quantum theory could be made equivalent to an alternative approach developed a few months before by Heisenberg, Born and Jordan.[8] He then developed his theory to show how it applied to the Stark effect.[9] Finally, he derived a second Schrödinger equation in which his wave function Ψ depended on time.[10]

Niels Bohr had not been successful at extending his own theory to atoms with more than one electron and to molecules.[11] Schrödinger himself did make unpublished attempts to do this with his own theory but without success. However, he was not aware of the new work being done by Uhlenbeck and Goudsmit which suggested that an individual electron could have different quantum states that were visualised as the particle spinning clockwise or anticlockwise.[12] When Schrödinger's wave functions were modified to include these spins the spectrum for the two-electron helium atom was explained,[13] as was the chemical bonding of the hydrogen molecule by Schrödinger's colleagues Heitler and London.[14] Indeed this progress, and subsequent work by many others, allowed the Nobel Prize-winning American theoretical chemist Linus Pauling to state: "His great discovery was based on an idea, the idea that the properties of atoms and molecules could be calculated by solution of a differential equation... The discovery of the Schrödinger wave equation was a great event for physics."[15]

Not long after the publication of the first paper by Schrödinger, the Cambridge theoretical physicist Paul Dirac published his own form of quantum mechanics which incorporated the treatment of Albert Einstein's theory of relativity and, most remarkably, predicted a new kind of particle which was called a positron.[16] This particle had the same properties as an electron except for an opposite charge and it was observed soon after its prediction. In 1929 Dirac stated: "The underlying physical laws necessary for the mathematical theory of a large part of physics and the whole of chemistry are thus completely known, and the difficulty lies only in the fact that the exact application of these laws leads to equations much too complicated to be soluble."[17]

It was easy to write down Schrödinger's equation for any number of electrons and protons but it was much harder to solve it, even in an approximate form, for an atom or molecule with more than one electron. However, the idea of the orbital for one electron at once found many uses in the qualitative description of chemical bonding and properties of molecules. Ultimately, it was only when electronic computers were developed and made generally available many years later that significant progress could be made in accurate calculations of Schrödinger's equation for larger atoms and molecules.

After 1926 Schrödinger did not hesitate to communicate his breakthrough in lectures to scientists around the world. His own personal biographical sketch of that time stated rather succinctly:

> Schrödinger is mainly a theoretical physicist. He had been interested in various domains of his science and published pretty good work on the statistical theory of matter, on colour-vision and on the theory of the atom although nothing outstanding until 1926. In 1926 he issued a rapid sequence of papers on wave-mechanics, which represented an important progress and made his name popular. The so-called Schrödinger wave equation has become the instrument for treating all problems in atomic and molecular physics.[18]

Boyhood

Erwin Rudolf Josef Alexander Schrödinger was born on August 12, 1887 in Vienna. His father Rudolf would have preferred a career in science but took over the family business as the owner of a small factory. His mother was Georgine (née Bauer) and her mother Emily (née Russell) was born in England. Georgine's sister Minnie helped bring Erwin up and he spoke English with her from an early age. Georgine's other sister Rhoda was also a doting aunt. Georgine took him to visit members of their family back in Leamington Spa in England. Minnie was married to Max Bamberger who was a Professor of General Chemistry at the Vienna Polytechnic (which became eventually the Technical University of Vienna). Thus, even as a child Schrödinger was exposed to science and to the English way of living.

Life was comfortable for the young Erwin in the late 1800s. He was an only child and his family had a large apartment in central Vienna with a view of St. Stephen's Cathedral. The Austro-Hungarian Empire was at its peak and Vienna was the centrepiece. Science, music and the arts were all flourishing with great names such as Boltzmann, Freud, Mahler, and Klimt producing their best work. This highly creative atmosphere was one Schrödinger never forgot and, when in positions abroad, he always longed to return to work in his home country, which indeed he did twice in his subsequent career.

Erwin did well in his studies and started at the Akademisches Gymnasium in Vienna in 1898. This was the oldest secondary school in Vienna with an emphasis on Latin, Greek and German Literature. Mathematics was studied as a minor component. There were many bible classes which gave Erwin a negative scepticism for organised religion throughout his life. The facilities were very traditional with poor lighting which gave eyesight problems for several children. Erwin had to wear strong glasses from the age of 12 and they became something of a trademark for him.

Erwin's father Rudolf continued his scientific interests and introduced his son to Darwin's theory of evolution which was then still quite novel. This was a biological interest that came out in an unexpected way for Schrödinger some 40 years later. He graduated from the Gymnasium top of his class and entered the University of Vienna in 1906. Through the work of physicists such as Doppler, Stefan and, most notably, Boltzmann, the University had gained an outstanding reputation for teaching and research in physics in the second half of the 19th century.

Four-year-old Erwin Schrödinger with parents Rudolf and Georgine, and Aunt Minnie (on the left).

Working first in Graz and later in Munich, Leipzig and Vienna, Ludwig Boltzmann had developed and applied statistical methods to atoms and molecules that enabled macroscopic quantities such as specific heats and entropy to be calculated from first principles. This was one of the major scientific breakthroughs of the 19th century and Boltzmann remains one of the great names of science to the present day. Sadly, he took his own life in 1906. He was the scientific hero of Schrödinger who always regretted not being able to work with Boltzmann. In due course, Schrödinger's own revolutionary work would give rigour to Boltzmann's ideas.

As is often the case with the most brilliant students, it is the youngest and most enthusiastic lecturers that catch their eye and get to know their students best. This was the case with Friedrich Hasenöhrl. He had studied

Erwin Schrödinger aged 17.

with Boltzmann and the great Dutch Nobel Prize winner Hendrik Lorentz and had done early work linking energy and mass that predated Einstein's theory of relativity. Schrödinger at once was inspired by the rigour and depth of Hasenöhrl's lectures. He later claimed that nobody outside his family had a stronger influence on him than Hasenöhrl.[19]

Hasenöhrl was invited to the 1st Solvay Conference in Brussels in 1911 which was also attended by Albert Einstein, Max Planck, Marie Curie and the British physicist Frederick Lindemann who recruited Schrödinger to the University of Oxford 22 years later. Tragically, Hasenöhrl was killed in the First World War in 1915. Also, through the mathematics course of Professor Wilhelm Wirtinger, Schrödinger mastered in depth the methods of differential equations which would be crucial for his later breakthrough in wave mechanics.

In the University of Vienna, Schrödinger met other promising students of physics. Hans Thirring was one year younger and he at once observed that Schrödinger was something special.[19] Thirring had a successful career in theoretical physics and subsequently corresponded with Schrödinger on a very regular basis. He played a key role in bringing Schrödinger back to Austria from Oxford some 30 years later in 1936 and again from Dublin in 1956.

Schrödinger also attended the lectures in practical physics given by the Director of the Physical Chemistry Institute Franz Exner. In Exner's department important work had been done on synthesizing radium chloride which was sent to Rutherford in the UK who used the material for great discoveries on the structure of the atom. Exner had also been involved in providing uranium compounds used by Marie and Pierre Curie for their groundbreaking work on radioactivity. He was, therefore, very well connected with the great experimental scientists of the time.

Exner trained many eminent physicists including Victor Hess who won the Nobel Prize for the discovery of cosmic radiation and later became a colleague of Schrödinger's after he had moved to Graz in Austria from Oxford. Schrödinger performed an experimental project under the supervision of Exner on the conduction of electricity on the surfaces of insulators in moist air. He spoke on this work at a meeting of the Academy of Sciences in Vienna in 1910 and his thesis enabled him to be granted the degree of Dr.phil. which is equivalent nowadays to a Master's Degree.

Military

Austria-Hungary at that time had a scheme in which young men who had gone through a significant education could volunteer for military training of one year. This enabled them to avoid the three years of compulsory military service. Therefore, in 1910 Schrödinger volunteered for service in the fortress artillery. Over Christmas he went on a ski trip with his friend Hans Thirring who broke his leg and had to leave the military service. This enabled Thirring to take up a vacant assistant post with Hasenöhrl that Schrödinger was hoping for.

Thus, when Schrödinger passed his military training a year later as a cadet officer and returned to the University, he could not work with Hasenöhrl and continued experimental research with Professor Exner and his assistant Fritz Kohlrausch. He, however, rapidly concluded that he was not ideally suited for experimental work and by this time the great experiments in physics were being done in other countries such as France, the UK and Germany. So he turned to doing individual theoretical work on problems of interest to experiment. This became a theme for much of his subsequent research.

The next qualification in his academic career was the Habilitation which would enable him to give his own lecture courses for which he was paid. To achieve the Habilitation he had to publish his own papers and give a public lecture on his research. He developed a theory for diamagnetism in which the electric currents associated with the motion of electrons produce an internal magnetic field. In this way Schrödinger was starting to think about the intricate properties of electrons. His theory, based on the velocity distribution of the Scottish scientist Maxwell, did not give good agreement with experiment. It was only his own form of wave mechanics that enabled this agreement to be improved several years later.

Another new theoretical study he initiated was on the problem of how a solid melts to form a liquid. He employed a recent advance by Peter Debye in treating molecules as dipoles with equal but opposite charges at either end. However, he ignored the important longer-range van der Waals forces between molecules. Yet again, it was only after his own form of quantum mechanics was developed that it became possible to calculate these forces accurately. However, even in the present day, it remains a challenge to explain from first principles all the quantitative details of a solid melting to form a liquid.

Professor Exner was keen for Schrödinger to do research directly relating to experiments being done in his department. Victor Hess had started an adventurous programme of balloon ascents observing radiation in the atmosphere. Accordingly, Schrödinger published a theory related to the observations of Hess and he mentioned that his theory could not include radiation from outer space.[20] The observations made by Hess led to the discovery of cosmic rays for which he was awarded the Nobel Prize in 1936. Little did Hess and Schrödinger realise that they would both be caught up with the Nazis in Graz in 1938. In the flourishing Vienna of 1912, the hardworking scientists were oblivious to the dark political scene that would emerge.

In the next year there was a great scientific congress in Vienna to which many of the leading scientists of the day were invited to lecture. As he had achieved his Habilitation, Schrödinger was allowed to attend but did not give a presentation. This was the first time he had seen the great Albert Einstein who spoke on his latest ideas on gravitation. This sparked a new interest for Schrödinger who just a few years later started an intense correspondence with Einstein that lasted throughout their careers. Furthermore, it was the subsequent realization by Einstein of the importance of the idea of de Broglie that an electron could have wave-like properties that convinced Schrödinger to develop his own wave equation.

The assassination of the Archduke Franz Ferdinand on 28 June 1914 in Sarajevo set in motion the fast sequence of events that led to the First World War. With Vienna as the epicentre of the Austro-Hungarian Empire, mobilization occurred at once. The great days of the city as a beacon for creativity, discovery and intellectual advances were over. Being aged 26 and a reserve in the army Schrödinger was called up at once as an artillery officer.

He was sent to the border with Italy which at that time was neutral in the war. Schrödinger was highly fortunate as many of his former student colleagues were sent to the great battles of the Russian front where the Austro-Hungarian army suffered badly. With his individual, creative and theoretical style of pursuing research he was able during this period to continue to think about scientific problems without the need for laboratories or libraries.

He even managed to complete some scientific papers. One was on the capillary pressure of gas bubbles and another was on the Brownian motion of particles, the theory of which had been pioneered by Einstein. Schrödinger developed a theory for Brownian motion that started a series of papers on applying statistical ideas to various problems in the spirit of his hero Boltzmann.[21] Given this interest, it does seem surprising that in later years he was stubborn in not accepting a statistical interpretation of his own wave function in contrast to so many other scientists.

Italy entered the war in May 1915 on the opposing side to Austria and Schrödinger was moved to Görz near Trieste. Here there was fierce fighting with the losses on both sides in the hundreds of thousands.

However, as he was commander of an artillery battery, he did not see very dangerous service himself. His efforts must have been useful as he received a military citation and was promoted to Oberleutnant, the highest lieutenant officer rank. It was around the time that his former teacher Professor Hasenöhrl sadly lost his life in an infantry charge in the Tyrol.

Schrödinger in uniform in 1915.

On 21 November 1916 Emperor Franz Josef died after 67 years as head of the Austro-Hungarian Empire and the atmosphere in Austria at once changed. Charles, his successor as Emperor, attempted to make peace with Italy but without success. Schrödinger was then moved back to Vienna to teach a course for anti-aircraft officers. His experience as an artillery officer inspired him to write a paper on the audibility of large explosions. He developed his own classical wave equation for this problem which would sow the seeds for his great work on another wave equation.[22] During this period he first met the future polymer scientist Herman Mark who he was to interact with in Vienna some 20 years later.

Around this time Schrödinger also published his first paper on quantum theory. The pioneer in quantum theory had been Max Planck. When working in Berlin and building on the ideas of Boltzmann, Planck proposed that the energy of electromagnetic radiation is quantized according to the formula $E = h\nu$ where h is Planck's constant and ν is the frequency of radiation. Schrödinger wrote a detailed review of the theory of specific heats to which Planck, Einstein, Debye and Born had all contributed.[23]

In due course, Schrödinger's own wave equation was to allow for the calculations of the quantized energies for the different degrees of freedom needed for the accurate predictions of specific heats for molecules and materials but he could not imagine this at the time. Many of the experimental results for specific heats had come from Walther Nernst in Berlin and his assistant Frederick Lindemann. Thus Schrödinger would have been familiar with the name and scientific contributions of Lindemann when he was recruited by the British professor to come to Oxford in 1933.

Schrödinger's review on specific heats was published in the new journal *Naturwissenschaften* which covered all the sciences.[23] It was edited by Arnold Berliner and was a German equivalent to the British journal *Nature*. In due course Schrödinger became a close friend of Berliner, submitted several more papers to his journal and they corresponded frequently.

Schrödinger studied Einstein's general theory of relativity when he was at the Italian front in 1916. This allowed him to write two short papers on the topic on his return to Vienna in 1917.[24,25] Einstein had produced a new cosmological constant in his equations and Schrödinger suggested they could be solved without the need for this constant. Throughout his subsequent scientific career, Schrödinger returned many times to relativity and cosmology, and his correspondence with Einstein on the topic was voluminous. However, he was not able to make the fundamental contribution to this subject in the way he achieved so powerfully with his breakthrough in wave mechanics in 1926.

Universities

Schrödinger was building a reputation for himself, and he was approached in 1918 by the Franz Josephs University of Czernowitz about an appointment there in succession to Josef Geitler von Armingen, an expert on electromagnetic waves. Schrödinger anticipated working on theoretical physics and also developing his philosophical interests. Czernowitz was a small, cultured city which had been ruled by several different nations and had been occupied by Austria since 1775. However, in the chaos at the end of the war the city was taken over first by Ukraine and then Romania. This made it impossible for Schrödinger to take up an appointment there. He would look back wistfully at this possibility throughout his life.

In the last year of the First World War the Austro-Hungarian Empire fell apart and life became very difficult in Vienna. Schrödinger had continued to live in the apartment of his parents and food became scarce. There was runaway inflation. His father's business had collapsed and when the war finished in November 1918 Schrödinger's military pay was stopped. His income from the University was hardly sufficient to support himself, let alone his family. This was the first time in his life that Schrödinger had experienced real hardship. Then in 1919 his father Rudolf died.

It has to be said that up to this point Schrödinger had received an excellent education which had allowed him to move eventually to a top university in which he could undertake his own research. He had developed his own particular style of finding a problem of interest to experimentalists, proposing an appropriate equation to describe the system and using rigorous mathematics to find analytical solutions. He did this work completely on his own and not in a research group or even guided by a more senior scientist. He had been very lucky that he had not seen very dangerous service in the First World War and had managed to continue to write research papers during that period.

Elsewhere, many of the most celebrated theoretical physicists of the time were emerging from large research groups headed by distinguished professors. For example, Arnold Sommerfeld in Munich supervised over 40 students or postdoctoral assistants who are well-known names in physics to the present day. This included seven Nobel Prize winners and such great names as Heisenberg and Pauli. In a large research group such as this the supervisor would suggest a problem to a student who could discuss it with colleagues every day.

The same was true with the research group supervised by Max Born in Göttingen. However, Schrödinger did his research on his own and this enabled him to tackle unique problems in his personal style. He had a good supply of research projects from his close link to the experimental department of Exner. There is a parallel to Paul Dirac at Cambridge University who developed his quantum theory entirely on his own but was still aware of the great experimental discoveries being made on the structure of the atom at the Cavendish Laboratory in Cambridge.

However, unlike Dirac, Schrödinger was not focused just on science and nothing else. He was an avid theatre goer and read very widely in philosophy. Schopenhauer was a particular favourite. His view that nature was a continuous chaotic struggle reflected Schrödinger's subsequent personal life rather well. He also read widely on the Hindu philosophy of Vedanta and on Spinoza. In his writings, and especially after he became famous with his wave mechanics, Schrödinger often referred to philosophies such as these.

Inspired by experimental work going on in Exner's department Schrödinger moved into the new area of research on colour theory. He published papers on topics such as the theory of pigments with high luminosity and colour measurement.[26] This research also linked to his philosophical interests. It is still cited by researchers working on colour theory in the present day.

As is true in universities throughout the world, the appointment to professorships can be complicated and can have political connotations. This was true at the University of Vienna where replacements for Hasenöhrl to the Chair of Theoretical Physics and Exner to the Chair of Experimental Physics were to be made in 1920. The appointments of

Schrödinger in 1920.

Gustav Jäger and Felix Ehrenhaft, respectively, were disappointing to Schrödinger as he did not consider them to be physicists in the top rank. Around this time he was offered an assistantship with Max Wien at Jena in Germany. The financial situation in Germany was better than Austria and the salary offered was quite attractive. So, even though the position in Jena was not a permanent one, Schrödinger took the major step in April 1920 of moving away from Vienna for the first time. He was always very open to offers and, from now on, he would be on the move many times in his career.

Just before his move to Jena, Schrödinger married Annemarie Bertel who was called Anny. She was aged 23 and Erwin 32. Anny was the daughter of a prosperous photographer in Salzburg and Schrödinger had known her since she was a teenager when she sometimes looked after the children of his colleague Fritz Kohlrausch. To cut a very long story short, Schrödinger had many affairs after his marriage but Anny remained with him right to the end and accompanied him to the many positions he took up in places all over Europe.

Despite the severe financial and political problems in the early 1920s, theoretical physicists had become much in demand in German universities following the great advances in quantum theory and relativity, and many positions were opening up. Almost as soon as the newly-wed couple arrived in Jena in April 1920, Schrödinger was offered a permanent associate professorship at the Hochschule für Technik in Stuttgart and he moved there in October. Erich Regener was the Professor of Experimental Physics in Stuttgart and had done accurate measurements on cosmic rays. He was also well acquainted with the great scientists of the day such as Einstein and Marie Curie. The post at Stuttgart, therefore, was an attractive one for Schrödinger.

At this time Erwin's mother had become very ill in Vienna with cancer. The value of his late father's investments and pension had been much reduced by inflation and she had been turned out of their comfortable apartment. This had a major influence on Schrödinger who later referred to these difficult circumstances in bitter terms. It did much to make him neurotic about financial security and he was always tempted by an apparently lucrative offer to a new position. So just six months later, Schrödinger was on the move again to a more senior professorship in Breslau, his third new appointment in 18 months. At this time he was somewhat annoyed that he had been passed over for two professorial appointments back in Austria at Graz for what seemed to be political reasons. Breslau was close to the border with Poland where there was political unrest but the work of other distinguished physicists there, such as the spectroscopist Rudolf Ladenburg, impressed him.

Schrödinger had become interested in the extensions made by Sommerfeld to the Bohr theory of the atom. In his first paper on the theory of atomic structure Schrödinger suggested that modified elliptical orbits of the electrons could explain the observed spectrum of the sodium atom.[27] His thinking was moving towards the detailed understanding of electron orbits.

Zurich

However, he was then approached by yet another university and this time it was a major appointment. Einstein, Max von Laue and Peter Debye had all held professorships at the University of Zurich and the officials there were looking to make a new appointment to continue this great tradition in theoretical physics. Schrödinger received a very strong reference from his recent colleague Regener who had been impressed by the clarity of Schrödinger's lectures even though he had moved on from Stuttgart all too quickly. In addition, the faculty at Zurich had noted the breadth of his papers covering many different areas of physics.

Switzerland had avoided the horrors of the First World War, there was no hyperinflation and the salary and pension offered were more attractive than those Schrödinger had been receiving at the German universities. At the relatively young age of 34 for a full professor, Zurich was an ideal appointment. He was well aware he had not yet made a research contribution at the very top level. This was Schrödinger's opportunity to settle down and produce his great work.

As soon as he started at the University of Zurich in October 1921 Schrödinger became ill with a serious bout of bronchitis. Five appointments in five different cities and three different countries in two and a half years were taking their toll. In addition, both his father and mother had died during this period. Tuberculosis was suspected and he needed to move to a high altitude to recuperate. Accordingly, the Schrödingers went to Arosa, a town in the Swiss Alps close to the now-famous ski resort of Davos.

In Arosa, Schrödinger wrote two papers that continued his progress with quantum theory.[28,29] The first was on the quantization of vibrations to calculate the specific heats of solids at high temperatures and the second was on the quantized Bohr orbits of an electron — a subject now very close to the wave mechanics Schrödinger would invent. In the second paper, by building on a theory developed by Hermann Weyl at the ETH Zurich, Schrödinger mentioned the possible need for the use of complex numbers in the description of the Bohr orbits. This phase factor was realized to be important many years later when quantum field theories for fundamental forces were developed. This also closely links with Schrödinger's subsequent realisation that his wave functions could be of a complex form.

After nine months Schrödinger was feeling well enough to return to Zurich. After this he often visited Arosa and it was the place where he discovered his equation four years later. Schrödinger was inspired by the mountains where he could concentrate without interruption of his work. His postings in the First World War in the Tyrol had also, most fortunately, allowed him this opportunity. He always worked very quietly on his own. As Anny stated:

> He never worked really with somebody. He was always alone. He liked to talk with people, but he never worked with them. Now in these days it's all team work, isn't it? He never would have got used to that.[30]

The University of Zurich was close to its illustrious counterpart, the ETH, also in Zurich. Not only was Peter Debye a professor there but also Hermann Weyl. Weyl was a mathematician who had studied under David Hilbert in Göttingen. He had derived fundamental theorems for the distribution of eigenvalues and was one of the first to realise the link between electromagnetism and gravity. Michael Atiyah, a Fields medalist, Master of Trinity College Cambridge and President of the Royal Society, stated in 2002 that Weyl was "one of the greatest mathematicians of the first half of the twentieth century, no other mathematician could claim to have initiated more of the theories that are now being explored."[31]

Joint seminars between the University and the ETH on theoretical physics were organized alternating between the institutions, and Schrödinger was an enthusiastic participant. This was essentially the first time he could attend regular seminars with other like-minded theoretical physicists and it was the perfect opportunity to discuss the fast-moving developments in the theory of atomic physics.

Zurich was not having the political problems that were arising elsewhere in central Europe and the atmosphere experienced by the Schrödingers was a free and easy one. Erwin and Anny, however, had no children and were starting to drift apart. Anny became very friendly with Hermann Weyl and Schrödinger was starting to look elsewhere. He would continue in this mode for the rest of his life and it caused complications.

As a full professor at the University of Zurich Schrödinger had to give an inaugural lecture. He chose the title "What is a Natural Law?", which resembles the grand title of his famous book *What is Life?* that he published some 20 years later. In his inaugural lecture he emphasised the statistical nature of matter, with the influence of Boltzmann and Exner being prominent. It remains strange that he strongly resisted a statistical interpretation of his own wave function in his later years when he had been a strong supporter of statistical ideas as a younger professor. Schrödinger delayed some six years in publishing his inaugural lecture and only had the confidence to do this after he had proved himself as a great scientist with the invention of wave mechanics.

His rising reputation allowed Schrödinger to be invited to his first Solvay Conference on Physics in 1924. These conferences, supported by the King of Belgium, had become the leading international gathering for the discussion of the very latest developments in physics. The subject was "The electrical conductivity of metals and related topics" and Schrödinger did not present a paper. The photograph for this 4th Solvay Conference shows Marie Curie, Ernest Rutherford, Paul Langevin and William H. Bragg in the front row with Schrödinger somewhat nervously appearing in the third row. It would not be very long before Schrödinger would be moved to the first row and indeed his wave mechanics eventually provided a rigorous theory for explaining electrical conductivity.

It should be noted that scientists based in Germany and Austria were not invited to this Solvay meeting due to the tensions still prevailing after the First World War. So there was no Planck, Einstein, Sommerfeld or Born. However, as Schrödinger was based in Switzerland he did receive an invitation and there was space for him due to the other absentees.

By 1924, ideas on identical particles had emerged in physics. A significant contribution came from Satyendra Bose who was a young Indian physicist working at the University of Dhaka in what is now Bangladesh. He wrote a paper which derived Planck's radiation law by using a new way of counting identical particles. He sent the paper to Einstein who was at once struck by the originality of the work. Einstein translated the work into German and sent it to be published in the *Zeitschrift für Physik*.[32] Bose had already interacted with Einstein by translating his work on General Relativity into English.

This work was then followed up by Einstein himself who predicted a new form of matter, now known as the Bose-Einstein condensate, in which all the identical particles are in the same quantum state at a temperature very close to absolute zero. In his work, Einstein also drew attention to the idea of de Broglie that an electron could have properties of both a particle and a wave.[33] This put seeds into Schrödinger's mind that would eventually lead to his wave equation just two years later.

Schrödinger also contributed to these developments by using the new Bose-Einstein statistics for the calculation of entropy. He sent his paper to Einstein who replied in favourable terms. Schrödinger then extended their ideas further and suggested to Einstein that they write a joint paper on the thermodynamic properties of the ideal gas. Einstein, however, declined. In those days, unlike the current time, the great scientists preferred to be the sole authors of papers and both Schrödinger and Einstein had hardly any joint papers with co-authors. This work eventually came out as a publication only in the name of Schrödinger just after his first great paper on wave mechanics was published and accordingly it was somewhat redundant.[34]

Schrödinger then wrote another paper on Einstein's gas theory in which he referenced de Broglie's work for the first time in using formulae for energy levels.[35] He speculated at the end of his paper "on the possibility of representing molecules or light quanta through the interference of plane waves." He was now getting so close to his wave equation. However, in his paper he also stated that the Bose-Einstein condensate could not form. This was not correct but it took over 70 years for the first Bose-Einstein condensate to be made and observed in the laboratory of Eric Cornell and Carl Wieman at the University of Colorado, Boulder in the USA.[36] They were awarded the 2001 Nobel Prize for Physics for this work together with Wolfgang Ketterle.

In 1925 a search was carried out for the Chair in Theoretical Physics at the University of Innsbruck in Austria. Schrödinger was the first-choice candidate with Arthur March, himself from Innsbruck, as the second choice. March had worked with Schrödinger's hero Hasenöhrl in Vienna where he had come to know Schrödinger well. He was tempted by a position at Innsbruck, back in his home country, in the mountains and quite close to Munich. However, the financial situation in Austria was still difficult and the salary offered initially was not competitive to Zurich.

Word of the offer was leaked and there was even a report in the local newspaper that a very expensive offer was to be made to Schrödinger that the university could hardly afford.

By this time, Schrödinger's research was developing quickly and he knew he would have major offers from elsewhere so he declined the approach from Innsbruck. He received many offers of appointments throughout his career from institutions in numerous countries. He frequently, and perhaps sometimes foolishly, gave the approaches serious consideration. Arthur March was appointed at Innsbruck and, as we shall see, will appear again more significantly in the Schrödinger story.

By October 1925 de Broglie's thesis had been published and Debye, who knew about Schrödinger's interest, suggested he give a talk on the subject at the ETH-Zurich University colloquium. Felix Bloch was then a research student at ETH who went on to a stellar research career discovering new effects in solid state physics and nuclear magnetic resonance for which he received the Nobel Prize in 1952. Bloch recalled:

> Schrödinger gave a beautifully clear account of how de Broglie associated a wave with a particle and how he could obtain the quantization rules of Niels Bohr and Sommerfeld by demanding that an integer number of waves should be fitted along a stationary orbit. When he had finished, Debye casually remarked that he thought this way of talking was rather childish. As a student of Sommerfeld he had learned that, to deal properly with waves, one had to have a wave equation. Then just a few weeks later he (Schrödinger) gave another talk in the colloquium in which he started by saying: "My colleague Debye suggested that one should have a wave equation, well, I have found one!"[6]

This great breakthrough of the Schrödinger equation occurred over Christmas 1925 in Arosa where he had previously convalesced just after he had moved to Zurich. It has been commented very widely that an unknown girlfriend accompanied him to Arosa and served as an inspiration. This point is discussed here in some detail as it has received substantial publicity.

In his comprehensive biography of Schrödinger, Walter Moore states: "Erwin wrote to an old girlfriend in Vienna to join him in Arosa while Anny remained in Zurich."[37] However, Moore was not able to identify the girlfriend. As the only evidence for his statement, Moore references an article published five years before by Ernst Peter Fischer who wrote:

> His wife was having an affair, and Schrödinger contacted an old girlfriend of his youth. They spent their Christmas holidays together in the Swiss skiing resort of Arosa.[38]

I contacted Professor Fischer about the origin of this statement and he very kindly replied on 13 December 2020 saying:

> When I wrote the paper in 1984 I was just changing from a laboratory worker in biophysics to a historian of science. I was working on a biography of Max Delbrück who is mentioned in Schrödinger's *What is Life?*. In the 1970s I had stayed in Delbrück's laboratory and he liked to tell stories about great men of science that he knew personally. I am sure that I heard of the dark lady in these days.

Max Delbrück was a German physicist who studied originally with Born in Göttingen and then in Berlin with Lise Meitner. He briefly overlapped with Schrödinger in Berlin for a few months in 1932 before going to the USA where he had a very distinguished career in molecular biology that led to the 1969 Nobel Prize in Physiology or Medicine. Schrödinger discussed Delbrück's work in biology in a book he wrote in Dublin in 1944 called *What is Life?*. All of his biographers use the word "gregarious" to describe Delbrück.[39] He was not a student in Zurich in 1925 and most likely would have only heard the story of Arosa through indirect gossip.

Several writers emphasise the statement from Weyl to the historian of science Abraham Pais — "Schrödinger did his great work during a late erotic outburst in his life" — as evidence linked to the unknown girlfriend.[40] However, this statement from Weyl does not refer directly to the time Schrödinger spent in Arosa.

The most concrete evidence of the lady in Arosa is in the recorded interview carried out by the historian of science Thomas Kuhn in 1963 with Anny Schrödinger. She states:

> We loved Arosa and in this quiet little Arosa came the first ideas about the wave mechanics… But in '25 when he started to work on that, we

> were in Arosa. I remember that he told me in Arosa about the paper of Einstein and de Broglie.[30]

This interview was conducted 38 years after 1925 but her replies were still clear and detailed. So it is quite likely that the "dark lady" of Arosa, the only witness to the moment of Schrödinger's great breakthrough, was just his wife Anny.

The first communication of Schrödinger on the discovery of his equation seems to be a handwritten letter sent from the Villa Herwig, Arosa to Wilhelm Wien in Munich on 27 December 1925. He stated:

> Just now a new atomic theory is niggling me. If only I knew more mathematics! I am very optimistic about this thing and hope that if only I can master the calculations, it will be very fine. I think I can provide a vibrational system in comparatively natural ways, not through ad hoc assumptions.[41(t)]

In his letter to Wien he also wrote down the formulae for the transitions between the electronic states of the hydrogen atom. The original letter, essentially the first mention of the new wave mechanics, sold for £22,500 at the auction house Christies in July 2020.

It seems that after returning to Zurich Schrödinger consulted with Weyl who was an expert on eigenvalues and who may have advised him on the analytical solution to the radial distance part of the Schrödinger equation in terms of a set of mathematical functions called associated Laguerre polynomials. By a coincidence, Weyl had just examined a thesis by an ETH student Walter Rotach on Hermite and Laguerre polynomials.[42] This may be related to the "more mathematics" that Schrödinger needed to complete his solution. Accordingly, the paper "Quantization as an Eigenvalue Problem" was received by the journal *Annalen der Physik* on 27 January 1926 and was published on 13 March.[2] The first page of the paper is shown here. The paper gives an acknowledgement to Weyl.

The paper, the first with the Schrödinger equation, rates amongst the greatest publications in the history of science and one of the most important discoveries of the 20th century. The new work very soon received acclaim from the most distinguished theoretical physicists. Max Planck on 2 April 1926 wrote from Berlin:

I read your article the way an inquisitive child listens in suspense to the solution of a puzzle that he has been bothered about for a long time, and I am delighted with the beauties which are evident to the eye.[41(r)]

361

3. *Quantisierung als Eigenwertproblem;* *von E. Schrödinger.*

(Erste Mitteilung.)

§ 1. In dieser Mitteilung möchte ich zunächst an dem einfachsten Fall des (nichtrelativistischen und ungestörten) Wasserstoffatoms zeigen, daß die übliche Quantisierungsvorschrift sich durch eine andere Forderung ersetzen läßt, in der kein Wort von „ganzen Zahlen" mehr vorkommt. Vielmehr ergibt sich die Ganzzahligkeit auf dieselbe natürliche Art, wie etwa die Ganzzahligkeit der *Knotenzahl* einer schwingenden Saite. Die neue Auffassung ist verallgemeinerungsfähig und rührt, wie ich glaube, sehr tief an das wahre Wesen der Quantenvorschriften.

Die übliche Form der letzteren knüpft an die Hamiltonsche partielle Differentialgleichung an:

$$H\left(q, \frac{\partial S}{\partial q}\right) = E \,. \tag{1}$$

Es wird von dieser Gleichung eine Lösung gesucht, welche sich darstellt als *Summe* von Funktionen je einer einzigen der unabhängigen Variablen q.

Wir führen nun für S eine neue unbekannte ψ ein derart, daß ψ als ein *Produkt* von eingriffigen Funktionen der einzelnen Koordinaten erscheinen würde. D. h. wir setzen

$$S = K \lg \psi \,. \tag{2}$$

Die Konstante K muß aus dimensionellen Gründen eingeführt werden, sie hat die Dimension einer *Wirkung*. Damit erhält man

$$H\left(q, \frac{K}{\psi}\frac{\partial \psi}{\partial q}\right) = E \,. \tag{1'}$$

Wir suchen nun *nicht* eine Lösung der Gleichung (1'), sondern wir stellen folgende Forderung. Gleichung (1') läßt sich bei Vernachlässigung der Massenveränderlichkeit stets, bei Berücksichtigung derselben wenigstens dann, wenn es sich um das *Ein*elektronenproblem handelt, auf die Gestalt bringen: quadratische

The front page of Schrödinger's first paper on wave mechanics.[2]

Planck also discussed the work with Einstein who wrote on 16 April 1926 to Schrödinger:

> Professor Planck pointed your theory out to me with well-justified enthusiasm, and then I studied it too with the greatest of interest… The idea of your article shows real genius.[41(t)]

This was to be the start of a very regular correspondence between Einstein and Schrödinger on the interpretation of wave mechanics that lasted the rest of their lives. Einstein wrote again ten days later on 26 April 1926:

> I am convinced you have made a decisive advance with your formulation of the quantum condition, just as I am equally convinced that the Heisenberg-Born route is off the track.[41(t)]

Heisenberg's theory, which was published in 1925 shortly before Schrödinger's, expanded quantities like the coordinates and momentum of an electron in terms of Fourier series in the energies for the transitions between states of the system.[43] Heisenberg found the unexpected result that the product of momentum multiplied by position does not equal the reverse product of position multiplied by momentum. Born then explained to Heisenberg that this result mapped on to the mathematics of matrices.[4]

Some of Schrödinger's handwritten notes survive from this time but there are only a few pages that appear to relate directly to his first paper on wave mechanics.[18] From these notes, and from subsequent comments he made, he first attempted to propose an equation for the electron in the hydrogen atom that included relativistic effects. However, this proved to be too complicated for an easy solution and the results did not seem to agree with experiment. It was left to Klein and Gordon to publish subsequently the relativistic equation.[44,45] Schrödinger was fortunate in taking his path as it was his non-relativistic equation and its generalization to systems with many electrons that has proved to be so powerful.

Following his first paper on wave mechanics, and encouraged by the acclaim he had received, Schrödinger went on to publish one new paper on his equation in nearly every month in 1926 until the summer.[7–10] This was the most intense period for research in his life. As he had proposed his equation in the first place he had a head start over his competitors who quickly took up his approach.

In his second paper he published the analytical solutions of his equation for the harmonic oscillator and for a rigid and non-rigid rotator.[7] These equations at once had applications to the vibrations and rotations of molecules. They allowed for the detailed interpretation of measurements of infrared and microwave spectra and for the extraction of important quantities such as the lengths of bonds between atoms in molecules.

He also cleverly showed how his own method of quantum mechanics could be made equivalent to the more complicated matrix approach that had been published the year before by Heisenberg and developed by Born and Jordan.[8] Born stated on Schrödinger's work:

> This was a new approach to quantum mechanics which, at first glance, had no connection with our own work. It was of fascinating power and elegance and as it used mathematical methods well known to every physicist (while our matrix method was not familiar to many of them) it quickly became the standard theory... From this time on wave mechanics was favoured by the overwhelming majority of physicists.[4]

Born at once set his research group on applying wave mechanics to the helium atom. This proved to be a more difficult problem than the one electron in the hydrogen atom due to the repulsion between the two electrons in helium. However, accurate results were soon obtained for the ionization energy of helium by Born's collaborators which were very close to the observed value.[13,46]

In addition, it quickly became clear that Schrödinger's theory gave an essentially complete description not only of the electronic structure of atoms with many electrons but also their properties such as the inert nature of rare gases like helium, neon and argon, and the colours and shapes of compounds containing the ions of transition metals such as copper and iron. Schrödinger's orbitals also gave a rigorous mathematical basis to the periodic table. The solutions of Schrödinger's equation directly gave integer quantum numbers, denoted by n, l and m, to specify the electronic orbitals which were expressed in terms of known mathematical functions: the associated Laguerre polynomials and the spherical harmonics.

A significant calculation[47,48] had been published by Wolfgang Pauli who had managed to use Born and Heisenberg's matrix theory to calculate the energy levels of the electron in the hydrogen atom only with "difficult and abstract work" whereas Schrödinger's treatment "could be understood without difficulty", in the words of Born.[4] Even Pauli himself, who was the most acerbic of theoretical physicists and who once critically described a work of a competitor as "not even wrong",[49] said of Schrödinger's papers: "I believe this work counts among the most important of what has been written recently."[50] This was praise indeed.

However, unlike the other theoreticians, Heisenberg did not find Schrödinger's theory so compelling. He attended a seminar that Schrödinger gave in Munich to members of Sommerfeld's department on his new theory. Some of those present were visibly annoyed when Heisenberg, still only 25 years old, said to Schrödinger: "Well, that cannot be true because when you make these assumptions, you cannot even explain Planck's law."[51]

Schrödinger brilliantly applied his theory to predict the spectrum of the hydrogen atom in the presence of an electric field — the so-called Stark effect.[9] He developed a form of perturbation theory for this purpose that still bears his name. He calculated analytically the intensities of the lines in the spectrum, drew them in his notebook[18] and published them.[9] His results agreed perfectly with the experiments. Nobody working in quantum mechanics had managed to do something quite like this before. Even Heisenberg acknowledged the importance of this work in being able to evaluate the matrix elements needed to compute spectra.[52] Behind the scenes of his published paper were many pages in Schrödinger's notebook deriving relationships between associated Laguerre polynomials that were need to calculate the spectral lines.[18]

In his final paper in the series, Schrödinger proposed a version of his equation in which his wave function depends on time.[10] This more general equation has also seen a huge number of applications including to processes when atoms or molecules absorb radiation or when molecules collide and react. It is the equation on his gravestone in Alpbach, Austria. However, this paper was his final burst of genius and Schrödinger did not again publish an original paper at this very top level. As we shall see later in this book, he did, nevertheless, publish some subsequent articles that had a significant influence on other sciences and scientists.

Theoretical physicists throughout the world immediately took up Schrödinger's methods. Max Born was interested in the collisions of an electron with an atom, the angular distributions of which were starting to be measured experimentally. Schrödinger's theory had worked perfectly for an electron bound to the nucleus of the hydrogen atom and it was of clear interest to examine if the theory could be extended to an unbound scattering of an electron from an atom. The theory for the scattering of waves was well established using classical mechanics and this seemed the right approach for adaptation using de Broglie's ideas.

Using Schrödinger's language, Born wrote down the initial wave and identified the squared amplitude of the scattered wave with the absolute square of the wave function $|\Psi|^2$. He then used some approximations to derive the probability for electrons being scattered by an atom into different angles.[53] This paper was the first of a huge number devoted to the wave mechanics of scattering processes. Extensions of this theory have enabled, for example, the properties of new particles to be predicted and discovered in accelerator machines and even the dynamics and rate constants for the chemical reactions of molecules can be calculated accurately from solutions of Schrödinger's equation.[54]

Another original aspect of Born's paper, for which he was awarded the Nobel Prize in 1954, was the identification of $|\Psi|^2$ as a representation of probability. Following Schrödinger's first great paper there was at once an intense debate on the interpretation of his wave function Ψ, and this discussion has gone on to the present day.

Born was the first to show that a probability interpretation of the wave function was needed to calculate scattering processes and this rapidly became the general and widely used interpretation of Ψ. Bohr himself supported this description and it became known as the Copenhagen Interpretation. However, most strangely, Schrödinger himself was never happy with this and neither was Einstein who produced his famous phrase "God does not play with dice".[55] Indeed, their concern with this interpretation led both Einstein and Schrödinger nearly ten years later to propose thought experiments to attempt to show the unsatisfactory aspect of the analysis. This led to the famous "Schrödinger's Cat" paper which is discussed in Chapter 3.

Heisenberg was not happy with Born and reproached him for "going over to the enemy camp" and said: "The more I ponder about the physical part of Schrödinger's theory, the more disgusting I find it."[56] Despite the fact that Born never worked directly with Schrödinger he kept up a very regular correspondence with him right up to Schrödinger's death in 1961. Born and Schrödinger both found themselves as refugees in the 1930s with dependents to support. They were first based, respectively, in Cambridge and Oxford, and then in Edinburgh and Dublin. Born also subsequently supported Schrödinger at important times when others in the scientific community had turned against him, and this included the nomination of Schrödinger to be a Foreign Member of the Royal Society as is discussed in Chapter 5.

Schrödinger was well aware that up until his first paper on wave mechanics he had not made an iconic contribution to physics. Accordingly, he worked very energetically to promote his new theory to audiences around the world. In the summer of 1926 he lectured in his old universities of Jena and Stuttgart, and also in Munich and Berlin. The reception to his lecture from the senior scientists in Berlin, including Planck, Einstein, Nernst and von Laue, was very favourable and Planck invited him to a reception in his house. This did no harm at all to Schrödinger's case when he was invited to a Chair in Berlin in the next year.

Schrödinger also received an invitation from the great man of Danish science Niels Bohr who had won the Nobel Prize four years before. Bohr met Schrödinger at Copenhagen railway station in October 1926 and could not leave the lecturer alone for several days in intense discussion of the new theory. In particular, Bohr was enthusiastic about the statistical interpretation of the wave function while Schrödinger would not relent. The strain was too much for him and he took to his bed in Bohr's house. It was the tradition in those days for the host to invite a distinguished speaker to stay at their own house — something that has continued in some scientific centres almost up to the present day. Bohr could not resist coming to Schrödinger's bedside to continue the discussion. Heisenberg very much sided with Bohr in this debate and produced his famous uncertainty principle a few months later which states that the more precise the momentum of a particle is determined the less precise is its position.[57]

Schrödinger was aware of the emerging importance of scientific research in the USA and arranged for the translation of his collected papers on wave mechanics into English.[58] This is a language which Schrödinger spoke perfectly because of his family links. At the end of 1926 he also published a paper in the *Physical Review*, an American journal which was fast evolving as a leading journal for all areas of physics. These publications were a break from essentially all of Schrödinger's previous articles which had been published in German journals and in the German language.

The *Physical Review* paper had the somewhat awkward title "An Undulatory Theory of the Mechanics of Atoms and Molecules".[59] It gave an emphasis to the work of de Broglie and a justification of the Schrödinger equation. In his discussion of the application of his theory to the Stark effect, Schrödinger also stated: "the charge of the electron is not concentrated in a point, but is spread out through the whole space, proportional to the quantity $|\Psi|^2$." Perhaps somewhat inadvertently he was using the language of the Copenhagen Interpretation to justify the formulas used for his calculations.

Just before Christmas in 1926, and almost exactly one year after he discovered his equation, Schrödinger and Anny set sail in a French liner to visit the USA for the first time.[60] He kept a rather amusing diary of the trip. The Atlantic crossing took ten days and, similar to his time a few years later as a Fellow of Magdalen College in Oxford, he did not like being thrown together at dinner with people he did not know well.

Unlike many people who travel to the USA for the first time Schrödinger was not so impressed initially. He did not take to New York and he criticized the country for "not having a uniform culture with a common historical view". He met the future Nobel Prize Winner John van Vleck in Minneapolis and thought he "acted smarter than he is".[60] In the Midwest he lectured in Madison, Minneapolis, Ann Arbor and Urbana. Everyone wanted to meet him and he was a celebrity. At one time he had to give three talks in a 30-hour period. He gave 57 lectures during this three-month visit. His trip forward and backward in many cities across the USA had some similarities to the extensive tours given today by modern rock stars.[60]

The Schrödingers spent several weeks in Madison at the University of Wisconsin and then travelled by train to the California Institute of Technology (Caltech), Pasadena, stopping off at the Grand Canyon on the way. At the front row for his lectures at Caltech were Robert Millikan, who had won the Nobel Prize for Physics just three years before for measuring the charge of the electron, and the grand old Dutch physicist Hendrik Lorentz who was also a visitor giving lectures on the same day as Schrödinger. Lorentz had won the Nobel Prize in 1902 and was one of Schrödinger's heroes. This was a relaxed opportunity to discuss the fundamentals of quantum mechanics a long way from home. Schrödinger liked Pasadena and the surrounding southern Californian area and wished he could have spent longer there.

After Pasadena he took the train back via Salt Lake City and over the Rocky Mountains to Denver, on to Chicago and then to MIT and Harvard. He also met the spectroscopist Robert Wood in Baltimore who had provided several experimental results on the spectra of atoms. His final lecture was at Columbia University in New York before taking the boat for the long cruise home. After his initial somewhat negative views, Schrödinger had begun to warm to the USA. On his visit he was offered positions in Madison and in Baltimore which were both financially attractive.[60] However, by this time he knew he was being considered for a major appointment in Berlin.

Berlin

The Friedrich Wilhelm University was the oldest university in Berlin. Max Planck had held the Chair of Theoretical Physics since 1889. He had won the Nobel Prize in 1918 for his discovery of energy quanta and was to retire in 1926. Max von Laue held a Chair there also and he had won the Nobel Prize in 1914 for his work on the diffraction of X-rays by crystals. In addition Albert Einstein had a non-teaching professorship being the Director of the Kaiser Wilhelm Institute for Physics in Berlin. He had moved from the ETH Zurich in 1914 and had been awarded the Nobel Prize in 1921. Being appointed to Planck's Chair was the most prestigious move for a theoretical physicist.

Several of the names already mentioned in this book were considered for the Chair including Born, Debye, Heisenberg, Sommerfeld and Schrödinger. Sommerfeld had been the most prolific at supervising early career theoretical physicists who had gone on to major academic careers and this made him first choice. However, he did not want to leave Munich. Debye's interests were thought to be more on the experimental side while Heisenberg, although clearly recognized as a genius, was considered to be too young for a major Chair at the age of just 25.

The appointments committee noted Schrödinger's ingenious invention of wave mechanics, his superb lecturing style and his "charming temperament of a South German".[37] He was also very highly rated by Planck who had been so impressed by his first paper on wave mechanics. As Schrödinger had most elegantly discovered a novel justification for Planck's quantum theory he was a highly appropriate replacement for the great scientist.

As is often the case with senior university appointments, the University of Zurich came back with a counter offer to attempt to retain their star physicist. Anny recalled:

> No, that was impossible because they didn't offer him enough. The Swiss are a bit careful and they wouldn't pay — especially the university. It has not as much money as the E.T.H. The offer in Berlin was very high from the first moment on. So Zurich tried to get a bit more, but it never came near the salary which Berlin offered. When it came to the end, then they offered him a double-professor at the E.T.H. and at the university so that both could pay, but even then he wouldn't have had as much as he got in Berlin. Berlin was in a very good position then and offered really a lot.[30]

The students in Zurich heard about the offer and organised a torchlight procession that ended in front of Schrödinger's house, and he and Anny found this very touching. Planck, however, was very persuasive and Schrödinger could not resist the excellent offer. So he and Anny moved to Berlin in August 1927. They rented a large apartment at 44 Cunostrasse in the pleasant area of Grunewald. Many subsequent letters from Schrödinger to the great scientists of the day were sent from that address.

Not long after his move to Berlin came the 5th Solvay Conference in Brussels.[61] The topic was "Electrons and Photons" and Schrödinger was asked to give one of the main lectures. The Conference did much to promote Schrödinger's breakthrough. However, as shown in the photograph, Schrödinger was still not considered senior enough to be placed in the front row of the photograph for the Conference where Planck, Marie Curie, Lorentz, Einstein and Langevin were seated.

The participants of the 1927 Solvay Conference. (From the left) **3rd row:** Piccard, Henriot, Ehrenfest, Herzen, de Donder, Schrödinger, Verschaffelt, Pauli, Heisenberg, Fowler, Brillouin. **2nd row:** Debye, Knudsen, W.L. Bragg, Kramers, Dirac, Compton, de Broglie, Born, Bohr. **1st row:** Langmuir, Planck, Madame Curie, Lorentz, Einstein, Langevin, Guye, Wilson, Richardson.

Essentially all of those names who had contributed to the recent development of the quantum theory were at this Solvay Conference including Bohr, Born, de Broglie, Dirac, Heisenberg, Pauli and Schrödinger. As many as 17 of the attendees had won the Nobel Prize or were to be awarded the Prize. The anti-German prejudice that had prevented Einstein and others from attending the Solvay conferences held after the First World War had melted away. The meeting was chaired by Lorentz who Schrödinger had recently met in Pasadena where Lorentz had mentioned to him that he would be invited to give a lecture at the 5th Solvay Conference. Sadly, Lorentz died a few months later.

This meeting is often considered the most celebrated of all the Solvay Conferences and the statistical interpretation of Schrödinger's wave function was a major discussion point. Both Einstein and Bohr were highly prominent in the debate. From this conference to the present day the alternative views of Einstein and Schrödinger have been marginalized and the statistical Copenhagen Interpretation of the wave function, as advocated by Bohr, Born and Heisenberg, has very largely prevailed. However, the alternative views have still been influential and stimulated subsequently some of the most radical new ideas on quantum mechanics from scientists such as David Bohm, Hugh Everett and John Bell that are very much discussed today.

Experimental evidence for the new quantum theory was also a topic at the Solvay Conference. The experiments of Clinton Davisson and Lester Germer, in which electrons scattered from the surface of nickel showed a diffraction pattern, and those of Arthur Compton, in which the frequencies of X-rays were changed when scattered by free electrons, supported de Broglie's ideas on wave-particle duality. Of these three American scientists, Compton had been able to make the long boat trip over the Atlantic for the meeting.

Schrödinger's talk was entitled "Wave Mechanics" and he started off by summarizing the main conclusions from his pioneering papers published in 1926. He also discussed some advances in his theory that had been made by others including the incorporation of relativistic effects and the extension to many-electron systems. He anticipated molecular orbital theory in stating that a many-electron wave function can be expanded as a linear combination of one-electron orbitals.

The published discussion following Schrödinger's paper is particularly fascinating as it gives a first glimpse into how Schrödinger's theory was to become such a powerful method for performing calculations on atoms and molecules.[61] Born pointed out that Kellner in his research group had extended Schrödinger's theory in an approximate way to the helium atom.[13] Lorentz, however, was alarmed at the extra computations that were needed and Heisenberg expressed his doubts that Schrödinger's theory would be extendable to systems with many dimensions. Schrödinger in response mentioned the work of Hartree in Cambridge that was making this extension.

There were interesting comments on the meeting.[61] The younger participants came away enthused. Heisenberg said: "Through the possibility of exchange between the representatives of different lines of research, this conference has contributed extraordinarily to the clarification of the physical foundations of the quantum theory; it forms so to speak the outward completion of the quantum theory." Born wrote to the organisers to say it was the most stimulating conference he had ever taken part in. However, some of the older guard in the front row of the conference photograph were not so impressed. Paul Langevin said: "The confusion of ideas reached its peak!" This was the conference, however, that promoted Schrödinger into the top division of great scientists.

Paul Dirac from Cambridge was the youngest person to attend the Solvay Meeting at the age of 25. He was still to produce his greatest work and spoke on his field theory of light. His view of the meeting was:

> I listened to the arguments, but I did not join in them, essentially because I was not very much interested. It seemed to me that the foundation of the work of a mathematical physicist is to get the correct equation, that the interpretation of these equations was only of secondary importance.[62]

Dirac had a background studying engineering and this underpinned his no-nonsense approach. Because Schrödinger had discovered the equation, he was always rated very highly by Dirac. However, Dirac had some after-dinner discussions on wider matters with the other young scientist at the meeting. Heisenberg, who was just one year older, recalled Dirac's comment: "I cannot understand why we idly discuss religion. If we are honest — and scientists have to be — we must admit that religion is a jumble of false assertions, with no basis in reality."[62] Dirac became famous for such statements.

On returning to Berlin, Schrödinger was thrown into a major course of lectures covering several areas of physics but with emphasis on electron theory. He was an eloquent and popular lecturer. At that time in Berlin there was a remarkable range of lectures that the fortunate students could attend. Planck was still lecturing, Lise Meitner gave a course on nuclear physics, Nernst on experimental physics, Fritz London on chemical bonding and Schrödinger's previous Breslau colleague Ladenburg on spectroscopy.

Fritz London overlapped several times with Schrödinger in his career. If he had not died in 1954 at the relatively young age of 54, he very likely would have won the Nobel Prize and he was nominated many times. He had studied with Sommerfeld and then taken a position with Ewald in Stuttgart. In 1926 he had become aware of Schrödinger's papers and, following Sommerfeld's encouragement and support, received a grant from the Rockefeller Foundation to work in Zurich with the discoverer of wave mechanics.

London penned a clever letter to Schrödinger suggesting in humorous terms that Schrödinger might have suggested de Broglie's ideas in his early paper of 1922. However, Schrödinger was not amused and responded rather sarcastically: "What you find in the murky material which has so far been discarded, you should attribute to you and not to me. If you extract something interesting, I shall be very pleased indeed."[63]

London arrived in Zurich early in April 1927 when Schrödinger was in the USA. Then, almost as soon as Schrödinger returned to Zurich, he was making the arrangements to go to Berlin. However, in Zurich London met Walter Heitler. He had done graduate studies in Munich and then was given a grant to work with Niels Bjerrum in Copenhagen. When Schrödinger's papers came out Heitler, like London, was very keen to learn more and transferred his grant to Zurich.

However, as has been noted several times in this book, Schrödinger's style was to carry out research on his own. In contrast to Sommerfeld and Born, who ran large research groups, Schrödinger was not interested in supervising the research of students or postdoctoral fellows. So both London and Heitler, finding themselves unsupervised in Zurich, decided instead to collaborate together on extending Schrödinger's theory. They chose a key problem in chemistry — the bonding of the hydrogen molecule. They wrote down an approximate form for the wave function of this two-electron system in terms of a product of the hydrogen atom wave functions (or orbitals) for the electrons centered on each atom. This "valence bond" form of the wave function gave some electron density in the region between the two atoms to produce a chemical bond. After some non-trivial calculations they were able to compute the strength of this chemical bond. This work was submitted for publication almost exactly one year after the submission of Schrödinger's first paper on wave mechanics.[14] It was the first advance in applying Schrödinger's quantum mechanics to a molecule with more than one electron and remains a major paper in quantum chemistry.

This work also illustrates well how the application of Schrödinger's theory to molecules with more than one electron, which is what made him so famous to the wider scientific world, was done by others and not by Schrödinger himself. However, he was aware of these important developments and took Fritz London as an assistant when he moved from Zurich to Berlin. He also arranged for Heitler to join his Institute for Advanced Studies in Dublin in 1941 where he eventually took over from Schrödinger as Director. Heitler took up Irish citizenship and was elected a Fellow of the Royal Society.[64] He even wrote the Biographical Memoir of Schrödinger for the Royal Society.[65]

Another young scientist who came to Zurich to work with Schrödinger was the American Linus Pauling. He had won a Guggenheim grant which enabled him to come to Europe to learn about the new ideas in quantum mechanics in several centres. He stopped off in Munich to see Sommerfeld, visited Born in Göttingen and then went to Zurich to learn wave mechanics.

Like Heitler and London, Pauling found Schrödinger elusive in Zurich but immediately realised the potential of the work of the two

Erwin Schrödinger (left) and Fritz London, 1928.

young physicists on the chemical bond. As a chemist he could see how the Heitler-London valence bond theory could be extended to calculate the structure and properties of many important molecules including those of interest in organic chemistry. In due course he used this theory to explain the tetrahedral shape of the methane molecule and the strong bonding in aromatic molecules like benzene. His book *The Nature of the Chemical Bond*[66] brought this research together and he was awarded the Nobel Prize for Chemistry in 1954. Pauling was also involved with movements to stop nuclear testing and in 1962 was awarded a second Nobel Prize, this time for Peace.

When Fritz London took his Rockefeller grant to Berlin with Schrödinger he found the appointment surprisingly congenial. He discovered that Schrödinger's attitude of not interfering with his research had the big advantage that he could work independently. London's work with Heitler on chemical bonding was becoming well known and they had received major compliments on it from no less than Heisenberg, von Laue and Born.[63] Taking quantum mechanics beyond the hydrogen molecule proved difficult at that time because of the computations involved but London found he could simplify the calculations using the powerful methods of group theory and this work became influential.

Fritz London, Michael Polanyi and Henry Eyring were also the first to calculate for a chemical reaction a potential energy surface which is the electronic energy of a molecule expressed in terms of the inter-atomic coordinates.[67,68] In modern times, potential energy surfaces underlie almost any simulation of a molecular system. London also applied a perturbation theory approach to Schrödinger's wave functions to provide a theoretical framework for the long-range dispersion forces between atoms.[69] These "London forces" are significant for the properties of liquids and gases and even for the structures taken up by biological molecules.

Tempting offers to Fritz London came from Born in Göttingen and Sommerfeld in Munich but he preferred the dynamic scientific scene then prevalent in Berlin which was at its peak in the late 1920s. He much enjoyed the seminars where the highly promising younger members like himself, Viktor Weisskopf, Max Delbrück, Eugene Wigner and Leo Szilard could present their latest ideas to the great senior physicists present including Planck, Einstein, Schrödinger, von Laue, and Nernst. With quantum mechanics starting to influence chemistry, brilliant researchers such as Michael Polanyi, Lise Meitner and Otto Hahn would also come over from the Kaiser Wilhelm Institute for Chemistry. After the seminar the participants would often adjourn to a local tavern to continue the discussions. Sometimes the Schrödingers would invite the group back to their apartment for Viennese sausage parties and the photograph on the next page shows that a good time was had by all.

As we have seen before, Anny Schrödinger summed things up rather well:

> Berlin was the most wonderful and absolutely unique atmosphere for all the scientists. They knew it all and they appreciated it all... one had a bit forgotten the first war and before the second, so it was absolutely a wonderful time. The theatre was at the height, the music was at the height and science with all the scientific institutes, the industry. And the most famous colloquium... the Berlin Academy had published lectures which were very famous too. There were lots of friends who came together, not on a special day; it was absolutely a very social life. My husband liked it very much indeed.[30]

Sausage party at the Schrödingers' apartment in Berlin in 1932. Schrödinger is in the right-hand corner.

It has to be said that Schrödinger continued to be much in demand for lectures as news of his great breakthrough with wave mechanics spread far and wide. Sir William Henry Bragg had won the Nobel Prize in 1915 with his son William Lawrence Bragg for his work on the diffraction of X-rays by crystals. He had attended the 5th Solvay Conference where he had heard several of the pioneers of the new quantum mechanics speak. He was now the Director of the Royal Institution, in a venerable building in the centre of London.

The Royal Institution had become famous not only through the pioneering experimental research of Humphry Davy and Michael Faraday but also through the communication of science from lectures to the public. This tradition has continued in great style to the present day. The Royal Institution is based in Albemarle Street at a walkable distance from Buckingham Palace and Downing Street, and members of the Royal family often attended the discourses given on Friday evenings. So when Bragg invited Schrödinger to give a course of lectures he was pleased to accept as it would, for the first time, enable him to speak on wave mechanics to a distinguished English audience. Bragg himself gave a lecture at the Royal Institution in advance of Schrödinger's arrival as a prologue to the talks on wave mechanics and emphasized the wave-particle duality of electrons.

Schrödinger gave four lectures in London on 5, 7, 12 and 14 March 1928 with the title "Wave Mechanics" and Anny accompanied him. The lectures were published in English and were dedicated to the memory of his teacher Friedrich Hasenöhrl.[58] The first two lectures largely described material from the great papers of 1926 while the third and fourth lectures presented some new ideas including the time dependence of the absorption of light by atoms and molecules leading to selection rules and the perturbation corrections needed to allow for the motion of the nucleus in the hydrogen atom. Schrödinger did not lecture at a Friday evening discourse. His lectures were too technical for the general public so they were presented on Mondays and Wednesdays.

In the same week as Schrödinger's lectures in March 1928 the Friday evening discourse at the Royal Institution was given by Edward Milne on "The Sun's Outer Atmosphere". Schrödinger subsequently wrote on Milne's work on relativity and the expanding universe. He had just been appointed to the Rouse Ball Professorship of Mathematics at Oxford University.

Between two of his lectures at the Royal Institution, Schrödinger visited Cambridge. He gave a lecture at the Kapitsa Club in Trinity College on Saturday, 10 March, on "The Physical Meaning of Quantum Mechanics".[70] The Russian physicist Pyotr Kapitsa had started this club to hear the latest developments in physics in an informal setting with blackboard and chalk. It was attended by the brightest physics students and research fellows in Cambridge. The talks were normally given on Tuesdays but a Saturday was a special case for Schrödinger between his lectures at the Royal Institution. The talk given the week before at the Kapitsa Club was by Douglas Hartree on "X-ray study of heat motion in NaCl crystal".[70] There had been many famous previous speakers at the Kapitsa Club including Bohr, Ehrenfest, Franck, Langevin and Landau. In 1925 Heisenberg had been a speaker and had briefly mentioned his new ideas on quantum mechanics which crucially caught the attention of Ralph Fowler and were subsequently developed by his research student Paul Dirac. However, it is unlikely that Dirac, who always attended the Kapitsa Club talks, would have been very excited by the title of Schrödinger's talk in 1928 as he was never impressed by the efforts to extract physical meaning from wave mechanics.

Just one month before Schrödinger visited England, Dirac had published his great paper "The quantum theory of the electron".[16] The work, which was built on the Klein-Gordon relativistic form of the Schrödinger equation, introduced a consistent way of dealing with both relativity and electron spin and gave the equation which is engraved on Dirac's memorial stone in Westminster Abbey. This stone is next to that of Isaac Newton, one of Dirac's illustrious predecessors holding the Lucasian Chair at the University of Cambridge.

From Schrödinger's letters of that time he seemed rather slow to pick up the significance of Dirac's great work. However, two years later, he did publish on the "trembling electron" arising from Dirac's theory.[71] Dirac was always a great supporter of Schrödinger and gave strong assistance when the latter was considered for Foreign Membership of the Royal Society in 1948 despite considerable opposition from other Fellows. He also wrote an obituary for Schrödinger in 1961 in *Nature* in which he stated:

> His great discovery, namely, Schrödinger's wave equation, as the basis of the description of atomic processes, was one of the most surprising of all the sudden advances that have occurred with the development of scientific knowledge.[72]

Dirac and Schrödinger were to meet famously in Stockholm in 1933 to be presented with their Nobel Prizes. They also got together several times in Dublin when Schrödinger was based there and they corresponded regularly.

It seems that Schrödinger had caught a cold during his visit to Cambridge and was confined to bed in the house of George Birtwistle, a Fellow in Mathematics at Pembroke College who had written an influential book on *The New Quantum Mechanics*. There Schrödinger was visited by the physicist George P. Thomson from Aberdeen University who had some exciting new experimental results on electron diffraction. Schrödinger recalled this visit in a letter to Thomson on 5 February 1945:

> After mentioning briefly the new theoretical ideas that came up in 1925/26, I wish to tell of my meeting you in Cambridge in 1927/28 (I think it was in 1928) and of the great impression the marvellous first interference photographs made on me, which you kindly brought to Mr Birtwistle's house, where I was confined with a cold. I remember particularly a fit of scepticism on my side ("And how do you know it is not the interference pattern of some secondary X-rays?") which you immediately met by a magnificent plate, showing the whole pattern turned aside by a magnetic field.[73]

Thomson's results demonstrated experimentally the wave nature of the electron and so were of particular interest to Schrödinger. George Thomson was the son of the great Cambridge physicist J.J. Thomson who was awarded the Nobel Prize for Physics in 1906 for his discovery of the electron. Following his father, George Thomson also won the Prize in 1937 together with Clinton Davisson.

Further details of Schrödinger's interesting visit to Cambridge in 1928 may be inferred from a letter he received from Professor Horace Lamb on 3 May 1931:

> Lord Rutherford has shown me your letter to Fowler who is at present in America, and has asked me to send you such recollections of Maxwell as I have. You may remember that I met you at Rutherford's house when you were in Cambridge, and there was some talk about Maxwell then.[18]

Lamb was a very senior Professor of Mathematics at Cambridge who was born in 1849. He had written classic texts on hydrodynamics and the theory of sound. In his letter to Schrödinger he mentions meeting him previously at Lord Rutherford's house in Cambridge. It is likely this was in 1928 as there appears to be no evidence of a further visit of Schrödinger to Cambridge before 1931.

Lamb mentioned that there had been some discussion at Lord Rutherford's house about the great Cambridge theoretical physicist James Clerk Maxwell who had established the famous equations for electromagnetic radiation. Maxwell's contribution had been described by Einstein, on the 100th anniversary of his birth, as the most profound and the most fruitful that physics has experienced since the time of Newton.

Lamb had been a colleague of Maxwell's for a few years in Cambridge before Maxwell's death in 1879. In his letter, he commented in some detail to Schrödinger about personal aspects of Maxwell, describing his lectures as "not very well arranged but delightful to listen to, not so much for the subject matter as for the humorous illustrations and odd items of thought which came out of them."[18] For Schrödinger, who had published the equation describing the motion of electrons, this must have been an historical moment to meet this human link with Maxwell's equations for electromagnetic radiation.

Despite all the busy lectures and meetings, Schrödinger was not so complementary about his visit to England and, after returning from Cambridge to London, complained to Paul Ehrenfest on 12 March 1928:

> What is so depressing me here in England is the watered-down, dispassionate gentlemanly atmosphere. You have to keep saying "I'm sorry" all the time and wrap up what you want to say because nudity is frowned upon here in every sense. I feel like I'm tied up. Add to that the uncomfortable rooms, the terrible cold in all rooms of the house, plus a murderous cold and cough.[41(t)]

Schrödinger finished his trip to England by visiting Dr. Frederick Pidduck at Corpus Christi College, Oxford on 15 March. This was his first visit to Oxford after the publication of his 1926 papers. After their meeting, Pidduck used the associated Laguerre polynomials that Schrödinger had introduced for the hydrogen atom to obtain an exact solution of Dirac's relativistic equation.[74] It is possible that this visit influenced Schrödinger to come to Oxford some five years later as he would have expected to be able to interact with such like-minded theoreticians.

After his visits to London, Cambridge and Oxford Schrödinger and Anny made their way to Leiden in the Netherlands. There Paul Ehrenfest, who had taken the Chair of Lorentz, held a lecture series to rival the Kapitsa Club in Cambridge. After a good dinner, Ehrenfest would fire probing questions at the visiting speaker who would be staying in the guest room of his house at 57 Witte Rozenstraat. The speaker would then write their signature on the wall in the hall just outside the guest room. Many of the great theoretical physicists had signed in this way including Einstein, Planck, Bohr and Dirac. Schrödinger added his signature on 21 March 1928 and, perhaps to emphasise her importance, Anny added her signature also.

Being so much in demand for lectures did not enhance Schrödinger's scientific productivity for new publications and he produced few papers of note during the six years he spent in Berlin. His six-month burst of creativity of 1926 was not to be repeated again. He only published two papers in 1928 with just some reviews and short essays in 1929. It is true his *annus mirabilis* of 1926 was an almost impossible act to follow but Schrödinger was reluctant to get into the details of extending his theory of wave mechanics in the more practical way achieved by others such as Born, London, Heitler and Pauling. In the very competitive scientific atmosphere of Berlin in the late 1920s, this lack of productivity was getting noticed. For example, Leo Szilard said: "Unfortunately Schrödinger is doing too much reading and not writing anything."[75]

He did, however, publish a novel idea in which he compared his Schrödinger equation to the diffusion equation for Brownian motion with an imaginary diffusion coefficient.[76] In the present day this suggestion has been turned into a very powerful "Diffusion Monte Carlo" algorithm as the diffusion equation can be solved with a simple random walk procedure which is very easy to apply on modern electronic computers. This approach is enabling the Schrödinger equation to be solved numerically for quite complicated multi-dimensional systems which are hard to tackle with alternative approaches.[77]

The Prussian Academy of Sciences was a highly prestigious scientific society based in Berlin. It was over 200 years old and had just 32 members in sciences, classics and philology. Being a member was a great honour. Einstein had first presented his general theory of relativity to the academy in 1915. There were just five physicists in the academy in 1929: Planck,

Einstein, von Laue, Warburg and Paschen. Schrödinger was elected that year and was then the youngest member aged 42. The Academy was very important to Planck who was its secretary. It was expected that members would write papers for the academy and Schrödinger published two papers in 1930 on Heisenberg's uncertainty relations and on relativistic wave mechanics. The Academy would come to play an important role in the political storm that would soon overwhelm German science.

In 1929 Schrödinger made a brief visit to Innsbruck to see Arthur March that turned out to be somewhat fateful. March had taken up the Chair that Schrödinger had declined and was preparing a book on the foundations of quantum mechanics. He had just been married to a pretty young lady Hildegunde. On his return to Berlin Schrödinger mentioned to Anny how he had been impressed by her charm.[37]

In 1931 Schrödinger was delighted to be elected a member of the Royal Irish Academy. He was well aware of the many distinguished mathematicians and theoretical physicists who had come from Ireland including George Boole, William Hamilton, Joseph Larmor, George Stokes and Lord Kelvin. Schrödinger's equation built on the work of Hamilton. He often found that the distinguished academies he was elected to would be useful to him in his subsequent career and this would certainly be true with the Royal Irish Academy.

Nazis

However, by this time the political atmosphere in Germany was changing at a rapid pace. Back in 1908 Adolf Hitler had moved to Vienna to try his luck as an artist. Things did not go well and he was homeless for a period. Schrödinger was just two years older than Hitler and during this period was studying hard nearby at the University of Vienna under Hasenöhrl. There is no evidence that these two Austrians ever met.

Then, in the summer of 1925, Hitler wrote his infamous book *Mein Kampf* in the Landsberg prison following his failed Beer Hall Putsch in Munich. In his book he espoused his pernicious anti-Semitic ideas and stated that they arose from his period of poverty in Vienna. Just a few months later Schrödinger published his equation early in 1926. So in this short period in the mid-1920s, one Austrian produced a publication which led to extraordinary destruction and another wrote an article which led to many hugely positive advances and eventually the new quantum age of the 21st century.

In October 1929 the Wall Street stock market crashed. There were very serious economic ramifications around the world and this included Germany. After the severe financial problems following the First World War, the situation had gradually improved there through the 1920s but this all changed in 1929. Unemployment rose fast and there were riots on the streets of Berlin. The political situation was becoming complicated with Soviet Russia continuing to expand its influence and right-wing movements becoming more popular throughout Europe. Hitler and his party of National Socialists took full advantage of this situation and he eventually became Chancellor of Germany on 30 January 1933.

In the run up to this momentous event Schrödinger had been working hard to communicate his breakthrough in quantum mechanics to the world and had not had any time to be concerned with the politics that was soon to have dramatic effects on his life. Other physicists in Germany, however, had become very involved with politics. A prominent example[78,79] was Johannes Stark who had been awarded the Nobel Prize in 1919 for his research partly on the "splitting of experimental lines in electric fields". Some seven years later Schrödinger's theory provided perfect agreement with the experimental results on the Stark Effect for the hydrogen atom.[9]

In 1915 Stark was hoping to take up a position in Göttingen but Peter Debye was appointed instead. In no uncertain terms, Stark blamed the "Jewish and pro-Semitic circle" of mathematicians and theoretical physicists and its "enterprising business manager" Sommerfeld.[78] Stark was appointed to a less prestigious post at Greifswald, which is on the coast of north-eastern Germany. Here a very conservative and nationalistic faculty and student body was present — an atmosphere that he found congenial.

After the end of the First World War several German cities were taken over by socialists groups and this was the case in Greifswald, much to Stark's alarm. This was the start of his interest in politics even before the emergence of the Nazis. Following the award of the Nobel Prize, Stark moved in 1920 to the University of Würzburg in his home state of Bavaria. Stark had become jealous of the more liberal physicists in Berlin and Munich who had dominated the German Physical Society and had alienated more conservative physicists such as himself who were based in the less fashionable cities.

In Würzburg, Stark at once attempted to start an alternative German Professional Community of University Physicists which he hoped would control research funds. However, Wilhelm Wien from Munich was then the President of the German Physical Society. He had won the Nobel Prize in 1911 for his work on the heat emission of black bodies which assisted greatly in the developments of Planck's ideas on quantum mechanics. Wien was greatly respected by the physicists in Germany and successfully rebuffed Stark's initiative.

By this time, the revolutionary new theories of Albert Einstein were the subject of intense debate and were not accepted by several of the old guard of German physics. Einstein himself was Jewish and had given up German citizenship when he moved to Switzerland in 1895 at the age of 16. His theory of relativity was spectacularly confirmed by the Cambridge scientist Eddington in 1919. He had observed the solar eclipse on the African island of Príncipe and had shown that Einstein's theory of relativity was needed to predict the shift of light rays by the gravitation of the sun. This observation at once propelled the name of Einstein to scientific stardom around the world and there was great jealousy from others such as Stark.

A non-scientist, Paul Weyland booked the huge Berlin Concert Hall for a meeting of the self-named "Association of German Natural Scientists for the Preservation of Pure Science" which Einstein himself dubbed as "Anti-Relativity Company Ltd." This meeting was widely advertised and reported in the German press. One of the physicists who spoke against Einstein's theory was Ludwig Glaser who was an advanced student being supervised by Stark. Glaser's anti-Einstein comments at the meeting were resented by more senior colleagues back in Würzburg. Thus, when Glaser came up for his Habilitation shortly afterwards, there was strong opposition.[78]

Stark was furious and resigned his Chair. As a recent Nobel Prize winner he at once expected offers for other appointments elsewhere but they did not arise. With Einstein now commenting widely on his support for world peace and the Weimar republic, Stark became increasingly bitter. He wrote a book in 1922 with the title *The Contemporary Crisis in German Physics* in which he attacked relativity and atomic physics. In the book, Einstein's theories were heavily criticised as was his promotion of his theories through newspaper articles and foreign lectures which had, in Stark's eyes, betrayed German nationalism. However, there were no direct anti-Semitic views expressed in the book. Von Laue, in particular, was very critical of Stark's book and the antagonism between these two Nobel Prize winners was never resolved.

Another influential Nobel Prize winner who had become very opposed to Einstein on personal and scientific grounds was Philipp Lenard. He had won the Nobel Prize in 1905 for his work on cathode rays and the photoelectric effect. Einstein himself used the results to show that quantum theory was needed to explain the photoelectric effect and received the Nobel Prize for this work in 1921 (but curiously not for his theory of relativity). Lenard did not accept Einstein's theory on the photoelectric effect and resented the publicity he was receiving about his work on relativity.[79]

Despite the fact that he had received the prestigious Rumford medal of the Royal Society, Lenard was highly critical of how ideas in German science had been "stolen" by English scientists, this view being related to the hatred that arose during the First World War. Lenard was also furious that it was English scientists who had verified Einstein's theory of relativity.

The antipathy between Lenard and Einstein had come to a head at the Annual Meeting of the German Natural Scientists in Bad Nauheim in September 1920 when they had bitter discussions that were widely reported in the press.[79] Not long after this, Stark and Lenard formed their own organisation, the nationalist Deutsche Physik. This movement was concerned with the rhetoric of science as opposed to science itself. It quickly moved to criticism of much of modern physics which they often dubbed "Jewish physics". Many of the scientists already mentioned in this book came from Jewish families and were to be caught up in the rapidly approaching crisis in which Stark and Lenard played such a negative role from the early days of national socialism.

In 1922 the liberal politician Walther Rathenau was assassinated and there were open threats also to Einstein's life. The German government ordered that flags should be flown at half-mast but Lenard ignored the order at his institute in Heidelberg and this led to a demonstration against him by socialist students. Then, when Hitler was sent to jail in 1924, Lenard and Stark published an open letter of support. They wrote: "Hitler and his comrades appear to us as gifts of God from a long darkened earlier time when races were still purer, persons still greater, spirits less fraudulent."[78] Their letter was influential as it provided intellectual support to Hitler's ideas from some of the leading scientists in Germany. Stark even wrote to Hitler to offer him the use of his own house for recuperation after he left prison, and he subsequently joined the National Socialist Party. Stark, Lenard and their colleagues had little interest in Schrödinger during their political activities of the 1920s and early 30s. They had bigger fish to fry.

German scientific leaders such as Max Planck stood back from the fray. Planck's main concern was how the emerging debates relating science and politics would harm the world-leading position achieved by German science. Due to the ugly atmosphere and threats, Einstein had understandably been spending more time abroad where he felt safe. In 1933 Planck wrote to Ehrenfest in Leiden, where Einstein was staying, in a slightly critical tone expressing the view that he hoped Einstein would retain his home in Berlin and give at least one lecture there a year. This ambivalent attitude would not make it easy for Planck in the coming years. One commentator was particularly critical:

> Flanked by swastikas, he praised his Führer to start speeches and executed the Nazi salute. When asked to bar Jewish students from his classrooms and then fire Jewish staff, he did as he was told. Planck's chosen path was working within the new system, however deranged, and he was trying to make a positive difference, however small.[80]

As soon as Hitler was appointed Chancellor by President Hindenburg in January 1933, Stark enthusiastically wrote to an acquaintance, the newly appointed Minister of the Interior Wilhelm Frick, saying that he and Lenard would be very pleased to give advice. Lenard went one step further and offered his services to Hitler. He said the entire university system in German was rotten and he would be willing to find and evaluate German physicists to take up positions. Their influence became apparent at once and, by July, von Laue was complaining that it was necessary to go through Lenard and Stark to make appointments.

Einstein was travelling back to Germany by boat after visiting the California Institute of Technology when he heard in March 1933 that his apartment in Berlin had been raided several times by the Nazis as had his summer house in Caputh. He was advised by friends that it was dangerous to return to Germany. He arrived in Antwerp and sought the assistance from the King of Belgium whom he had got to know at the Solvay Conferences. Einstein renounced his German citizenship and resigned from the Prussian Academy of Sciences.

The threats against Einstein were horrendous and he was public enemy number one in Germany. One anti-Semitic publication authorised by Goebbels ran a photograph of Einstein with the caption "Bis Jetzt Ungehaengt" or "not yet hanged". An associate of Einstein, the German-Jewish philosopher Theodor Lessing, received similar publicity and was shot by a Nazi in Czechoslovakia. In the summer of 1933 Einstein secretly moved to a small house in the coastal town of Cromer in Norfolk, England where he had armed guards. A few months later he appeared at a full Albert Hall in London where he said: "Without the European freedom of mind there would have been no Shakespeare, no Goethe, no Newton, no Faraday, no Pasteur and no Lister."[81]

Einstein had a major influence on the British attempts to assist the many refugee scientists who left Germany in the 1930s. He had become a close friend of Schrödinger in the late 1920s as a colleague in Berlin. They often went sailing together and Schrödinger stayed at Einstein's vacation house in Caputh. Schrödinger always acknowledged that Einstein's realisation of the importance of de Broglie's ideas gave him the inspiration for the Schrödinger equation. They kept up a close correspondence after Einstein was forced to leave Germany.

On 12 August 1933 Schrödinger wrote to Einstein in affectionate terms:

> I needed to give you a sign of life after the long time since I last said goodbye to you in the beautiful Caputh. I hope and believe I do not need to fear that the love and warm friendship with which you set me in your thoughts has been clouded.[41(t)]

Schrödinger had held Einstein's Chair in Zurich and had, in a certain sense, followed him also to Berlin. As Einstein had, out of necessity, moved abroad Schrödinger was now very seriously thinking of doing the same.

The Prussian Academy of Sciences published a statement critical of Einstein. The minutes of the Academy of 11 May 1933, which had been written by Planck, stated:

> The overwhelming majority of physicists realised that Einstein's work could be compared with importance only with Kepler or Newton. Therefore it is deeply to be regretted that Einstein by his own political behaviour made his continuation in the Academy impossible.[82]

Von Laue called an extraordinary meeting of the Academy and insisted this statement must be withdrawn. However, he obtained only minimal support and his proposal was not accepted. Schrödinger was not involved in any of these Academy meetings.

On 23 March 1933 the Reichstag met in Berlin to vote on the enabling act that would give dictatorial power to Hitler. The Brownshirts and the SS arrested all the possible opponents they could find and surrounded the voting chamber to ensure the act was passed. Things were moving very quickly. On 6 April Hitler gave a speech to the professional association of German doctors when he said he would be quickly eliminating the excess of Jewish intellectuals from the cultural and intellectual life of Germany.

This was followed on 7 April with the Law for the Restoration of the Professional Civil Service. This forbade "non-Aryans" and "politically unreliables" from holding positions in the Civil Service. Shortly afterwards a "non-Aryan" was defined as someone with a parent or grandparent of Jewish blood or religion. This covered a large number of German scientists and many of the theoretical physicists already mentioned in this book. Initially, there were some exceptions such as those who had been involved in the First World War or were in post at the start of the war.

In cities throughout Germany there were torchlight processions of Nazi supporters who took the opportunity to loot shops run by Jewish people or beat up people on the street who objected. On 10 May students from the University of Berlin, who had set up "The Committee for Fighting the Un-German Spirit", burnt over 20,000 books and periodicals from the public and private libraries. Among the authors whose publications were burnt was Albert Einstein. The propaganda minister Joseph Goebbels delivered an address to the crowd, which was also broadcast on the radio, in which he referred to the books being burned as "intellectual filth" and "Jewish asphalt literati". There followed similar events in towns throughout Germany.

The naïve theoretical physicists in cities such as Berlin, Göttingen and Munich, who had been busily working on their scientific research, were not prepared for this onslaught. There is a story, attributed to Anny Schrödinger, that her husband tried to stop the looting of a Jewish store and was set upon by the stormtroopers, but was saved by a physicist who recognised him.[37]

What now followed was the mass exodus of many of the most brilliant scientists from Germany. In the coming Second World War, the leading scientists and mathematicians working in the UK and the USA played a crucial role in winning the war through many advances including the invention of radar, the cavity magnetron, the decoding of secret messages, and ultimately the nuclear bomb. Through the expulsion of many of its top scientists, Germany did much to lose the Second World War.

However, not all the physicists wanted to leave Germany. As has already been discussed, the role of Max Planck was ambiguous. As President of the Kaiser Wilhelm Society and Secretary of the Prussian

Academy of Sciences in Berlin he was in the most influential position. Following the startling announcement by the Jewish chemist Fritz Haber that he was leaving Germany, Planck requested and received an interview with Hitler himself in May 1933.[82] He told Hitler: "The forced immigration of Jews would kill German science and Jews could be good Germans." Hitler replied: "But we don't have anything against the Jews, only against communists." Planck felt that "this was the worst possible reply he could get because it took from him every basis for further negotiation".[82]

Even though he was aged 75 by this time, Planck was a firm German patriot and stayed on in his influential positions. He hoped that, through his influence, the disastrous policies of the Nazis on German science could be softened. In this he was unsuccessful.

The Second World War was to be defined ultimately through advances in Physics and Engineering whereas Chemistry was the dominant science of the First World War through munitions and poison gas. Together with Haber, Carl Bosch had developed a high-pressure catalytic method to fix nitrogen to yield ammonia and this revolutionised the production of fertilisers and chemicals. Both were to win the Nobel Prize in Chemistry for this work (Haber in 1918 and Bosch in 1931). Bosch was the first head of the chemical company I.G. Farben which had become one of the biggest corporations in the world and was hugely influential in Germany as an employer and supplier of vital chemicals.

Shortly after Hitler became Chancellor he had a meeting with Bosch to discuss more government investment in the industrial process to produce synthetic oil, a procedure that would be crucial for the German war effort. In a similar tone to Planck, Bosch informed Hitler that the removal of Jewish scientists from their research positions in universities and institutes would put physics and chemistry in Germany back by 100 years. Hitler at once replied: "Then we will work 100 years without physics and chemistry!"[83] After this, Hitler refused to meet with Bosch again and the Nazis took over I.G. Farben which became a major and infamous component of the German war machine.

Heisenberg played an even more dubious role than Planck. He was also highly patriotic and wrote to his former mentor Born to say that following Planck's discussion with Hitler it was clear that "only the very least are affected by the law, you and Franck certainly not, nor Courant — the political revolution could take place without any damage to Göttingen physics".[82] This naïve view could not be further from the truth. Indeed, Heisenberg went on to play a leading role in research on a possible nuclear bomb in Germany in the Second World War.

James Franck in Göttingen, who had won the Nobel Prize in 1925 with Gustav Hertz "for their discovery of the laws governing the impact of an electron upon an atom", suggested to Born and the mathematician Richard Courant that they all should resign their professorial positions at once even though they came under the exceptions to the new Civil Service Law. Franck had served in the First World War and had been awarded the Iron Cross for his work with Haber on poison gas. Initially, Born and Courant declined to resign but Franck went ahead with a courageous letter to the press that received considerable publicity.

Like several of the Jewish physicists, Franck first moved to Bohr's laboratory in Copenhagen on a temporary basis and, in due course, found a position in the USA at Johns Hopkins University, Baltimore. He did important work in the Manhattan Project on the chemistry of new radioactive isotopes. He finished his career at the University of Chicago where the James Franck Institute is still named after him. Franck was subsequently followed abroad by many other physicists.

Another physicist from Göttingen who took a very different path was Pascual Jordan. He had published important fundamental papers with Born and Heisenberg on the theory behind the matrix form of quantum mechanics. He joined the National Socialist party in 1933 and was even part of a stormtrooping unit. However, he did actively support Einstein in opposition to Lenard and Stark. During the Second World War he joined the Luftwaffe and worked at Peenemünde where the V2 rockets were built.

However, the great majority of physicists in Germany were opposed to the Nazi regime. All of those classified as "non-Aryans" had to find a way to leave Germany as a matter of urgency and a small number of others who were bitterly upset by the developments left as well. This latter group included Schrödinger.

Shortly after the Nazis came to power the Academic Assistance Council was set up in Britain by several notable people including Lord Rutherford who had been awarded the Nobel Prize in Chemistry for his experimental work on the structure of the atom. This Council over the next few years helped over 2,600 academic refugees to come to Britain. As many as 16 later won Nobel Prizes, 74 became Fellows of the Royal Society and 34 Fellows of the British Academy. They became a significant component of a thriving scientific and academic community in Britain that has continued to the present day. Furthermore, after being rescued by Britain, several of these scientific refugees moved on to the USA where they had a similar influence.[84]

Frederick Lindemann had been the Dr Lee's Professor of Experimental Physics at Oxford University since 1919. He had done important experimental work with Nernst in Berlin on specific heats at low temperatures that was known to Einstein and Schrödinger. He had also proposed a mechanism for unimolecular chemical reactions that still bears his name today. He was a regular attender at the Solvay Conferences and was well aware of the lead that Germany had taken in physics. He was a flamboyant character and was known throughout Oxford as "Prof". He was a wealthy man and had a Rolls Royce and a chauffeur. He was a supporter of the Academic Assistance Council and in 1936 was instrumental in its move to be called the Society for the Protection of Science and Learning.

Lindemann was also a close friend of Harry McGowan, the Chairman of Imperial Chemical Industries (ICI) through whom he got the promise of substantial funding for several temporary research fellowships for established scientists from Germany to come and work in the UK. Under McGowan, ICI had grown to be the leading industrial company in the UK and was competing with DuPont in the USA and I.G. Farben in Germany.

Accordingly, Lindemann's chauffer took him in his Rolls Royce to Germany in 1933 to see if he could encourage some of the most promising but demoralised physicists to come to Oxford. Lindemann arrived in Berlin on Easter 1933 and went to see his old research supervisor Nernst. There he met Professor Franz ("Francis") Simon and realised at once that here was an opportunity to build up a new research effort in low temperature physics at Oxford with Simon leading the team.

Francis Simon had worked previously with Nernst and was now leading an excellent group doing research on low temperature physics in Breslau. Kurt Mendelssohn was the cousin of Simon and was collaborating with him. He had already visited Lindemann's Clarendon laboratory in Oxford to install a helium liquefier apparatus. In addition, Nicholas Kurti, who came from Hungary, was also working with Simon as was Heinz London, the brother of Fritz London. All of these scientists came from Jewish families.

ICI had major interests in the properties of gases and Lindemann anticipated he could persuade them to support the work of this team led by Francis Simon. He then made the arrangements for them to move to Oxford. In due course, Simon, Mendelssohn, Heinz London and Kurti were all elected Fellows of the Royal Society. During the Second World War, Simon devised the diffusion method of separating uranium isotopes that produced the material used in some nuclear bombs. He was knighted and eventually took over Lindemann's Chair at Oxford in 1956 but sadly died just a month later.

Lindemann, however, did not stop with Simon's team. He had heard about the brilliant work of Fritz London who had nominally been Schrödinger's assistant. As a Jewish Privatdozent working in Berlin in 1933, London's prospects were bleak. Like several others he visited Bohr in Denmark and Ehrenfest in Leiden asking for advice. He wrote to the Academic Assistance Council in London and said he could get good references from top physicists such as von Laue, Schrödinger, Born and Franck and mentioned that Slater and Pauling had taken up his ideas in the USA. Through his brother Heinz, he also knew that Kurt Mendelssohn had been working in Oxford and asked him to put in a good word to Lindemann.[63]

Although he was not a theoretician, Lindemann realised the important role theoretical physics had played in driving forward new developments in physics in Germany. He consulted with Sommerfeld who said that Hans Bethe and Fritz London were the two outstanding younger theoreticians in Germany. Bethe had already had an offer from Manchester

University. Einstein had given his positive view on London when he had been a Visiting Fellow at Christ Church Oxford, a post which had been arranged by Lindemann.

Lindemann also appreciated London's more practical approach to theory and anticipated he could work on low temperature problems with Simon's experimental group that included his brother Heinz. This line of research turned out brilliantly and London subsequently developed the theory for both superfluidity and superconductivity.

During his stay in Berlin in Easter 1933 Lindemann was invited to meet with Schrödinger in his apartment at the 44 Cunostrasse. Wanting to impress Lindemann in an English way, Schrödinger offered him tea. Schrödinger explained he was very unhappy about what was happening in Germany. Lindemann mentioned he had made an offer to Fritz London to come to Oxford. Schrödinger then enquired if he had accepted and Lindemann said he was still considering the offer. Schrödinger said that he did not understand why London had not accepted the offer but if he did not want to go to Oxford then Schrödinger himself was interested.[1,37] Lindemann was taken aback and asked Schrödinger if he was serious and he said he was. Having spent vacations in England as a child and speaking perfect English, Schrödinger felt comfortable with this move.

Schrödinger was not a "non-Aryan" and was not going to be dismissed from his Professorship in Berlin. However, he was well aware by now that several of his close colleagues and associates were looking for posts overseas. In due course many of the leading scientists mentioned in this book including Bethe, Bloch, Born, Debye, Delbrück, Einstein, Franck, Frisch, Haber, Heitler, Hess, Infeld, London, Mark, Meitner, Peierls, Polanyi, Simon, Stern, Szilard, Teller, Weisskopf, Weyl and Wigner all became refugees. Together with Schrödinger, as many as 12 of these scientists had won or would win the Nobel Prize. German science was to be destroyed by the Nazis and it would never recover its world-leading position.

Schrödinger had been particularly concerned by the treatment of his close friend and colleague Einstein who was the first to be forced overseas. He had also been upset when he and Anny applied to visit her mother in

Austria. At that time the government in Austria was opposed to the Nazis and payment of a fee of 1,000 marks was required to cross the border from Germany. This was the last straw for Schrödinger and he was determined to move to England. Lindemann was a shrewd operator and he realised that someone of Schrödinger's standing, even if he was supported by ICI funding, would need the prestige of a Fellowship at one of the Oxford colleges. Lindemann's own college Christ Church had already provided a Visiting Fellowship for Einstein, which had had its complications, so he approached Professor George Gordon, the President of Magdalen College, Oxford.

CHAPTER TWO

To Oxford and the Nobel Prize

Oxford

The University of Oxford is the oldest English-speaking university in the world and there is evidence that teaching existed there in 1096.[1] Situated just 60 miles west of London the University has played a major role in the history of the United Kingdom. In 1933 there were 24 colleges associated with the University which also had departments in many academic areas. The colleges are self-governing and some have considerable endowments. They control their own membership and have their own internal governance structures, buildings and grounds. Each college elects its own head who chairs the Governing Body of Fellows, which agrees on the main decisions of the college. Until quite recently one of the college heads would also become Vice-Chancellor of the University of Oxford. Many of the colleges were founded originally by bishops or royalty.

Oxford, as the University is frequently known, was traditionally very strong in the arts and humanities, and particularly in subjects such as classics, ancient history and theology. To the present day, Oxford has provided through its students most of the UK Prime Ministers and a large number of prominent politicians. Many of the members of the establishment professions such as the Civil Service, the Law and the Church of England were also educated in Oxford. Up until the mid-19th century, the Fellows in many of the colleges had to be in holy orders and women were only first accepted to take degrees at Oxford in 1920.[1]

The traditional colleges were slow to include women as students and Fellows, and this only happened at Magdalen College in 1979.

Some great scientists such as Robert Hooke, Robert Boyle and Christopher Wren had attended the University of Oxford and were early Fellows of the Royal Society which was founded in 1660. Until recent times, however, Oxford did not have a significant reputation in scientific research. In the early part of the 20th century, the small number of science professors were heads of departments and were playing a leading role in building up their subjects.

By 1933, only two Oxford scientists had won the Nobel Prize. These were Frederick Soddy, who had worked with Rutherford and won the Chemistry Prize in 1921 for his work on isotopes, and Charles Sherrington, who was a Fellow of Magdalen College and won the Prize for Physiology or Medicine in 1932 for his discoveries on the nervous system. By contrast, the University of Cambridge had won 14 Nobel Prizes by this time and some of these were great discoveries such as the electron by J.J. Thomson and the structure of the atom by Rutherford.

In 1933, as is the case in the present day, tutorial teaching of undergraduates in Physics was conducted in the Oxford colleges while lectures and laboratory work were largely carried out in departmental buildings. Mathematics was taught in the colleges in the style of the humanities. In the early years of the 20th century, Chemistry had become a major subject for research in Oxford with generous funding from the thriving chemical industry. Medical sciences and some areas of biological sciences were also developing quickly.

However, research in Physics at Oxford was in the doldrums in 1933. John Townsend had been the Wykeham Professor of Physics since 1900. He had done important work on the conductance of gases but was not sympathetic to revolutionary advances such as quantum mechanics and relativity. Consequently he was not popular with the research students in Physics at Oxford in the 1930s. By this time, the research of the other Physics professor, Frederick Lindemann, had also stagnated. There was no Professor of Theoretical Physics and the limited research in this area was very largely conducted in colleges by mathematicians. Lindemann was well aware of this dire situation and that is why he went to Germany with the aim of recruiting top physicists to Oxford.

Magdalen College, Oxford was founded in 1458 by William Waynflete, the Bishop of Winchester. He provided a significant endowment for the College and built some fine buildings set in beautiful grounds. Not long after its foundation, the Magdalen Great Tower was built when Thomas Wolsey was Bursar and this has been one of the most visible features of Oxford to the present day. In the 19th century, the University established four Chairs associated with Magdalen and these were called the Waynflete Professorships after the founder. The subjects were Mathematics, Chemistry, Physiology and Philosophy. They put Magdalen at the forefront of science in Oxford.[2]

In 1933 Sir Charles Sherrington was the Waynflete Professor of Physiology and just one year before had been awarded his Nobel Prize. He was brought up in Ipswich, Suffolk. Sherrington had initiated a dramatic improvement in the reputation of Medical Sciences in Oxford and two of his research students at Magdalen, Howard Florey and John Eccles, went on to win the Nobel Prize for themselves. Sherrington had been President of the Royal Society, the leading position in science in Britain and its Dominions. He had also been knighted and awarded the Order of Merit by the King. In later years Sherrington and Eccles had a lively correspondence with Schrödinger on the theories of the mind. Sherrington and Schrödinger were foundation members of the Pontifical Academy of Sciences in 1936 — a body that was soon to provide a helpful refuge for Schrödinger.

Robert Robinson, the Waynflete Professor of Chemistry, was one of the most distinguished organic chemists in the world. He was born in Chesterfield, Derbyshire. He went on to win the Nobel Prize for Chemistry in 1947 for his research on the synthesis of organic molecules of biological importance such as penicillin and morphine. His own samples of those compounds he made still exist in the Chemistry Department at Oxford today. He would in due course be knighted, become President of the Royal Society and be appointed to the Order of Merit — a similar career track to Sherrington. Robinson was a shrewd and well-travelled chemist who had realised that, through the work of Pauling and others, Schrödinger's wave mechanics was starting to provide a theoretical framework for his subject in explaining and predicting the bonding and structures of organic molecules. He was a strong supporter of bringing Schrödinger to Oxford and assisted him in his subsequent career.

Arthur Dixon was the Waynflete Professor of Pure Mathematics at Magdalen. He had been elected a Fellow of the Royal Society in 1912. He had written a small number of papers on algebra and geometry and also on Bessel functions in the early 1930s. This should have given some academic overlap with Schrödinger.

Magdalen College has also been home to the Sherardian Chair of Botany since 1734. Arthur Tansley was the holder of this Professorship from 1927–37. He was a pioneer in the science of ecology and, through detailed studies including Wicken Fen near Cambridge, coined the term ecosystem. A Fellow of the Royal Society, in due course he was knighted for his contributions to ecology. In addition to botany, Tansley had also undertaken research on psychoanalysis and had even worked with Schrödinger's famous co-patriot Sigmund Freud in Vienna. There has been a distinguished line of holders of the Sherardian Chair including John Sibthorp, Humphry Sibthorp and Charles Daubeny. Cyril Darlington was appointed to this Chair in 1953 and his discoveries on chromosomes were to influence Schrödinger who presented his own thoughts on this topic when in Dublin in the 1940s.

With a few exceptions such as Thomas Wolsey, who became the Chancellor to King Henry VIII, Magdalen College had not produced many notables before the 20th century. In the 19th century, the brilliant playwright Oscar Wilde, who was an undergraduate in the College, was another exception. However, the situation was changing rapidly in the 20th century. In the Presidency of Sir Herbert Warren and then Professor George Gordon several well-known names associated with the College had become highly prominent. Edward, the Prince of Wales, was a student in the College from 1912–14 and went on to become King in 1936 but abdicated shortly afterwards and was not crowned. In the 1930s Magdalen men were also Editor of the *Times* (Geoffrey Dawson) and Archbishop of Canterbury (Cosmo Lang), and the college was at the heart of the establishment.

George Gordon had been President of Magdalen since 1928 and had been quietly working to build up a Fellowship of distinguished scholars in many subjects. He was born in Scotland and had been a Fellow at Magdalen from 1907–15. Like many of his generation, he fought in the First World War and was wounded in France. This affected his health for the rest of his life. He then became Professor of English Literature at Leeds University and, after that, Merton Professor of English Literature at Oxford from 1922–28.

In the early 1930s Magdalen College did not have a very strong reputation in Physics but, through its Waynflete Professorships of Physiology, Chemistry and Mathematics, and the Sherardian Chair of Botany, had played a leading role in Oxford in some of the other sciences. Lindemann was aware of this situation and thus considered Magdalen to be an ideal College to appoint Schrödinger to a Fellowship to enhance significantly the reputation for research in Physics in Oxford.

However, there was a complication. Schrödinger informed Professor Lindemann that he would like to bring an assistant to Oxford. This was not so unusual in Germany where Fritz London had moved with him from Zurich to Berlin. He had in mind Arthur March who was contemplating writing a book with Schrödinger on wave mechanics. Schrödinger had visited March on more than one occasion in Innsbruck and had been very much charmed by his wife Hilde.

Lindemann calculated that the ICI funding he had arranged for scholars from Germany could be used to help provide some of the significant salary that Schrödinger would require and might also stretch to supporting Arthur March. Lindemann was aware that Magdalen had been quite advanced in funding competitive Research Fellowships for able young scholars (then and now called "Fellowships by Examination") so he asked President Gordon if Magdalen could provide a Fellowship and some additional funds to support Schrödinger. Finding these resources over the next few years was going to be a significant problem for Gordon.

The colleges in Oxford have a democratic structure with all major decisions being made by the Governing Body of Fellows at a College Meeting. At Magdalen College, there is always a Meeting on, or close to, St Mary Magdalen Day of 22 July. In 1933 this was on a Saturday so the

Fellows met on the morning of the previous day. President Gordon then wrote on 21 July 1933 to Lindemann:

> The Schrödinger matter went through quite well this morning, that is to say the question of his election has been set down for the College Meeting on October 3rd, and will, I hope, be favourable. The proposal is that he be elected to a supernumerary non-stipendiary Fellowship, and that we make him annually an appropriate grant. I named as the figures £200 or £250. His Fellowship, if he is elected, will run for 5 years, and of course he can be re-elected. He would be in the same class of Fellowship as Benecke, our Senior Fellow, and would be a member of the Governing Body. I very much hope it goes through. I urged on the Fellows present the advisability of treating all our proceedings today as confidential, but you must know how difficult it is to secure that: men tell their wives, *etc.*[3]

George Gordon worked very hard to assist Schrödinger over the next five years. As President of Magdalen he wanted to do the best he could for his most distinguished new appointment. The role of Gordon is not as well known as the more flamboyant characters who assisted Schrödinger after he left Germany such as Lindemann and subsequently Éamon de Valera in Ireland. President Gordon wrote dozens of letters to Schrödinger, Lindemann, the Rockefeller Foundation, ICI, the Academic Assistance Council, and many academics in Magdalen College, the University of Oxford and elsewhere trying patiently to assist Schrödinger. He did this every year from 1933 until 1938. His letters, some in typescript and others more informal and handwritten, largely tell the non-scientific part of the Schrödinger story during this period. Many of these letters are reproduced and discussed in this book.

Tyrol

In the meantime Max Born, who also recurs many times in the Schrödinger story, had made a firm decision to leave Göttingen and Germany. He was from a Jewish family but had hesitated after James Franck had decided to emigrate and had been urged by Heisenberg not to leave. Having served in a cavalry regiment in the First World War he was not forced immediately to resign his Professorship. However, he was becoming increasing concerned about stories in the newspapers of intellectuals in Germany being arrested for possessing banned books. He was receiving several offers of positions elsewhere and arranged to have a vacation at Selva in the South Tyrol, away from all the turmoil in Germany in May 1933, to consider his future. By a coincidence, at the start of the trip he witnessed the burning of the books in Göttingen and this made him even more determined to leave.[4]

The South Tyrol then became the centre in the summer of 1933 for an extraordinary set of meetings of many of the refugee physicists and their families, and also several other important scientists. Born had received many offers of appointments from places including Columbus USA, Paris, Belgrade, Brussels, Istanbul and Cambridge. Einstein had even suggested an appointment in Jerusalem. Born wrote back to thank him for the idea but the political uncertainty in Palestine would make it difficult for him, particularly with a young family.[4]

Born had worked before at Cambridge as a student with J.J. Thomson and, while in the South Tyrol, made a visit to nearby Zurich to discuss the Cambridge position with the British physicist Patrick

Max Born and his son Gustav in the 1920s.

Blackett.[5] Blackett had started as a student at Magdalene College in Cambridge and had worked with Rutherford. He had invented a cloud chamber for detecting charged particles and this would win him the Nobel Prize in 1948. He was a major player in British science and eventually would become President of the Royal Society.

The proposed details of Born's appointment as a lecturer for three years in the Department of Mathematics at Cambridge were explained to him by Blackett, as was an attractive Fellowship at St John's College where Paul Dirac was a Fellow. Born's wife Hedi at once travelled to Cambridge, had tea with Lord and Lady Rutherford, organised a house to rent at 246 Hills Road and arranged for their son Gustav to be accepted at the Perse School. This was remarkable efficiency and things were all set for the Borns to move to Cambridge.

Meanwhile in nearby north Italy, Schrödinger and Anny had arrived, as had Wolfgang Pauli and Hermann Weyl.[6] Max Planck was also staying in the vicinity and he met with Born and Schrödinger in a final attempt to try to persuade them to stay in Germany. This was, however, to no avail as they had already been making their arrangements to move to England.

Also staying in Bressanone in the Tyrol were Arthur and Hilde March. Schrödinger went for a long bicycle ride with Hilde.[7] Almost exactly nine months after this, she had a baby girl Ruth in Oxford. Some 75 years later, over breakfast in the President's Lodgings at Magdalen College in Oxford with Ruth and the author of this book, Gustav Born, the son of Max Born, recalled numerous details of this "last waltz" gathering of the great theoretical physicists in the Tyrol.

In the meantime, Lindemann had written to Schrödinger on 17 August 1933 from Minard Castle, Argyll where he was staying:

> Just a line in great haste to thank you for your letter which was forwarded to me here and to say how pleased I am to learn that all is settled, though for the moment of course nothing is announced.
>
> I am still hoping to come abroad at the end of this month. If I can manage it I should be delighted to come and see you at Bressanone, Provincia Babrano on the way… I wonder whether you could send me a line to Oxford telling me how long you will be there… If, however, you and Born are not going to be there after 15th September I will make a special effort to visit you on my way out.[3]

Schrödinger then replied in English on 23 August 1933 from Solda, Alto Adige:

> You are sure to find me at Solda, if you choose to, on your way to Northern Italy. I should greatly appreciate to know of your visit three or four days ahead in order to advertise it to the March's and to my wife, who will at that time be somewhere about the Alto-Adige and who would like to join us, if possible.
>
> For the latter reason the middle of September at Malcesine would suit us all best since (1) my wife, who is now at Selva in Gardena with the Borns, wrote me that she would like to stay there till the middle of September and (2) Arthur March, who spends his summer in Bressanone, wrote me that he intends to make a trip to Innsbruck and Vienna in the first week of September in order to verhandeln about his leave… I do hope that we'll manage this meeting and then I'm sure it will be enjoyful to all of us.[3]

Schrödinger, in letters and lectures in English, often liked to use a German word to add some colour and character and here "verhandeln" means "negotiate".

So Lindemann finally joined the motley crew of theoretical physicists in the Tyrol in his Rolls Royce. Perhaps encouraged by his success with Schrödinger, he also made an attractive offer for Born to come to Oxford. Born, however, was never impressed by Lindemann who he considered to be "a mediocre physicist, a remarkable personality and a remarkable snob"[5] and he explained he was already committed to Cambridge. In later years, Born was also highly critical of Lindemann's role in the Second World War as Winston Churchill's scientific advisor. Born met Francis Simon in Zurich and, despite his reservations on Lindemann, convinced him that it would be sensible to accept the offer from Oxford, given the worsening political situation back in Germany. As refugee scientists in Britain, Born and Simon were to have a very frequent correspondence.

Lindemann finally met with Schrödinger in September at Lake Garda. He communicated the positive details of the letter from George Gordon arranging a Fellowship at Magdalen College with every prospect of tenure together with a research grant.[7] He was also able to give more details of the financial support from ICI which he said would be equivalent to a Professorship. There were no formal duties apart from research on which a report would need to be written for ICI every year and it would be necessary to live in Oxford. Everything was falling into place for the move of the Schrödingers to the UK.

Schrödinger had kept up a regular, informal and personal correspondence with Paul Ehrenfest in Leiden. Ehrenfest was born in Russia but moved to Austria after the Russian Revolution. He had studied under Boltzmann so he had a special relationship with Schrödinger. He had done important work relating Boltzmann's statistical mechanics to the new wave mechanics and also on the time dependence of Schrödinger's wave function. Leiden was a good place for Schrödinger to visit on journeys back and forth from central Europe to Britain, Belgium or France. He had done this in March 1928 on his way back from England when he and Anny put their signatures on the wall outside Ehrenfest's guest room. On 26 September 1933, Schrödinger wrote from Malcesine, Lake Garda to Ehrenfest to say:

> I will be in Brussels at the Solvay Conference from 22–29 October and I would like to visit you in Leiden beforehand. I am looking for quarters in Oxford. I won't go back to Berlin, I have given up my position there. I was offered and accepted a preliminary hideout in Oxford. Basically verbally — or you can guess it.[8(t)]

Very sadly, this letter was not read by Ehrenfest. He was suffering badly from depression and had taken his own life by gunshot just the day before Schrödinger's letter was written. His son had Down's syndrome and Ehrenfest had also shot him just before shooting himself. Schrödinger first heard the tragic news from Léon Brillouin who wrote in French on 2 October 1933 to say:

> I had learned a few days ago, with deep emotion, the suicide of Ehrenfest. It gave me infinite pain. I cannot guess the real causes of this act. I had had such a previous affection and I cannot get used to the idea of his disappearance.[8]

The death of Ehrenfest was felt deeply throughout the community of theoretical physicists. He was described by Einstein as "the best teacher I have ever known".[9] Paul Dirac was also very much disturbed by the tragedy. Shortly before the incident he had several scientific discussions with Ehrenfest at a conference in Bohr's institute in Copenhagen. On leaving the meeting Ehrenfest said to him, "What you have said, coming from a young man like you means very much to me because maybe a man such as I feels he has no force to live." Not normally an emotional man, Dirac wrote a four-page letter to Bohr describing his last discussions with Ehrenfest and blaming himself for the tragic event.[10]

Magdalen

Meanwhile, back in Oxford, George Gordon was getting nervous about the election of Schrödinger to a Fellowship. He received a visit about this from Patrick Johnson, a young Physics Fellow at Magdalen. On 1 October 1933, Gordon wrote to Lindemann:

> Pat Johnson came to see me tonight and asked me if I knew (1) that it was being said in Cambridge that Schrödinger was coming to Magdalen, and (2) that Schrödinger's wife had written recently to a lady in Oxford — the wife of a science don I think in St John's — asking her to help them in finding a house, because her husband had been offered and had accepted a post at Oxford — at any rate how it reached Johnson…
>
> I only hope that this gossip does not reach our Fellows before Tuesday morning for we need a 3/4 majority. It would be easy to misinterpret these reports as meaning (a) that our decision has been anticipated and (b) that S. is coming to Oxford in any case and has, therefore, already been provided for and that anything we may do for him will be in the nature of a luxury… What I want is an answer (as specific as you can make it) to the suggestion that someone may report that S. is assured for Oxford whether we elect him or not.[3]

Even nowadays, getting a positive vote for a Fellowship election with a three-quarter majority can be quite a challenge for a Magdalen President, as the writer of this book knows all too well. Gordon needed to prepare very carefully for the College Meeting with the Fellows of Magdalen on 3 October 1933. A very persuasive paper making the case had to be sent out seven days in advance and it would be read very carefully by all the Fellows present. It stated:

> Question of election to a Supernumerary Non-Stipendiary Fellowship under Statute IV, Clause 47. This proposal has arisen from the movement among the English Universities to help disposed or threatened German scholars by providing them for a time with such means and conditions of life as may enable them to carry on their work. Letters from the Hebdomadal Council and from the Academic Assistance Council (of which Professor Robinson is a representative member) have been read at recent College Meetings.
>
> Shortly before the College Meeting of July 21st information was received that an attempt was being made to provide at Oxford for Professor Erwin Schrödinger of Berlin, who was then living in daily fear of prison or the concentration camp, and found it impossible in these circumstances either to teach or to research. It is understood that laboratory facilities can be assured for Professor Schrödinger and a certain amount of money has been guaranteed for his living expenses here. But there was still the problem of giving so distinguished a man, if he should come to Oxford, some appropriate academic status. The money also fell short of what was required by some £200 or £250 per annum. It was in these circumstances that the President was approached and asked if he thought it possible that the College would be willing to consider Professor Schrödinger for election to a Fellowship...
>
> It was generally agreed that the type of Fellowship most appropriate to such a case was a Supernumerary Non-Stipendiary Fellowship, for this among other reasons, that it would leave the College free, if he were elected, to deal with the question of financial provision from year to year and by way of Grant.[11]

As we have seen already, Schrödinger was not, at least at that time, "living in daily fear of prison or the concentration camp" but President Gordon wanted to impress on the Fellows the urgency of the appointment. It is clear also from this statement that the arrangements which were being made to provide for Schrödinger and for his Fellowship were constructed with no permanent financial provision in place for an extended period. This would cause major uncertainties for Schrödinger in his three short years in Oxford. Statutory Professorships at the University of Oxford were given full tenure but there were no such suitable positions available for Schrödinger at that time so he was provided with an ad hoc appointment of a temporary nature.

In his papers for the key College Meeting, Gordon also included a brief Curriculum Vitae of Schrödinger. This stated:

> Dr Erwin Schrödinger, Professor of Theoretical Physics at the Friedrich Wilhelm University, Berlin since 1927. Born Vienna 1887. Not a Jew.
>
> Father Rudolf Schrödinger, of Vienna, a botanist, Mother, English. Educated in Vienna, Privatdozent, Vienna 1914. Professor at Stuttgart 1920; Breslau 1921; Member of Academies of Science in Berlin, Vienna, Leningrad, and Dublin. Fellow of the American Physical Society. Mateucci Medalist. Nobel Prizeman.
>
> Has published numerous scientific papers in physical journals and in the Sitzungsberichte of the Academies of Science of Vienna and Berlin, has collected papers on wave mechanics (English translation from the 2nd German edition 1928). He is a theoretical physicist and has dealt much with the philosophical side of Mathematical Physics, especially with the Philosophy of Causality. Married, has lectured in England and speaks English perfectly.[11]

There are some errors in this statement. It was not Schrödinger's mother who was English but his grandmother. Also, at least on 3 October 1933, Schrödinger was not a Nobel Prizeman. The paper also states explicitly that Schrödinger is "Not a Jew". This was not an anti-Semitic statement but was probably included to make clear that Schrödinger was unusual as most of the scientific refugees who were forced to leave Germany at that time were Jewish.

Magdalen College was certainly not anti-Semitic. It gave scholarships to several Jewish students who had become refugees from Germany. A good example is Heinz Koeppler. He was born in Germany in 1912 and was brought up in a Jewish family in Berlin. He came to Magdalen as a student in October 1933 to study history and was made a Demy (the

elected a Fellow of the British Academy. A different, more complicated, character was Cesare Foligno. He was Serena Professor of Italian and an expert on Dante. An Italian national he was an active supporter of Mussolini and had given talks to groups of Italian fascist supporters based in Britain.[12] He attended Magdalen College Meetings right up to the announcement of war on Britain by Italy on 10 June 1940.[11] He then very quickly escaped to Italy and spent the rest of the war in Naples. After the war, he translated into English Curzio Malaparte's extraordinary book *Kaputt* which vividly portrayed the excesses of Nazi leaders.[13] Foligno never returned to Magdalen.

As soon as he had been informed by the delighted President Gordon of the election, Lindemann wrote in English to Schrödinger on 3 October 1933:

> Cher Ami. I am glad to tell you that you were today officially elected a Fellow of Magdalen. The bells have been ringing and all the proper steps have been taken to indicate their pleasure. Owing to one or two formalities it will not be published in the *Times* for a few days, but at least the matter is settled. You will no doubt receive an official letter from the President in a few days and I take it you will reply in the ordinary way. There is no need to bother about anything he might say about the financial aspect. I will explain all this when we meet. In your reply it would be best not to refer to it at all I think.
>
> You need not worry in the slightest about the permit. Simon, Kurti, Mendelssohn and London are all here and have had no difficulties whatever. I am also glad to be able to tell you there should be no trouble bringing your belongings to England… I am making enquiries about your car and will let Mrs Schrödinger know when she comes here.[3]

The day after Schrödinger's election to a Fellowship at Magdalen College, Gordon wrote to Professor C.S. Gibson, the Secretary of the Academic Assistance Council which was based at the Royal Society at Burlington House in London. Gordon said on 4 October 1933:

> In your letter to me of 5 July about academic assistance to distressed German scholars you asked that you should be informed if my College

> was able to take any action on the matter. I am glad to be able to tell you that, as the *Times* announced this morning, we yesterday elected to a Supernumerary Fellowship Dr. Erwin Schrödinger who has recently resigned his Professorship of Theoretical Physics at the Friedrich-Wilhelm Universität. We expect him in Oxford towards the end of the month.[14]

However, Gordon was already starting to have concerns about the financial arrangements for Schrödinger, which would become even more significant in the next three years despite Lindemann's verbal assurances. On 6 October 1933 Gordon wrote to Lindemann to say:

> Is it legitimate for me to know what sum, and from what sources, you have procured for Schrödinger? And if that transaction is now complete, except for what Magdalen can do? I should like to know before I meet our Grants Committee on October 11, even if you label certain parts of the information as confidential and for myself alone.
>
> I have written to Schrödinger conveying the good will of the College and asking him when he may be expected in Oxford. With luck I may be able to have him matriculated on October 21. The question of a degree for him can wait a little? Or what do you think?[3]

It should be noted that to be a member of the University of Oxford matriculation is required and to be a senior member a degree of M.A. is needed. Nearly all the academics at that time already had the appropriate degree (and Oxford even allowed a degree from Cambridge to count but no other University) but Schrödinger's position as a scientist from outside the UK was unusual and needed some attention.

On 7 October 1933 the *Times* also reported in its University News on German Scholars at Oxford:

> The election to a Supernumerary Fellowship at Magdalen College of Dr. Schrödinger, the pioneer in wave mechanics, is to be followed by further appointments to assist Jewish Scholars exiled from Germany. Balliol College has agreed to give hospitality to Professor I. Scheftelowitz, late of the University of Cologne, the Sanskrit scholar. All Souls is to receive Professor Cassirer the philosopher.[15]

This report seems to imply somewhat ambiguously that Schrödinger is a Jewish exile. Of the other names mentioned, sadly Scheftelowitz died just a few months later while Cassirer stayed one year and then went to Sweden where he corresponded with Schrödinger on philosophical aspects of wave mechanics.

Before Schrödinger could join the scholars at Magdalen College he attended his third Solvay Conference in Brussels held from October 22–29, 1933. This was the first time he was placed in the front row of the Solvay photograph together with Niels Bohr, Marie Curie, Lord Rutherford, Paul Langevin and Louis De Broglie. Heisenberg and Dirac attended, but sadly no Einstein. There were few German scientists as so many were being displaced at that time. New particles and isotopes had recently been discovered, including the neutron, the positron and deuterium, and this was the main topic of the meeting. The conference set the scene for many more dramatic discoveries in nuclear physics in the following decade which would lead eventually to nuclear energy and the nuclear bomb.

Schrödinger was essentially an interested observer at this Solvay Conference and had other things on his mind. He wrote in English from the Club de la Fondation Universitaire in Brussels on 22 October 1933 to thank Lindemann for his letters and congratulations on the Magdalen Fellowship:

> I feel very guilty in not answering your kind letters, the first announcing my election in Magdalen College, the second outlining the two letters for me and my wife which will make the entrance easier.
>
> By this time you will have received my wife's letter from Paris telling you that she finally decided not to go to Oxford before me, since the time had run short and we did not think her travel to be of so much use any more as to justify the expense. We shall hurry to Oxford as soon as this Conseil is over — by hurrying I mean that we shall come as quick as we can considering the difficulties which will originate from the "left-hand" in England and which render it almost indispensable to avoid passing London. Paris, although right-handed, gave enough difficulties to my wife, but she was very plucky (I don't dare to touch the volant in these places, not even in Bruxelles).

I should think that it will be possible to arrive in Oxford either the 31 Oct or the 1 Nov. Once more, please excuse my not answering to you for so long a time — of course I wrote to President Gordon immediately.[3]

The newspaper *Zeitung Berlin* was quick to report Schrödinger's exit from Germany. On 24 October 1933, under the heading "Loss to German Science" it stated (translated from German to English):

The German scholarly world is threatened with a grave loss. Prof. Erwin Schrödinger, the successor to Planck in the Chair of Theoretical Physics at the University of Berlin, has been appointed to the University of Oxford. Schrödinger had announced lectures in Berlin for the winter but the English University did not impose any teaching obligation on him, apparently not without his consent. It must be expected that Schrödinger, a significant scholar who has worked in Berlin since 1927, has left Germany. It is also regretted that Prof. Hermann Weyl, the Professor of Mathematics at the University of Göttingen, has accepted an appointment at the American Princeton University.[16]

In the meantime there had been more, somewhat frantic, letters between Gordon and Lindemann on the financial aspects of Schrödinger's new position in Oxford. This is illustrated in the letter Gordon wrote by hand to Lindemann on 24 October 1933 with the usual secretive comments on the financial arrangements for Schrödinger's appointment:

I read the letter from ICI to our Grants Committee this afternoon where they have agreed in the first instance that the College make a grant to S. of £150 for 1934. I was also asked to follow a suggestion of the Academic Assistance Council…to write to the Rockefeller Foundation to see whether they might be prepared to supplement this sum. It was further minuted that if the Rockefeller people return a negative it is then recommended that the College grant be automatically altered from £150 to £200. It must be said that my Committee was impressed by the hole which say £200 a year to Schrödinger is going to make on our annual grants budget. It may mean we must cancel a number of subscriptions such as the British School of Rome, Athens and Palestine. At

> a pinch they are prepared to do this I think in the name of Schrödinger but they naturally don't like it.
>
> Between ourselves, if Schrödinger is as brilliantly agreeable as I know him to be there will probably be a move in a year hence to transfer him into another category of Fellowship which would ensure him a superior fixed stipend for a period of years. But that is pure guesswork of mine and depends on many things: not to be talked of or commented on just now at any rate.[3]

The writer of this book can confirm that subscriptions to the British Schools of Rome and Athens are still kept up by Magdalen College today, although on occasion there is some discussion on this topic at the Fellowship Committee.

Nobel

Both Gordon and Lindemann had assumed erroneously that Schrödinger already had won the Nobel Prize before he arrived in Oxford, such was his reputation in the academic world. Ever since the creation of these prizes in Physics, Chemistry, Physiology or Medicine, Literature and Peace by Alfred Nobel in his will of 1896 they have captured the imagination of the scientific community and the public around the world.

In Physics the Nobel Prize is given to the person who made "the most important discovery or invention" in the field. Being awarded by the Royal Swedish Academy of Sciences and presented in a lavish ceremony by the King of Sweden, the Nobel Prizes have a special independence and kudos. A Nobel Prize cannot be awarded to more than three individuals and this has often caused complications when several scientists have been involved in a breakthrough. This was the case in quantum mechanics.

At its foundation, the financial award of the Nobel Prize was significant but it is the prestige that then, and now, is its feature. It is interesting to note that in modern times several new prizes have been introduced in particular scientific fields, such as the Breakthrough and the Kavli Prizes, which provide a financial award more significant than that for the Nobel Prize. These Prizes have become important but they still do not rank in prestige with the Nobel Prize. Indeed, once a scientist has won the Nobel Prize it puts them above other scientists in the public eye and they are often then asked for their learned opinion on topics well away from their own area of expertise. This is something that Schrödinger enjoyed but he found later it could sometimes have negative implications.

Schrödinger would have been well aware that many of the physicists he had interacted with had already won the Nobel Prize in Physics. This includes some remarkable names mentioned in this book including Lorentz, the Braggs and Curies, Lenard, Thomson, Wien, von Laue, Planck, Stark, Einstein, Bohr, Millikan, Franck, Hertz, Compton, Richardson, de Broglie and Raman. However, up until 1933 there had been no prize awarded for the new quantum mechanics.

The most influential theoretical physicist on the Nobel Committee for the Physics Prize was Carl Oseen who was a specialist in hydrodynamics, a classical area of physics. It took him a long time to come round to supporting a Prize for quantum mechanics which he had first considered in 1927. Indeed the Nobel Committee for Physics failed to agree on the award of a Prize in both 1931 and 1932. Due to the large number of nominations which had been accumulated, it was, however, very much expected that the prize for 1933 would go to quantum mechanics and this would be announced early in November of that year. Therefore, the tension was rising for Schrödinger and his colleagues as he prepared to arrive in Oxford.

In the preliminaries before the award of the Nobel Prize the Chair of the award committee writes to previous prize winners asking for their opinion on who is most deserving of the award. In addition, they write to other notable scientists such as presidents of academies or heads of major departments in the subject of the award. The committee always looks to see who wrote the very first papers that led to a "most important discovery or invention", in the words of Nobel. The responsibility on the Nobel Committee of Swedish physicists was huge and few members would have detailed knowledge of a particular field that was nominated. Letters from Einstein had a significant influence on the Nobel Prize Committee. In a letter to the Committee of 25 September 1928, Einstein had mentioned Schrödinger as deserving of the Prize although he considered de Broglie to have precedence. De Broglie was awarded the Prize in 1929. By 1933 Schrödinger had received as many as 41 nominations for the Prize including from Einstein, Planck, Bohr, and de Broglie. Einstein's letter to the Nobel Committee dated 30 September 1931 was written in German from his summer house in Caputh and was of particular significance:

> I nominate the founders of wave or quantum mechanics — E. Schrödinger of Berlin and W. Heisenberg of Leipzig. In my opinion, this theory contains without doubt a piece of the ultimate truth. The achievements of both men are independent of each other and so significant that it would not be appropriate to divide a Nobel Prize between them. The question of who should get the prize first is hard to answer. Personally, I assess Schrödinger's achievement as the greater one, since I have the impression that the concepts created by him will carry further than those of Heisenberg. This, however, is only my opinion, which may be wrong. On the other hand, the first important publication by Heisenberg precedes the one by Schrödinger. If I had to decide, I would give the prize first to Schrödinger.[17]

Einstein was to emphasise Schrödinger again in a letter to the Committee on 29 September 1932. However, although wave mechanics was a highly elegant theory which explained the spectrum of the hydrogen atom perfectly, the Nobel Committee was hesitating as a new discovery or invention arising from the theory was needed under the strict interpretation of Nobel's will. Heisenberg had received 21 nominations by 1933 and Niels Bohr had placed him first preference with Schrödinger second. Moreover, Heisenberg had used his theory to predict in 1927 that the ortho form of the hydrogen molecule (with parallel nuclear spins) would be three times more abundant than the para form (with opposite nuclear spins).[18] This was experimentally verified in 1929 by Harteck and Bonhoeffer and provided clear evidence of a new discovery arising from a quantum mechanical prediction.[19]

Also, by 1933 many predictions of the spectra for diatomic molecules had been made by Mulliken, Hund, Lennard-Jones and others using a molecular orbital theory based on Schrödinger's wave mechanics, and several of these predictions had been verified experimentally. As early as 1929, a Discussion of the Faraday Society in London had been devoted to this topic.[20] Therefore, the Nobel Committee was anticipating giving the prize for 1932, retrospectively, to Heisenberg and that for 1933 to Schrödinger. However, while the Nobel Committee was still deliberating in the spring of 1933, a startling development was made in an observation using a cloud chamber by Carl Anderson at Caltech of the positron, a particle which is essentially identical to the electron but opposite in charge.[21] This new particle had been predicted previously by Paul Dirac with his own relativistic form of quantum mechanics.[22] Just three years later Anderson himself would receive the Nobel Prize in Physics for this experimental work.

Dirac had only three nominations for the Nobel Prize, although one was from the influential William H. Bragg and that was enough. As Heisenberg's first paper[23] in 1925 had preceded Schrödinger's, who had himself shown the two theories to be equivalent, the Nobel Committee decided to give the 1932 Prize retrospectively to Heisenberg while splitting the prize for 1933 equally between Schrödinger and Dirac.

There was also the strong claim of Max Born who was the supervisor of Heisenberg for his Habilitation. Born had demonstrated how Heisenberg's theory could be interpreted using the properties of matrices and had also shown mathematically how commuting coordinates and momenta in the new theory link to the classical theory. Another strong candidate was Wolfgang Pauli. Like Schrödinger, he had been born in Austria. He had done his PhD with Sommerfeld on applying the old quantum mechanics to the hydrogen molecular ion. He had discovered the exclusion principle which states that no two electrons can have the same quantum numbers, or in mathematical terms, a wave function must change sign when the coordinates of any two electrons are interchanged. The exclusion principle was crucial in the successful extension of Schrödinger's wave mechanics to atoms and molecules with more than one electron which, by 1933, had been applied by several researchers. However, Einstein, who was the most influential nominator, had not mentioned Born or Pauli in his letter to the Awards Committee. Pauli was awarded the Prize in 1945, after support this time from Einstein, and

Born also eventually became a Nobel Laureate in 1954 for his statistical interpretation of Schrödinger's wave function.

The unlucky pioneer in quantum mechanics was Arnold Sommerfeld. There were several fundamental discoveries to his name including magnetic quantum numbers and the fine structure constant. He also did some of the earliest quantum mechanical predictions on the electronic structure of metals and solids that ultimately led to modern electronic devices. He was a research supervisor *par excellence* and seven of his own students or postdoctoral assistants went on to win the Nobel Prize including Heisenberg and Pauli. He was nominated a record 82 times but was never awarded the Nobel Prize.[17]

The Nobel Awards Committee for Physics made its final recommendation on 23 September 1933 and rumours on the decision then started to spread. The recommendation from the committee needed to be approved at the meeting of the Royal Swedish Academy of Sciences of 9 November before any official announcement could be made.

Just before this date, on 4 November 1933, the Schrödingers had arrived in Oxford having driven from Brussels. They stayed at the undistinguished Isis hotel on Iffley Road which was close to Magdalen College. President Gordon met with Schrödinger for the first time on 7 November. On the following day Gordon wrote to Charles Cruttwell, the Principal of Hertford College. Cruttwell was a member of the University Hebdomadal Council and was a delegate to Oxford University Press. He was mercilessly parodied by his former pupil Evelyn Waugh in several celebrated novels. He had enquired about Schrödinger's position at Magdalen. On 8 November 1933, Gordon wrote:

> Thank you for your note about Schrödinger. It may not be known to the *Times* office that his name occurs in Who's Who from which the main facts of his life can be derived. I have also a descriptive note here, which I could make available, as the work has chiefly secured him his reputation. He has only just arrived in Oxford, and I saw him in fact for the first time yesterday morning. He had heard nothing of a Nobel award and is very anxious that there should be nothing in the press about it unless it is absolutely assured. He tells me that he has been more than once spoken of it in the Berlin press as having received or being about to receive

a Nobel Prize, and he would not like this to occur again in England. The moment, however, that the *Times* can tell you the thing is certain I will pass through this note, which would be useful as an addition to the Who's Who particulars.[11]

An implication of this letter was that the rumour of an upcoming Nobel Prize for Schrödinger had by now reached several people in Oxford including Gordon. Furthermore, this letter implies that Gordon had prepared a document already that could be used in a press release once the official announcement had been made.

Presidents of Magdalen College spend a lot of time fundraising and Gordon was still working to find an appropriate financial arrangement for Schrödinger. The Rockefeller Foundation had used its funds to rebuild science in Germany after the First World War and was now making small grants to displaced scholars. Accordingly, Gordon wrote on 8 November to Professor Lauder Jones at the Rockefeller Foundation in Paris saying:

I am very glad to know that you feel able to submit a recommendation to your Committee in Paris supporting an appropriation of £100 for 1934 towards Dr Schrödinger's research expenses. There are one or two points on which I am afraid I have not made myself sufficiently clear in my former letters. The name of Dr Schrödinger was first proposed to us for election to a Fellowship after the sum of £400 had been ensured from outside sources. We were asked if we could do something for him academically (in other words, give him the standing of a Fellow), and also supplement the £400 so far as our resources permitted and our Statutes allow.

The Fellowship to which we have elected Dr Schrödinger is for a period of 5 years, no longer period being allowed under the appropriate Statutes. I shall look forward to hearing from you after the meeting of your Committee this month. The guarantee of another £100 would be most welcome to me personally, and would give Dr Schrödinger that feeling of modest financial security which is so necessary to the continued prosecution of his remarkable work.[11]

On the next day, 9 November 1933, as is also described on the first page of Chapter 1 of this book, Schrödinger came to Magdalen College and was formally admitted as a Fellow by President Gordon with the other Fellows present. The President and Fellows then all processed to the College Hall for a celebratory dinner in Schrödinger's honour.

After the dinner Schrödinger was weighed on the college scales in the Senior Common Room. He came out as 10 stone 9 lbs (68 kg). This is a quaint tradition at Magdalen College on special occasions or when the Fellows are feeling especially happy, perhaps after some fine wine at dinner from the voluminous College cellar. It is a tradition that still occurs today and the records of the weights are kept in a special book. There are four records in the book of Schrödinger being weighed in this way — the other three are in 1934, 1938 and 1948 and his weight hardly changed over this 15-year period.

Other notables to have their weight recorded in this book include many of the Magdalen Fellows mentioned earlier in this Chapter and also T.E. Lawrence of Arabia in 1912 and 1914. He was given a grant as a Magdalen Senior Demy for archaeology research in Syria and Jordan, and he then became famously involved in other highly significant activities in that region. Other signatures include the writer J.B. Priestley (1930), the Nobel Prize winner and penicillin pioneer Howard Florey (1935), the future Prime Minister of Great Britain Harold Wilson (1938), J.R.R. Tolkien (1953), Picasso's lover Françoise Gilot (1960), Magdalen's second Nobel Prize winner in Physics Tony Leggett (1963) and the first British female Nobel Prize winner Dorothy Hodgkin (1968).

In the meantime, on the same day of 9 November 1933, the Royal Swedish Academy of Sciences had met and decided formally to award the 1932 Nobel Prize in Physics to Werner Heisenberg "for the creation of quantum mechanics, the application of which has, *inter alia*, led to the discovery of the allotropic forms of hydrogen" and the 1933 Prize jointly to Erwin Schrödinger and Paul Dirac for "the discovery of new productive forms of atomic theory". An attempt to communicate the news to the prize winners was not made until the evening.

On 13 January 1943, when Schrödinger was in Dublin, he wrote in English to Max Born on how he heard about his Nobel Prize:

> I have an amusing recollection of which I must tell you. On the 9th of November 1933 dear George Gordon, President of Magdalen, called

> me to his office to tell me that the *Times* had said I would be among that year's prize winners. And in his chevaleresque and witty manner he added: "I think you may believe it as the *Times* would not say such a thing unless they really know. As for me I was truly astonished as I thought you had won the prize."[24]

So this was how Schrödinger learnt he had won the Nobel Prize. It is clear from the previous letters mentioned here that President Gordon had thought Schrödinger had already been awarded the Prize when he made the case for a Fellowship at the College Meeting of 3 October, but it is also clear that he shortly realised afterwards that this was incorrect.

A telegram had been sent by H. Pleijel, Secretary of the Royal Swedish Academy of Sciences, to Professor E. Schrödinger, University of Oxford, arriving at 8.37 pm on 9 November 1933 at the Oxford Telegraph Office. The telegram stated (translated into English from German):

> I have the honour to inform you that the Royal Swedish Academy of Science has decided to award you and Professor Dirac the Nobel Prize in Physics 1933. Letter follows.[25]

A telegram with essentially the same message, but in English, was sent to Professor P.A.M. Dirac, University of Cambridge, and arrived at 8.25 pm at the Cambridge Telegraph Office.[26] Anny Schrödinger heard about the good news independently. She subsequently wrote:

> On November 9, 1933, I was sitting in a small hotel in Oxford when I was called to the telephone at 9 pm. It was the *London Times* informing me that my husband was among the Prize winners whose names had just been cabled from Stockholm. An hour later the first reporter appeared at the hotel and at midnight the *London Times* called again to speak to my husband so that first interviews could be printed the next morning.[6]

On the next day, 10 November 1933, Gordon wrote to Schrödinger:

> May I send you my most cordial congratulations on this morning's splendid news? I wish you might have obtained an undivided prize, because the larger sum would have been so particularly useful to you at the present time. But the glory is undiminished, and even £4,500 is a

> magnificent gift. You will see that the *Times* made use of the statement which you were so good as to write out for me. For the sentence "he had to leave Germany" I have no responsibility, it is the work of the *Times* office and I wish it had been more accurately worded. I heard when I came back from dinner about 11 last night that the *Times* had rung me up, and they may have wished to ask me about this point. The call, however, was not repeated.
>
> With my renewed congratulations, in which all your new colleagues here join warmly, I am sure. Yours sincerely, George Gordon.[29]

The new Nobel Prizes were reported in many major newspapers around the world. The *Times* of Friday, 10 November 1933, was in its usual style of starting with notices of obituaries and legal statements, personal notes, advertisements, letters to the editor, stock prices, public school sports and then home news. The Foreign News section had the highly topical headlines "Herr Hitler at Munich, Tenth Anniversary of Putsch" and "Reichstag Trial", both written in non-celebratory language. Then there was the article "Nobel Prize in Physics, Professors Dirac and Schrödinger".[15] These headlines clearly put the Nobel Prize in the context of the political developments taking place in Germany.

The Nobel Prize article in the *Times* stated that Schrödinger "had to leave Germany owing to the Nazi revolution, but found hospitality at Magdalen College, Oxford, where last month he was elected to a supernumerary Fellowship." It also stated: "Schrödinger found out the proper wave equation… He also generalised his equation and the study of his waves from a single electron or a single mass point to quite an arbitrary mechanical system." The article correctly emphasised the crucial aspect of Schrödinger's equation that it is not just applicable to the hydrogen atom but to all atoms and molecules. The article also mentioned Dirac's prediction of the positron and, somewhat briefly, the award to Heisenberg.

President Gordon had done well to slip in the statement that Schrödinger was now a Fellow at Magdalen College, Oxford. This was strictly true as he had been formally admitted as a Fellow just hours before the announcement of the Royal Swedish Academy of Sciences. Indeed, quite correctly, Magdalen today claims Schrödinger as one of its own ten Nobel Prize winners. However, the statement made by the *Times* that

Schrödinger "had to leave Germany owing to the Nazi revolution" was clearly of concern to Gordon and might have contributed to later problems that Schrödinger had with the Nazi regime.

Also on the same page of the *Times* was a report on "Austrian Nazi Celebrations" that included the statements:

> The Nazis of Graz, the capital of Styria, celebrated the tenth Anniversary of the Munich Putsch last night by floating huge swastikas, outlined in burning lamps, down the River Mur, on which the city is situated. Hundreds of people occupying the bridges applauded the sight and resisted the efforts of the police to disperse them.[15]

Presumably, Schrödinger did not read this article although he did himself keep the *Times* report of the Nobel Prize which was published on the same page. If he had appreciated the implications of the article, perhaps he would have made a different decision when he moved to Graz three years later.

On the same day the *Daily Telegraph* also reported on the Nobel Prize.[15] The article was not as detailed as that in the *Times* but had the competitive headline "Nobel Prize in Physics. Oxford and Cambridge Divide the Award". It stated that Schrödinger was formerly Professor of Theoretical Physics at the Friedrich Wilhelm University Berlin, and recently elected to a "Research Fellowship" at Magdalen College, Oxford. It also said that Schrödinger had reached England "only on Sunday last. He dined in the college for the first time, and was admitted to a Fellowship."

The *Daily Mail* also reported on the award on 10 November 1933. This has always been a conservative British newspaper which was owned in the 1930s by Lord Rothermere. He was a supporter and friend of Hitler and Mussolini, and often took the opportunity to bring them praise. On its foreign news page the first article in the *Mail* wrote in celebratory terms on the tenth anniversary of Hitler's Putsch in Munich:

> Herr Adolf Hitler led veterans of the National Socialists movement this morning from the Burger Beer Hall to the Odeon Platz along the street through which they passed on November 9, 1923 when their attempt

to seize power in Germany failed. Ten years ago he led a forlorn hope. His followers were easily put to flight by the police and 16 of them were killed. Today the founder of the National Socialist Party, who once more led his staunchest disciples through the streets, is supreme ruler of Germany.[15]

The next article was on the Nobel Awards and concentrated on Dirac:

The Nobel Prize for Physics for 1933 has been divided between Professor P.A.M. Dirac of Cambridge University and Professor Erwin Schrödinger of Austria who is at present working in Oxford "in connection with the atomic theory". Dr Dirac, who is only 31 years old, was the youngest professor in Cambridge at the time of his appointment to the Lucasian Professorship last year… He is a shy man of few words, and has been called "the silent celebrity". A report issued this summer by the Nobel Prize Foundation stated that each prize this year would be worth £9,465.[15]

The press in Germany also commented on the award, although the emphasis here was on Heisenberg. The *Berliner Morgenpost* said that Schrödinger was "currently on vacation in Oxford".[15] It also gave extensive reports to the march led by Hitler in Munich on the tenth anniversary of the Beer Hall Putsch and the major elections to be held in just two days which would be the "moment of judgement" for the Führer. The *Wiener Zeitung* announced that an Austrian Dr E. Schrödinger had won the prize and was currently working in Oxford, but with no more detail. The *Svenska Dagbladet* from Stockholm in Sweden just stated that Schrödinger was from Berlin.[15]

In America the *Washington Post* gave more prominence to the Nobel Prize in Literature which had also been announced at the same time.[15] This had gone to Ivan Bunin who was reported as a Czarist poet, had opposed the communist regime in the Soviet Union and had emigrated. Even then, the Literature Nobel Prize had a political angle. The *New York Times* described Schrödinger as "succeeding Max Planck as Professor of Physics at Berlin University. Before the current semester opened he gave up that Chair to accept a call to Oxford." It was also stated, somewhat parochially: "The names of the winning physicists came as a big surprise in ordinarily well-informed circles here. It was expected that last year's prize would go to Percy W. Bridgman of Harvard."[15]

This comment from the *New York Times* is interesting as the Nobel records show that Bridgman did receive the largest number of nominations in 1933 (ten) but all were from professors in the USA who were not very well known.[17] An American cartel seemed to be in operation for the Nobel nominations that year but was not successful. Schrödinger received nine nominations in 1933 but this included de Broglie, Bohr and Einstein. Bridgman, however, did not have to wait long and was awarded the 1946 Nobel Prize for his research on high pressure physics.

From elsewhere round the globe, the *Times of India* referred briefly to the Nobel award next to a report on "Mr Gandhi to be boycotted". The *China Press* also mentioned the Prize alongside an article expressing concern of an outbreak of bubonic plague in China.[15]

Schrödinger himself was not so happy with some of the descriptions reported in the press. He communicated this to President Gordon who wrote back to Schrödinger on 13 November 1933 in his usual friendly tone:

> Many thanks for your letter of Saturday last. I have not yet seen the apparently indiscreet exaggeration of the Oxford correspondent you speak of: no one had the right to exaggerate in that way. But fame brings its penalties, and not the least, as you should know, is this sort of mangled talk.
>
> I am glad that you have been able to leave your hotel for 12 Northmoor Rd, and hope that when you are comfortably settled and at rest here, we may have the pleasure of seeing both Madam Schrödinger and you at our table here. I have just told my wife of your new address, and she will be calling no doubt at the first opportunity.
>
> I have asked the Vice-Chancellor when he can matriculate you as a member of the University and you will probably hear about this soon from Mr Weldon, one of our Fellows, who looks after these things. Presently when this is over we must get you a degree! I thought the *Times* photograph was excellent. Dirac looked as if he had been stunned and was just recovering consciousness.[29]

On 15 November 1933, the *Times* published an additional report implying they had overlooked the priority of Heisenberg in their previous article. It was stated: "The parts played by Schrödinger and Dirac were reported on Friday. Heisenberg's contribution may perhaps be regarded as the primary one."[15] Many leading scientists would dispute somewhat the tone of this statement. Indeed, Einstein himself stated in his letter to the Nobel Award Committee: "Personally, I assess Schrödinger's achievement as the greater one, since I have the impression that the concepts created by him will carry further than those of Heisenberg."[17] It was Schrödinger who discovered the practical way of performing calculations on atoms and molecules through his equation and also invented the wave function (or orbital) which allows for experimentally observed properties to be predicted.

Serious scientific publications also acknowledged the award. In November 1933 *Science*, the publication of the American Association for the Advancement of Science, reported that Professor Erwin Schrödinger "formerly of Berlin and now of Cambridge" had been awarded the Prize. However, this error was corrected later in the same article:

> The three Nobel Prize men, two Germans and an Englishman with a French name, all have picturesque personalities and their erudite adventures in physics, pursued in both European and American centers of research, have been accompanied by a happy companionship among the trio and with other students of physics…
>
> Although until recently the Vienna-born Professor Schrödinger held the chair of theoretical physics at the University of Berlin, to which he was appointed in 1927, he now finds working at Oxford more congenial than the present atmosphere of Berlin. He made an enforced departure from Germany due to the present political regime, and last month he was elected to a Fellowship in Magdalen College, Oxford.[27]

The implication here that the Nobel winners had done some of their work in America is an exaggeration. Schrödinger had carried out a lecture tour across the USA in 1927 but nothing more significant. Dirac had done a similar first tour in 1929 where he met up with Heisenberg in

Yellowstone National Park and they both gave lectures at the University of California, Berkeley.

The other leading English-language journal *Nature* also gave a detailed report and correctly linked Schrödinger to Magdalen College:

> The Nobel Prize for Physics for 1933 has been awarded jointly to Prof. P.A.M. Dirac and Prof. E. Schrödinger, both of whom have earned international reputations for their work on the quantum theory. The prize for 1932 has been awarded to Prof. W. Heisenberg, of the University of Leipzig. Prof. Schrödinger who, at the time of the publication of his paper on "Quantisation as a Problem of Proper Values" was at the University of Zurich, has since occupied a chair in the University of Berlin and is now at Magdalen College, Oxford.[28]

In addition, a description was given on Schrödinger's earlier papers before his discovery of wave mechanics:

> He had previously written a number of papers on various subjects in physics. Amongst them are works on the theory of pigments and on colour measurement, while one, "On the Coherence of Wide Bundles of Rays", is of an experimental character. One of his contributions in 1914 is of particular interest in that it seems to foreshadow his important work. This is on "The Dynamics of Elastically Coupled Systems", in which he considers a system of mass points which, in the limit, gives the partial differential equation of a vibrating string and in which considerations of group velocity play a part. Influenced by the ideas of de Broglie, Schrödinger developed wave mechanics as a coherent theory, the centre of which was the wave equation. The nature of the atomic problem appeared in a new guise, showing mathematically more resemblance to the vibrating string than to the planetary system.
>
> The methods of the new theory were classical in character, and gave the hope that the quantum theory might finally be absorbed into the classical doctrine. Schrödinger seems to have been imbued with this idea. The theory met at once with an enthusiastic welcome and made rapid strides. An important step was made when Schrödinger showed the relation of his theory to that of Heisenberg, and henceforth it was

possible for the two theories to draw from each other in spite of the difference in their philosophic outlook.[28]

Congratulations

After the announcement of the Nobel Prize, a huge number of congratulations poured in to Oxford, and Schrödinger carefully filed and kept the telegrams, letters and cards, several of which are described below.[29] Many of these letters came from scientists from Germany or Austria who were forced by the Nazis to leave their positions in 1933 or later. Schrödinger had been quick off the mark to congratulate Paul Dirac who wrote back on 14 November from St John's College, Cambridge with his own congratulations. He also stated: "I am very glad indeed to be in the company of you and Heisenberg and hope we shall have a good time together in Stockholm."[29] As many physicists know, the introverted Dirac was the last person you would expect to have a good time with. As is often the case with scientists who jointly win the Nobel Prize, there was a special respect between Dirac and Schrödinger that now lasted for all of their lives.

The first congratulatory telegram arrived from Niels Bohr from Copenhagen. Another arrived from Max Born, saying briefly: "Heartiest congratulations." Max Born had no reservations about awarding the prize to Schrödinger but he was somewhat hurt that he had not shared the prize with Heisenberg. To make things worse Hermann Weyl wrote to Born to say: "Don't make yourself sick, dear Born, that the golden Nobel Prize passed you over! Heisenberg got there first."[4]

Heisenberg, however, wrote back to Born after receiving his congratulations to say:

> The fact that I am to receive the Nobel Prize, for work done in Göttingen in collaboration — you, Jordan and I — this fact depresses me and I hardly know what to write to you. I am, of course, glad that our common efforts are now appreciated, and I enjoy the recollection of the beautiful time of collaboration. I also believe that all good physicists know how great was your and Jordan's contribution to the structure of quantum mechanics — and this remains unchanged by a wrong decision from

> outside. Yet I myself can do nothing but thank you again for all the fine collaboration, and feel a little ashamed.[4]

In due course, things were put right to some extent in 1954 when Born was awarded the Nobel Prize for "his fundamental research on quantum mechanics, especially his statistical interpretation of the wave function".

Fritz Haber, who had overlapped with Schrödinger in Berlin and was now a refugee living in Cambridge, telegrammed to Schrödinger to say: "Congratulate you most heartily." Haber had been offered a position in Rehovot, Israel by the chemist Chaim Weizmann who was also the first President of Israel. He departed from Cambridge two months later. Very sadly, the great physical chemist died of a heart attack in a Basel hotel on 29 January 1934 when he was on his way to Israel.

Another of Schrödinger's Berlin colleagues, Max Planck, still hoped he would return to Berlin, and wrote on 10 November 1933:

> The news that the Nobel Prize has been awarded to you and Heisenberg gave me enormous joy which was even more enhanced by the thought that this decision by the Swedish Academy of Sciences also means an honour to academia which, given the current circumstances, is so warranted and absolutely necessary and from which all your fellow academics will benefit.
>
> Of course, this honour, which has been bestowed on you, makes the pain of missing you even greater. To exaggerate this is absolutely impossible. Should really all bridges be destroyed?
>
> Would it be not be possible for you to convert your sabbatical year retrospectively into a year of leave? That way you might have one or two years to explore your future research plans without committing yourself totally and we, who are left behind, could, in the meantime, nourish ourselves with the hope that one day you may return.[29(t)]

Others to write to Schrödinger included William L. Bragg from Manchester who said: "I was delighted to see that you had been awarded the Nobel Prize. We all feel that this is most appropriate and send you our warmest congratulations." Another British Nobel Prize winner was Owen

Richardson from King's College London who had won the Prize in 1928 for his experimental work on thermionic emission. He had measured a high-resolution spectrum for the hydrogen molecule which needed Schrödinger's wave mechanics for its interpretation. He had also met Schrödinger at the 1927 Solvay Conference. His letter included the message:

> My hearty congratulations on the Nobel Prize. I had heard a whisper of it but the whispers are not always confirmed at Stockholm. It will be very welcome to you in the exceptional circumstances in which you are situated at present. I for one hope this will be instrumental in causing you to make England your permanent home. Mrs Schrödinger, I expect, is overjoyed. You will have a busy time getting ready for Stockholm and a wonderful welcome there.[29]

Louis de Broglie from Paris wrote to say that he had been proposing Schrödinger for the Nobel Prize since 1929 and had had considerable pleasure reading his magnificent works. Egil Hylleraas wrote from Bergen in Norway. He had carried out an important application of wave mechanics in 1929 showing it gave highly accurate results for the energy levels of the helium atom. He had done this in a *tour-de-force* calculation by including the distance between the two electrons explicitly in the wave function. He wrote to say he was writing a special article to summarise the work of the three new Nobelists.[29]

Arthur March sent a telegram from Innsbruck with his congratulations. He had not yet travelled to Oxford. Hermann Weyl, Schrödinger's former colleague in Zurich and close friend of Anny, had also had to leave Germany as his wife was Jewish. As had been mentioned in the *Zeitung Berlin*, he had taken up a position in Princeton at the new Institute for Advanced Study where Einstein had also been appointed. His telegram from the USA said, somewhat cheekily: "Hail Anny and Erwin the Nobel Prize winners. What about the Princeton invitation. Happily yours. Peter." Peter was the affectionate name Weyl used with the Schrödingers and Erwin had already been invited to visit the University of Princeton to consider a position there. Arnold Berliner, who had published several of Schrödinger's works in *Naturwissenschaften*, also sent his congratulations and he was soon to publish another key paper from Schrödinger written in Oxford.

Felix Ehrenhaft had taken the Chair of Exner, Schrödinger's old professor. He wrote to the former student of his University of Vienna with pride saying:

> What I had long expected and hoped for has finally happened. 24 hours before the award ceremony, I said to colleague Schweidler: you will see this time Schrödinger will be the Nobel Prize winner and it has happened. You can imagine how happy I am for you. I wish you a lot of success in your new sphere of activity and I remain filled with great pride for our Viennese physicist.[29(t)]

Ehrenhaft was Jewish and was soon to be a refugee himself in England. He had regular correspondence with Einstein who helped him emigrate eventually to the USA. Both Einstein and Schrödinger considered Ehrenhaft to be a "maverick physicist". Egon Schweidler, who is mentioned in Ehrenhaft's letter, had taught Schrödinger and was acknowledged in his Dr.phil. thesis of 1910. He was to stay in Austria during the Second World War.

Another physicist from Schrödinger's early days in Vienna was Stefan Meyer. He had worked with Exner on radiochemistry and had supplied the Curies and Lord Rutherford with supplies of radium. He wrote with pride to say:

> Today the newspaper brings the news that the Nobel Prize has been awarded to you, and we are all especially pleased about it. Congratulations towards you and your dear wife! I know Lindemann and Soddy from Oxford, but, as far as I know, they don't harmonize very well with one another...
>
> According to all I have heard, Oxford must be a very pleasant place to stay. Also, the individual English centres can be reached so quickly among each other that all friends in other places in England would also be easily at hand. You are sure to get used to it quickly: don't forget your old friends too! How much we cling to you in friendship and how often we commemorate you.[29(t)]

Meyer was Director of the Institute of Radium Research in Vienna and had been an assistant to the great Ludwig Boltzmann. After Boltzmann's suicide in 1906 he also temporarily became director of his institute. Like many of the scientists mentioned in this book he came from a Jewish family and just one month after the Anschluss in Austria in 1938 he was forced to retire from his post in Vienna. He went to live quietly in the Austrian countryside and somehow managed to survive the war. His brother Hans was a distinguished chemist who had a professorship at the German University in Prague. He was murdered in the Theresienstadt concentration camp in 1942. Stefan Meyer returned to his former position in Vienna after the war. The Institute for Subatomic Physics of the Austrian Academy of Sciences is now named after Stefan Meyer.

Friedrich Paneth wrote from Imperial College in London to congratulate Schrödinger. He had worked as Meyer's assistant in Vienna and eventually had become Professor at Königsberg. His parents were of Jewish descent and he left Germany in 1933 to take up a position at Imperial College. He was nominated several times for the Nobel Prize in Chemistry but narrowly missed out. He took British nationality and was elected a Fellow of the Royal Society. Unlike Schrödinger, he was successful in becoming part of the British scientific establishment. He won the Liversidge Award of the Faraday Society and, in the war, became director of the important British-Canadian atomic energy programme in Montreal. After the war he was invited back to Mainz to become the Director of the Max Planck Institute for Cosmochemistry.

Maurice Goldhaber wrote from Cambridge. He had been a student in Berlin where he had met Schrödinger and hoped to do research under his supervision. He was Jewish and had to leave Germany in August 1933. Schrödinger wrote a strong supporting letter to Rutherford and so Goldhaber moved to Cambridge where he worked in Chadwick's team which had just discovered the neutron. He was told he needed to find a College in Cambridge and enquired at Magdalene College where Patrick Blackett had been a Fellow. The Senior Tutor Vernon S. Vernon-Jones told him: "Ah, you are a refugee, I suppose we ought to have one." And then said, very generously: "I suppose you have no money, we had better give you a hundred pounds."[30] Goldhaber then emigrated to America where he in due course became Director of the Brookhaven National Laboratory and won the National Medal for Science and the Wolf Prize. In his congratulatory letter Goldhaber informed Schrödinger that he was settled at the other Magdalene and was looking forward to meeting Born, Dirac and Lennard-Jones in Cambridge.

The Vice-President of Magdalen College, Stephen Lee, wrote to congratulate the new Nobel Prize winner and also to give some very important extra information on the gown he should wear at High Table dinner in the College Hall:

> May I first of all congratulate you very sincerely on your Nobel Prize. It has come, I hope, as a good augury for a happy holiday sojourn in Oxford. Magdalen College is extremely fortunate to have elected as a Fellow one who has thus attained the highest honour in the world of learning.
>
> Now to small matters. I have consulted various experts on the question of what gown you should use when dining in Hall. As you are "on the Foundation of the College" but have not at the moment an Oxford degree it seems that it will be correct if you wear the gown of a scholar, or as we call them in Magdalen a "demy", when dining in Hall. I have therefore instructed the Porters at the College Lodge to have a demy's gown available to you when you wish to use it.
>
> If by any chance you wished to attend the College Chapel services on Sundays (9.30 am and 6 pm) as a Fellow you wear a surplice. I am glad to hear you have got a house now. My wife hopes to give herself the pleasure of calling upon Mrs Schrödinger when you are settled in.[29]

Schrödinger, however, was not inclined to attend a college chapel nor to wear a surplice. His new Magdalen colleague Sir Henry Miers also wrote in a very friendly style:

> Though I have not had the pleasure of meeting you since you became a Fellow of Magdalen College, I am glad as one of your colleagues to take the opportunity of writing afforded by the award of the Nobel Prize. Let me congratulate you heartily upon this great and well-deserved honour. I know the pleasure and pride which I feel is shared by all of our colleagues, and especially those who have been associated with the

scientific work of the University... I look forward to meeting you in Oxford before too long.[29]

In addition, the Magdalen Tutorial Fellow in Modern History J.M. Thompson wrote on 10 November 1933:

After our meeting last night I should like to be one of the first to congratulate you on the news which I heard by "wireless" when I got home and saw in the paper this morning. How delighted your wife will be too![29]

Professor Nevil Sidgwick was an Oxford chemist who had studied with Ostwald in Leipzig. He had used Schrödinger's orbitals to derive some simple rules for the chemical bonding of coordination compounds which are still used today. He wrote from Lincoln College to say: "May I congratulate you most heartily on your Nobel Prize. I am glad to learn that your connection with Oxford begins with such a moment."[29] He would soon be corresponding with Schrödinger encouraging him to apply for a vacant Chair in Edinburgh.

The other refugee recently arrived from Germany to join Magdalen College was the young Heinz Koeppler. He wrote: "As a German student studying here at Magdalen College, I am excited to be in Nobel company and I congratulate you most sincerely."[29] An additional and interesting letter came from Hamburg written by Albrecht Mendelssohn Bartholdy:

Let me tell you how I am delighted with the news about the Nobel Prize, and also that you have reached Magdalen. I have been to Balliol several times through my friendship with Lindsay. Magdalen is my oldest Oxford association; Paul Benecke is from our family, of all of us a bright cousin, and I also know Weldon well. When I was back in Oxford for the first time after the war, I was in Magdalen frequently.[29(t)]

He was the grandson of the great composer Mendelssohn. He founded the Hamburg Institute for Foreign Policy and was one of Germany's representatives at the Paris treaty negotiations after the First

World War. He was the leading expert on Anglo-Saxon Law in Germany. Being from a large and prominent Jewish family, and whose grandfather's famous compositions had been banned by the Nazis, he was to be expelled from his professorship in Hamburg in January 1934 and escaped to Oxford via Switzerland. In Oxford, Bartholdy had a Fellowship at Balliol College until his death in 1936. In his letter, he mentions his cousin Paul Benecke, the Senior Fellow at Magdalen College, whose mother was the composer Mendelssohn's daughter. Kurt Mendelssohn, who came with Simon to work in the Physics Department at Oxford, was also a distant relative.

Michael and Magda Polanyi were also by now refugees in England — in their case in Manchester where Polanyi had been appointed to a Chair in Physical Chemistry. They sent congratulations to Schrödinger and Anny by telegram saying they hoped to see them again soon. Michael Polanyi came from a Jewish-Hungarian family. He had done important work in Fritz Haber's Physical Chemistry Institute in Berlin which included calculating the first potential energy surface for a chemical reaction using approximations to Schrödinger's wave mechanics.

Polanyi continued fruitful physical chemistry research in Manchester building up an outstanding department. He was elected a Fellow of the Royal Society in 1944. In the late 1940s, he got to know Alan Turing who had been appointed to the Mathematics Department also at Manchester University. They took part in a famous symposium in 1949 on "The Mind and the Computing Machine" together with Douglas Hartree and John Young, who had been a Tutorial Fellow in Zoology at Magdalen College, Oxford and overlapped with Schrödinger.[31] Some of the modern approaches to artificial intelligence derive from this symposium.

Like Schrödinger, Polanyi became interested in philosophy and was, quite late in life, elected to a prestigious Senior Fellowship to undertake research in that field at Merton College, Oxford. The Polanyi Prize of the Gas Kinetics Section of the Royal Society of Chemistry bears his name. His son John, who was a student in Manchester and still holds a Chair in Toronto, won the Nobel Prize for Chemistry in 1986 for work on chemical reaction dynamics. Another son George studied economics at

Magdalen College, Oxford and worked on counter-intelligence in Europe in the war.

Many other congratulations came in including from Schrödinger's closest student-friend Hans Thirring from Vienna and from Patrick Blackett who had met with Max Born in the summer in Zurich to help arrange his appointment in Cambridge. Anny's mother Imgard sent a telegram saying: "Congratulations to our dear Nobel award winner with the utmost joy." Josefine Jünger, who was Anny Schrödinger's godmother, sent her warm congratulations to the Nobel Prize winner without any malice. Her twin daughters Ithi and Roswitha also both wrote letters. Ithi said: "I just read in the coffeehouse that a so-called Christmas wish has now been fulfilled for you. Congratulations on how happy I am that this great, great honour was given to you."[29(t)] While Roswitha wrote: "My stupid writing cannot possibly reflect the joy I felt when I myself heard that you have received such a great honour."[29(t)] The biography by Moore makes assumptions on Schrödinger's relationships with the Jüngers, but reference to these letters is not made there.[7]

In the meantime, Frederick Lindemann had been continuing to be concerned about the financial arrangements for Schrödinger. He wrote on 10 November 1933 to William Rintoul at the Imperial Chemical Laboratories to say:

> I am rather shocked to find that the maximum salary mentioned was £900. As you will remember, Sir Harry gave verbal authority for the thousand in exceptional cases at our meeting, so I take it that no difficulty will arise there.
>
> I am amazed this morning to see that Schrödinger has been awarded the Nobel Prize. I made sure it had been given to him when it was definitely in the papers. However, it is all right it has arrived in time to coincide with your first payment. I hope to be able to bring him up to lunch next Friday.[3]

So Lindemann, just like President Gordon when he had written the first paper for the Magdalen College Meeting on 3 October, had assumed all along that Schrödinger had already won the Nobel Prize before coming to Oxford.

Lindemann and Schrödinger in the garden of Francis Simon in Oxford, ~1934.

Stockholm

With the first announcement of the Nobel Prize on 9 November 1933 there was only one month before the presentation of the award in Stockholm and much urgent preparation was needed. This included buying clothes for Anny, preparing lectures and booking tickets. Only on November 18 did the promised confirmatory letter from Henning Pleijel, the Permanent Secretary of the Royal Swedish Academy of Science, arrive. He stated:

> I have herewith the honour to confirm my telegram by informing you that the Royal Swedish Academy of Science in its meeting of November 9th has decided on this year's Nobel Prize for Physics being awarded to you together with Professor Paul Dirac as a reward for the discovery of new fertile forms of the atomic theory.
>
> On behalf of the Academy I also have the honour to invite you to the solemnity to be held at Stockholm on the anniversary of the decease of Alfred Nobel (10th of December), where the amount of the prize as well as the diploma and the medal in gold will be handed over to

> you. Immediately after the festival a banquet will follow at Grand Hotel Royal in this town.
>
> If on your presumptive visit to Sweden you should be accompanied by a friend or by members of your family, they of course are also included in that invitation. In this case you would oblige me very much by informing me kindly of their names in order that cards may be duly issued to them.
>
> I also would be very much obliged to you if you would let me know, as soon as possible, whether you intend to give, on this occasion, the lecture which is incumbent on each Prize winner within six months of the Founder's day at which the prize was won, and what your wishes may be with regard to that lecture which has to treat the subject to which the prize has been awarded.
>
> Finally, as I presume that you will arrive in Stockholm some day before the 10th and that it could be of some interest to you to become acquainted, before the festival, with some of the leading persons in our Academy, Mrs Pleijel and I would feel really honoured if you would partake together with some scientists of this town in a quiet dinner at our home in the Academy at the 9th of December (Saturday evening).[29(t)]

Not all prominent physicists back in Germany sent congratulations to the 1933 Nobel Prize winners for Physics. We have already seen how Johannes Stark, the winner of the 1919 Nobel Prize for his observation of the splitting of spectral lines in electric fields, had become an outspoken National Socialist who railed against the new theories, especially the work of Einstein. Following the 1933 award he turned his attention against quantum mechanics. Heisenberg's matrix mechanics and Schrödinger's wave mechanics were all dismissed by him "as opaque and formalistic".[32] Heisenberg wrote a counter article but Stark responded by saying Heisenberg was still advocating "Jewish physics". With Schrödinger out of the way in Oxford and Einstein at Princeton, Stark's hatred now turned against Heisenberg.

Following on from his congratulatory message, Max Planck continued to be persistent in attempting to persuade Schrödinger to return to Berlin. On 19 November 1933 he wrote:

With all the understanding I feel for you, who is beset by floods of letters, I cannot resist the urge to say at least a few short words which your so kind letter of the 14th of the month puts in my mouth.

First of all, I understand perfectly that you are not approaching the Ministry of your own accord to convert your dismissal into a vacation. Each of the reasons you have given is quite obvious to me. But I must urgently ask you for something else. If the Ministry itself should contact you, for example with a suggestion for two years' vacation or something like that, then don't say no straight away, but set your conditions as far as you see fit and necessary, but always with the thought that a positive result will remain possible, and believe me that no one would be more pleased about such a thing than your Berliners.[8(t)]

Around this time Planck and Heisenberg had been discussing the importance for German science of trying to persuade Schrödinger to return to work in Germany. Heisenberg was particularly annoyed by Schrödinger's departure "since he was neither Jewish or otherwise endangered".[33] Planck wrote to von Laue: "I regard Schrödinger's resignation as a deep wound to our Berlin physics, which we must endure with all of the energy available to us."[33]

It seems Schrödinger was too busy with the preparations for the visit to Stockholm to reply in a positive way to Planck. With all the excitement of the initial announcement abating, the Schrödingers moved initially from the small Isis hotel to a temporary accommodation at 12 Northmoor Road in Oxford and then a few months later to a more spacious house at 24 Northmoor Road. This was just two houses away from the famous writer J.R.R. Tolkien who was a great friend of C.S. Lewis, Schrödinger's colleague at Magdalen College. Perhaps somewhat surprisingly, there is no record of Tolkien and Schrödinger ever meeting. It seems that Anny and Erwin were not so keen on keeping up the garden in their new house. Their close friend Hansi Bauer-Böhm said:

The house in Oxford had a big garden which was quite neglected and not at all like other English gardens, full of dandelions — "dents de lions" as Erwin called them, thrilled by their colour and the linguistic association.

The neighbours were less thrilled but appalled at such negligence and the weeds that were spreading everywhere.[6]

Hansi Bauer-Böhm was an artist who was born in Austria and had been very friendly with Anny Schrödinger from a young age. Anny had been the secretary to her father Fritz Bauer who was the General Director of the Phoenix Insurance Company. Hansi visited the Schrödingers in Oxford and settled in England in 1939 after escaping from Austria to Switzerland. She eventually lived in Walberswick, Suffolk and her paintings and sculptures are still sold today.

Schrödinger received strict instructions for his visit to Stockholm from the Nobel Foundation who were organising the celebrations. For a report that would be published, he was asked to send a high-quality photograph of himself and a short biographical note of 1–2 pages containing the most important dates in his life, a summary of his most important work and especially that for which the Nobel Prize was awarded. He was also asked to send a copy of the short speech he was to give at the Nobel banquet after the award ceremony and also the Nobel lecture and any associated photographs.

The invitation from the Royal Swedish Academy included the possibility of bringing family members to the celebrations in December and Anny joined her husband. Schrödinger did not leave many recollections of the visit to Stockholm but Dirac's mother Florence left extensive notes.[26] Dirac was not married at that time so his mother was his guest. Heisenberg also brought his mother Annie.

All the Laureates arrived with their female guests at Stockholm central station on the morning of 9 December. There were many cameramen and a famous photograph was taken as shown here. Schrödinger looks very happy in his somewhat casual plus-fours while the others look on rather sheepishly. They were taken to the Royal Grand Hotel for breakfast and were then given a tour of Stockholm. In the evening they attended the dinner hosted by Professor Pleijel, the permanent secretary of the Royal Swedish Academy of Science. A total of 32 Nobel laureates "and ladies" were present and the dinner went on until 12 am.[26]

The 10th of December was the big day for the presentation of the Prizes on the anniversary of the death of Alfred Nobel. In the morning the Laureate parties were taken to the Swedish Museum. In the afternoon they were taken on a procession through the crowded streets to the grand ballroom of Stockholm Concert Hall. The atmosphere was one of a fairy-tale in the cold darkness with the fir trees and electric lights.[26] The Nobel Laureates were to be seated on the platform opposite King Gustaf V in the front row.

The year of 1933 was the centenary of the birth of Nobel so it was a special occasion and several members of his family were present together with numerous members of the Swedish Royal Family. There were German, Austrian, British and Swedish flags. At 5 pm the trumpets played and the Laureates, each escorted by a host, were taken to their velveted seats on the platform. Schrödinger and Heisenberg were dressed in white tie dinner suits while Dirac wore a drab-looking dress suit.

Annie Heisenberg, Anny Schrödinger, Florence Dirac, Paul Dirac, Werner Heisenberg, and Erwin Schrödinger at Stockholm Railway Station, 9 December 1933.

Also present was Ivan Bunin who won the Prize for Literature. It was a smaller set of Laureates than normal as Thomas Hunt Morgan from the California Institute of Technology, who had won the Prize for Physiology or Medicine for explaining the role of the chromosome in heredity, was unable to attend and there was no Prize for Chemistry that year.

The Chairman of the Nobel Foundation Dag Hammarskjöld then gave a welcoming speech. In due course, he would become one of the most famous of all Swedish citizens and was awarded the Nobel Peace Prize posthumously in 1961. Then there was a speech of reminiscences on Alfred Nobel by Ragnar Sohlman who had been the executor of Nobel's will and had created the Nobel Foundation. This was followed by orchestral music.

The Laureates were called in turn with Heisenberg first, followed by Schrödinger and then Dirac. Before the presentation of the Nobel Prizes by the King, Henning Pleijel gave a summary of the work of each Laureate that had led to the Nobel Prize. By now they had got to know him well through the initial telegram, the confirmatory letter and a splendid dinner the evening before. He started off in some detail by summarising the research of previous Nobel Prize winners on whose discoveries the new developments in quantum mechanics were based. This included Planck, Einstein, Rutherford, Bohr and de Broglie. He then summarised the work of Heisenberg who was presented with his Nobel Prize by the King. After this it was Schrödinger's turn and Pleijel said:

> Professor Schrödinger. Through a study of the wave properties of matter you have succeeded in establishing a new system of mechanics which also holds good for motion within atoms and molecules. With the aid of this so-called wave mechanics you have found the solution to a number of problems in atomic physics. Your theory provides a simple and convenient method for the study of the properties of atoms and molecules under various external conditions and it has become a great aid to the development of physics.
>
> For your discovery of new fruitful forms of atomic physics and the application of these, the Royal Academy of Sciences has decided to award you the Nobel Prize. I request you to receive this from the hands of His Majesty the King.[17]

As is shown in the photograph below, Schrödinger then bowed to the King who presented him with a heavy blue and gold leather case. This contained the 23-carat gold Nobel Prize medal struck by the Swedish Mint. On one side, the medal portrayed the face of Alfred Nobel and his birth and death dates in Latin numerals. The reverse had engraved the name E. Schrödinger, the year 1933, and an inscription from the Aeneid: "Inventas vitam iuvat excoluisse per artes" which translated is "It is beneficial to have improved (human) life through discovered arts". There was also a portrayal of the goddess Isis, which is very appropriate for a laureate from Oxford. This medal would subsequently have an interesting story as explained in Chapter 5. The case for the medal contained a cheque made out to E. Schrödinger for 85,000 Swedish Kroner.

The King also presented a unique certificate designed by the Swedish artist Elsa Örtengren. This included rays of light shining on physics equipment and the words (in Swedish): "To award half of the prize to the one in the field of physics who has made the most important discovery or

The King of Sweden presents Schrödinger with his Nobel Prize.

invention. To Erwin Schrödinger for the discovery of new fruitful forms of atomic theory".[17] After the final award for Literature was made to Ivan Bunin the Swedish National Anthem was played. The Laureates and their guests were then taken to the Nobel Banquet held back at the Grand Hotel where they were staying. There were 300 people attending the dinner who were the great and good of Stockholm. Together with several members of the Royal Family this included politicians, ambassadors, bankers, professors, and members of the Nobel family and Foundation. Schrödinger was sat next to the Duchess of Östergötland while Anny was next to the Crown Prince of Sweden.

During the dinner, an orchestra played music by the composers Chopin, Strauss, Wallace, Friml, Söderman, Nevin, Coleridge, Tchaikovsky, Atterberg and Wagner, largely chosen to be of the different nationalities of the Nobelists. The dinner was served on silver dishes with a splendid menu consisting of Consommé des Gourmets, Filets de Sole Maire, Poularde de Sandemar Poêlée à la Renaissance, and Bombe Glacée Nesselrode. This was followed by Corbeilles de Petits Fours, Fruits and Friandises. The wines were Schloss Böckelheimer 1929, Ch. Lagrange 1924 and Louis Roederer, Grand Vin Sec.

After the dinner, the Nobel Prize winners were invited to give a short speech, which for them must have been the most nerve-racking part of the day, especially after several glasses of fine wine. Dirac gave an extraordinary speech on how physics theories could be applied to economics. It is always a risk when scientists move outside their expert field and the reactions from those present, and the newspapers next day, were somewhat puzzled. Heisenberg just gave a simple vote of thanks. Schrödinger's speech was warm and affectionate:

> Your Royal Highnesses! Ladies and gentlemen! There are things in life that cannot be learned through experience, but must be met right the first time. And if you don't hit it, you don't have the opportunity to correct the mistake the second time. I find myself in such a situation today. Never in my life have I had to say thank you for the Nobel Prize in a solemn speech and, if I don't get it, I will not be able to use the exercise I have gained on any other such occasion. Because the occasion is unique. Don't judge me too severely...

> And what I have to say isn't too difficult either. I have to thank you for all the beautiful, love and good that I can experience here. And it is said: Whatever the heart is full of, the mouth goes over. My heart is full, so full that I hardly know where to start. The first are probably the feelings of gratitude in memory of the well-known generous founder Alfred Nobel, who brought this institution into being through his princely gift to the science of the world. Next to him, the first and most heartfelt thanks go to His Majesty and the Royal Family, who, through their cooperation, their warm and active participation, make this festival a unique one, something that will not be repeated anywhere in the world, an honour for scientific work without equal, which, if we let our gaze wander through history, has a parallel at most in the solemn coronations of poets in medieval Rome...
>
> Let me close with the utterly selfish wish: I hope to return here very soon — and not just once, but often — to return. It will probably not be celebrations in halls decorated with flags and not with so much party clothes in the suitcase, but with two long boards on the shoulder and a backpack on the back.
>
> My toast goes to the everlasting happiness of Sweden and the happy, kind people who inhabit it![17]

Everything at the events was taken in by the Swedish press who reported on all the details and took many photographs. In the *Dagens Nyheter* next day Schrödinger was pictured reading a menu card and was also shown entering the banquet arm in arm with the Duchess of Östergötland. The Nobel Prize celebrations were an opportunity for Stockholm to be seen in all its glittering glory.

The next evening there was dinner at the Royal Palace hosted by the Crown Prince with other princes and princesses. Florence Dirac wrote a vivid description of the events.[26] Anny Schrödinger wore a long pale pink and silver gown. This time the instruction was black evening dress for men with medals and ribbons. Soldiers and sentries saluted as the Nobel Laureates and their guests entered the Palace. Anny was bowled over by the celebrations:

> The Banquet stuck out for its exceedingly happy, casual and vibrant atmosphere. The Swedish princes were the most charming hosts and

Anny and Erwin decided to spend New Year's Eve with Max and Hedi Born in Cambridge. It was an opportunity to go over the most extraordinary year they had all ever had. Not everyone got on with Schrödinger but Max Born did. He described him as:

> A most lovable person, independent, amusing, temperamental, kind and generous, and he had a most perfect and efficient brain. There was never a steady flow of letters between us. But if one of us informed the other about some new scientific idea, there followed an explosion of correspondence; often I got a long letter every day, full of mathematical and critical remarks about colleagues and the general situation in theoretical physics.[5]

For sure, 1933 had been one of the most dramatic years in the history of Germany. Back in Berlin, however, many people were still unaware of Schrödinger's new position in Oxford or would not accept it. Schrödinger himself wrote: "I did not go back to Berlin after the summer, but instead handed in my resignation, which remained unanswered for a long time. In fact they then denied ever having received it, and when they learnt I had been awarded the Nobel Prize for Physics they flatly refused to accept it."[35]

Some 73 years later, Schrödinger's daughter Ruth showed the writer of this book a German magazine published at the end of December 1933 in which the "Men of the Year" were declared as Adolf Hitler, Rudolf Hess, Werner Heisenberg and Erwin Schrödinger. Not completely the company Schrödinger would choose to be in.

CHAPTER THREE

Life and Work in Oxford

College

Early in 1934, George Gordon, President of Magdalen College, wrote to Schrödinger in friendly but slightly worried terms:

> My dear Schrödinger. I proposed on Monday last, and to the Council of the University, that you should be granted the degree of M.A. by decree and this, I should hardly say, was agreed to. It will be confirmed by the University on Tuesday next, and then the authorities will communicate with you about one more point I wish to raise. The fee involved (£12) is being paid by this College, so don't please pay it yourself! That would be dealing too generously with the University.
>
> I keep hoping that I may meet you some Sunday night in College at dinner, but so far have not been fortunate. That is the night when we have the largest attendance at Dinner in Hall and in the Common Room afterwards and when one has the most chance of meeting with one's colleagues.
>
> I hope you are well and happy. I was glad to see that you have gone on lecturing, and hope you can find an audience capable of following your teaching![1]

This was a typical letter from a President of Magdalen, starting off with somewhat minor details about the M.A. degree and then going on to the real point of the letter about Dinner in Hall on Sunday nights. We have already seen that the Magdalen Fellowship at that time was very distinguished. Apart from Schrödinger, it included one Fellow who had won the Nobel Prize (Sherrington), another soon to win it (Robinson) and one more to receive the prize in due course (Eccles). There were also four Fellows of the Royal Society and world-renowned experts in a wide range of subjects from Medieval English, to Middle Eastern Languages, to the French Revolution and to the Philosophy of Kant. Gordon wanted Schrödinger to take part and contribute to this distinguished intellectual society in a highly convivial setting. This was the opportunity for Schrödinger to interact in intellectual terms with his college community.

The traditional process then in use at the start of the High Table dinner at Magdalen College was that the Fellows, after the President, processed from the Common Room into the College Hall in order of seniority of election to their Fellowships. Schrödinger being the most recently elected Fellow in 1933 would be last in the procession. On Sundays, the dress would be black tie and all those dining at that time would have been male. Then and now all Fellows wear black gowns. At the start of the dinner on Sundays the world-renowned College Choir would sing a spine-tingling ancient hymn as the Grace from the gallery in the Hall. The cutlery at dinner would be silver and each Fellow would have a silver tankard, called a tun, for water. The lighting would be by candles. With each course, there would be ample wines of the very top quality from the voluminous college cellar and poured by the Fellows' Butler.

After dinner the President and Fellows processed to the candlelit Common Room for dessert where they would sit in a horseshoe formation around a great fire. Wines were then poured for each Fellow by the two most recently appointed (Junior) Fellows present and delicacies such as exotic fruits and chocolates were passed around. Towards the end of the dessert a Fellow, often with medical interests, brought round the silver snuff box for further refreshment of the mind. In 1933, the very best cigars would also be available and the atmosphere would be one of smoke. Somewhat remarkably the event still occurs in this style, apart from the cigars, on nearly every evening in the University Term.

The High Table and Common Room of Fellows at Magdalen can be forbidding for newcomers. The distinguished English writer and playwright Alan Bennett, who was a junior lecturer at Magdalen in the late 1950s, recalled his first High Table Dinner when his gown got stuck under his chair and he did not have the courage to release it. Accordingly, his arms were trapped and he could not eat any of the delicious courses that were served to him, much to the chagrin of the Fellows' Butler.[2] The atmosphere can be a challenge even in the 21st century. Quite recently, a candidate for a Fellowship was brought into the Common Room after Dinner and at once passed out when asked to sit next to a well-known Philosophy Fellow.

Many academics still find these quaint traditions enjoyable. After a hard day researching in the laboratory, library or archive, or following concentrated teaching to small groups of students, it is a pleasant way to relax with interesting colleagues who are expert in a range of diverse subjects. There are some reports that Schrödinger did not enjoy this opportunity. On college dinners, Max Born stated that Schrödinger said:

> You never know who your neighbour might be. You talk to him in your natural manner, and then it turns out he is an archbishop or a general, or even a former prime minister – huh![3]

However, John Eccles had a different view:

> In the same year Schrödinger and I were elected as Fellows of Magdalen. He was a Guest Professor and I was an Official Fellow transferring after seven years as a Fellow of Exeter College, Oxford. Thus we were considerably senior in academic status to many of the younger Fellows. Every Sunday evening in Term Time there was a formal dinner for Fellows who were clad in dinner jackets with black tie. After the dinner at High Table we retired to the Senior Common Room for Port and Madeira, with a fine selection of fruit and nuts and other dessert dishes.
>
> No servants were in attendance, the custom being that the two most Junior Fellows of the College would serve the Fellows seated in a crescent at small tables with their plates and utensils. Because of our recent election, Schrödinger and I had to participate in this menial procedure, so we set upon a strategy whereby we could recover from our disability in having last choice of all the luscious fruits, such as peaches in mid-winter. So to each Fellow sitting in the dim candle light we found that

by anticipatory moving of the fruit dish we could confuse each recipient as he put his hand to grasp his choice. We were almost always successful in this innocent game and rejoiced at the end as we each ate the originally chosen luscious fruit saved from our dish. We kept our clever secret and only now I divulge it so as to illustrate that my good friend Erwin Schrödinger had a good sense of humour and enjoyment of a game plan!

We had several Sunday evenings together in this way, but soon, because of new fellowship elections, we were no longer the Junior Fellows. However, we had learned not to be fooled in our choice of the best fruit![4]

It is clear that Schrödinger's colleagues at Magdalen worked hard to bring him into the culture of the College. Magdalen had a Philosophy Club which brought together scientists such as Sherrington, Tansley and Young with philosophers including Weldon, Wolfenden and James Alexander Smith, the Waynflete Professor of Metaphysical Philosophy.[5] Papers discussing the interpretation of science were often written in typescript and presented at the Club over a dinner. A year before Schrödinger arrived at Magdalen, the Club had met on 2 November 1932 to discuss a paper by the physicist Patrick Johnson on "The Basic Assumptions of Physics".[6] This paper emphasised the recent developments in the quantum theory including Schrödinger's equation which gave the energy levels of atoms. Those present discussed whether a mathematical formula used in physics was "simply a translation of a statement into symbolic language or whether it represented a radically different mode of thinking".[6] Several of the scientists present (Johnson, Young, Tansley and the Tutor in Medicine Malcolm MacKeith) took the former view while the philosopher Weldon did not agree. Through discussions such as this, some of the Magdalen Fellows had become aware of Schrödinger's important work. This would have done no harm when his election to a Fellowship was first raised in the College some nine months later.

After Schrödinger's arrival at Magdalen, Arthur Tansley, the Professor of Botany, organised a dinner to introduce Schrödinger to some colleagues from the Magdalen Philosophy Club.[5] This included the philosopher Harry Weldon and also Peter Medawar, who then was a young undergraduate prodigy studying Zoology at Magdalen. The young South

African Solly Zuckerman, a close friend of Tansley, was invited as a guest and recalled meeting Schrödinger at this dinner in his autobiography.[7] Zuckerman did research in Anatomy at Oxford. He went on to work closely with Weldon and Patrick Blackett on developing the British bombing strategy in the Second World War. In due course he became the Chief Scientific Advisor to the British Government in the 1960s.[7]

In Berlin, Schrödinger had attended the remarkable seminars in theoretical physics together with world-leading stars such as Einstein, Planck, Nernst and von Laue and with highly promising younger colleagues or research students including Delbrück, Szilard, Wigner and Weisskopf. In Zurich, Schrödinger had also attended seminars together with brilliant ETH professors such as Debye and Weyl, and this had inspired him to dream up his Schrödinger equation. There was no such seminar, however, of this quality at Oxford, even with the introduction of several excellent refugee physicists from Germany associated with Francis Simon. Schrödinger's Viennese friend, the philosopher of science Karl Popper, wrote:

> In Oxford I met Schrödinger, and had long conversations with him. He was very unhappy in Oxford. He had come there from Berlin where he had presided over a seminar for theoretical physics which was probably unique in the history of science... In Oxford he had been very hospitably received. He could not of course expect a seminar of giants: but what he did miss was the passionate interest in theoretical physics, among students and teachers alike.[8]

Schrödinger had been elected to a prestigious Fellowship at Magdalen College for five years. Few other refugee academics at Oxford had the privilege of being elected to College Fellowships at that time. He was not a theoretical physicist who had no other interests apart from mathematics. He was very knowledgeable on classical literature and had a deep interest in several aspects of philosophy, both areas very strong in Oxford and significant at Magdalen College. However, it seems he did not take up the opportunity to engage with many of his colleagues at Oxford in a major way despite the strong encouragement from George Gordon and several Magdalen Fellows.

As usual his dependable wife Anny summarised things rather succinctly:

> In Oxford he was not so happy because Oxford is no scientific centre. He really was paid for nothing – just because he was Schrödinger, of course, they gave him a high salary, but he had no duties whatsoever. He couldn't even give a lecture because the lectures are all made out. The scientific centre was Cambridge, of course, and not Oxford. He always called his high salary in Oxford, "I feel like a charity case."[9]

There are several other reasons why Schrödinger seemed to find it hard to settle in Oxford. He was an independent and informal character. He did not like traditions, rules and formal dress, which are still prevalent in Oxford. He was a lone scientist and not a collaborator. Furthermore, as a Nobel Prize winner he was at once distracted by many invitations to visit departments overseas and was always receiving job offers which he, sometimes rather foolishly, often took all too seriously. In addition, at Oxford, where the humanities still largely dominated, his great work was not known by many academics. The emphasis at Oxford was also very much teaching and lectures to undergraduates followed up by written examinations. Schrödinger was not used to teaching in this way and preferred high-level lectures on the latest research.

William J.M. Mackenzie was a Fellow teaching Politics at Magdalen who was subsequently to visit Schrödinger in Austria when there was considerable uncertainty about his safety there. Mackenzie summarised the changing atmosphere during Schrödinger's time at Magdalen:

> Magdalen was then being manoeuvred into a new period in which it sought intellectual rather than social distinction; what had been the softest and easiest College in Oxford, Gibbon's college, that of Oscar Wilde and of Edward VIII as Prince of Wales, became before anyone had noticed a focus of what were then radical politics and leftist talk. The heart of the matter was to use the College's wealth radically: and this meant for us primarily more scholarships, more graduate work, more research Fellowships, with selection at each stage on intellectual promise without regard to College connection, athletic talent or clubbability.

> We made ourselves rather unpopular at Eton and among old Magdalen rowing men, and our radicalism now looks rather old-fashioned. But we moved the place, and we got good students.
>
> There was of course an old guard in college, but fading through efflux of time and the strength of new appointments. There were also two great names, Sherrington and Schrödinger, but I knew neither well. Then there was what might be called a young guard, in particular C.S. Lewis and Bruce McFarlane, men of immense ability and energy but not radicals. But those who talked and drank beer in Harry Weldon's room would certainly include John Young and Peter Medawar in biology, Pat Johnson and James Griffiths in physics, John Morris in law, John Austin (but a little austerely) in philosophy.[10]

There was still the question of an Oxford degree for Schrödinger. As had been explained to him by Magdalen Vice-President Stephen Lee, he had to wear a demy's (scholar's) gown when dining in the Hall until he was admitted to an M.A. by the University. This must have been somewhat undignified for an academic of Schrödinger's status. Then on 30 January 1934 it was announced in the *Times*: "the degree of M.A. was conferred by decree upon Dr E. Schrödinger, Fellow of Magdalen College."[11] The photograph on the next page shows Schrödinger with his M.A. gown and mortarboard next to the Chapel entrance in St. John's Quad at Magdalen. He was looking pleased and was now an official member of the University.

Schrödinger and Anny had eventually found reasonably comfortable accommodation in a pleasant house at 24 Northmoor Road in Oxford. This was, however, only just in walking distance from Magdalen where many of the Fellows lived in rooms in the college taking up their entitlement under the Magdalen statutes. It was not easy for him just to drop in for dinner in his college. In addition, his personal life was becoming more complicated.

The ICI financial support was also used to bring Arthur March to Oxford on leave from Innsbruck. March retained a junior professorial position at the University of Innsbruck but he would need extra funds to live in Oxford. The justification for funding March was that he would collaborate with Schrödinger on a book on wave mechanics and would do research relevant to the interests of ICI. March also brought his wife

Schrödinger at Magdalen College in his M.A. gown, 1934.

Hilde who was expecting a baby in May 1934, some nine months after her bicycle ride in the Tyrol with Schrödinger. Arthur and Hilde lived at 86 Victoria Road which was a comfortable house just a few minutes' walk from the Schrödingers' house at Northmoor Road.

Patrick Blackett had become one of Schrödinger's closest friends in England and would have some of the most personal correspondence with him in the coming years. They had met with their wives at the last Solvay Meeting in Brussels and Schrödinger wrote to Blackett in English on 17 February 1934 expressing some concern about his housing situation in Oxford:

> You sent me a telegram something like the 12th or 15th of November. Which I have not yet answered because: (1) I did not know your address in that moment, and, (2) I had to answer about 5 or 6 inches of congratulations of which I first answered, (3) about 4½ inches of these, which I was liable to forget, unless I got rid of them fairly soon. Hence forgive me. For I know I should not forget you.
>
> I was so glad to meet you and your wife in Bruxelles, and the only thing I regretted was that you were in London and I was in Oxford. I must apologise for that. For I feel sure that you do not regret to be in London instead of being in Oxford.
>
> We have not yet settled down here. We have been absorbed in hunting for a house to live in, for weeks and months. It is only now, that we are seriously hoping to obtain one in two months from now, about the middle of April. After this I do hope that the distance between Oxford and London will not be prohibitive and that we shall see you either here or there, as often as possible.
>
> In the early days of March I am leaving to go to America for 5 or 6 weeks, my wife is going to Austria for that time. After which we'll be back in April to take possession of the "promised land" (das gelable Land, as Kanaan is called in the bible).[12]

On 30 May 1934 a daughter Ruth was born to Hilde March at the Acland hospital in Oxford. The father was registered as Arthur March. Then on 1 December 1934 Ruth was baptised in the chapel of Magdalen College by Adam Fox, the Dean of Divinity. Her father was recorded as Arthur March, University Professor, Innsbrück.[13] Although there was no mention of this at that time, and Ruth was not told herself until some 17 years later, her father was really Erwin Schrödinger. It should be noted that Adam Fox was one of the Inkling group with his colleague C.S. Lewis and J.R.R. Tolkien. He became Professor of Poetry at Oxford and then a Canon at Westminster Abbey where his ashes are buried in Poet's Corner very close to a monument recently dedicated to C.S. Lewis and quite close to one commemorating Paul Dirac.

Much has been written about the relationship between Hilde March and Schrödinger in Oxford, Austria and Dublin.[14] However, this was never hinted at in the many letters written by Gordon and Lindemann who were the people most responsible for bringing them to Oxford. Gordon worked very hard to enable Schrödinger to return to Oxford after he had left for Austria in 1936 and found himself in serious difficulties with the authorities there two years later. It would be hard to understand that Gordon would have done this if he had concerns about Schrödinger's private life or, indeed, had heard troubling talk on this topic in Oxford. Due to the High Table dining and excellent wine cellars there is much gossip at Oxford.

For most of his research career Arthur March explored a possible fundamental principle of quantum mechanics that there exists a minimum quantity of time called a chronon, just as Planck's constant is a minimum quantity of action (with units of energy multiplied by time). Heisenberg himself also published several papers on the chronon. March published nearly all his papers in German in journals such as the *Zeitschrift für Physik*. However, just before coming to Oxford he published a paper on a mathematical theory on regulation according to particle shape that continues to be highly cited to the present day in fields ranging from geology to materials science.[15]

The relationship between Arthur March and Schrödinger was always very cordial and March acted as a doting father to Ruth. In his three-year period in Oxford from 1933–6 March did not have a college position and also did not have a formal position in the Physics Department. He did publish one scientific paper in 1935 in the *Transactions of the Faraday Society*.[16] This paper, written in English, had the title "On the Absorption Theory of the Electrokinetic Potential" and describes the forces between colloidal particles. Somewhat unusually, he did not provide an institution or address with his paper and this may have been intentional due to the ambiguity of his position at this time. At the end of his paper he did express the acknowledgement: "I have much pleasure in thanking the Imperial Chemical Industries, whose generosity enabled me to carry out this investigation", a phrase which was also used by Schrödinger for some of his papers published in Oxford during this period.

The field of this paper by Arthur March was in colloids and was relevant to understanding the separation of mixtures. This area of research was of commercial interest to ICI. However, when March returned to

Innsbruck in 1936 with his family he did not publish again on colloids, reverting to his more fundamental theoretical work. He stayed in Innsbruck for the rest of his career leading a strong department in theoretical physics.

Einstein

Albert Einstein had also spent a small amount of time at Oxford in the early 1930s with his visits organised by Lindemann. Following the confirmation in 1919 of his General Theory of Relativity in astronomical observations by the Cambridge astrophysicist Arthur Eddington, Einstein had become a scientific superstar known the world over. His informal appearance and regular comments to the press added to his fame. Lindemann had arranged funding from the Rhodes Trust to support a visit by Einstein and in May 1931 collected him from a boat arriving at Southampton docks. His chauffeur drove Einstein in Lindemann's Rolls Royce to Oxford to be received with great acclaim. He stayed for nearly one month and gave a lecture to a packed Rhodes House of over 500 students and faculty. The lecture on "The Theory of Relativity" was given in German, without notes and with much use of the blackboard. The *Oxford Times* noted: "The front of the hall was filled by college principals, with younger members of the university seated at the back and in the gallery."[17]

Einstein gave two more lectures to smaller audiences and, after the third lecture, the *Oxford Times* reported that he was "gesticulating helpfully in curves with the chalk to explain it and turning repeatedly from his audience to the board and back".[18] It seems that very few of those attending understood a word of his lectures but they were still pleased to experience the great Einstein describing his work. Indeed, it seems that the Dean of Christ Church Henry White, a biblical scholar, slept throughout his final lecture.[19] One of Einstein's blackboards was kept after the lecture by Robert Gunther, a Fellow of Magdalen College who had founded the Museum of the History of Science in Oxford. This blackboard is still there today. The *Times*, *Nature* and the *Oxford Magazine* all reported on Einstein's lectures. The *Spectator*'s News of the Week column said: "while welcoming the benefit of Einstein's lectures for Anglo-German relations...to indicate the drift of his argument is frankly beyond our power."[18]

William Golding, who won the Nobel Prize for Literature in 1983 and wrote the famous novel *Lord of the Flies*, was an undergraduate studying natural sciences at Brasenose College in 1931. He recalled meeting Einstein in the grounds of Magdalen College:

> I was looking over a small bridge in Magdalen Deer Park, and a tiny moustachioed and hatted figure came and stood by my side. He was a German who had just fled from the Nazis to Oxford as a temporary refuge. His name was Einstein. But Professor Einstein knew no English at that time and I knew only two words of German. I beamed at him, trying wordlessly to convey by my bearing all the affection and respect that the English felt for him. It is possible – and I have to make the admission – that I felt here were two grade-one thinkers standing side by side; yet I doubt if my face conveyed more than a formless awe. I would have given my Greek and Latin and French and a good slice of my English for enough German to communicate. But we were divided; he was as inscrutable as my headmaster.
>
> For perhaps five minutes we stood together on the bridge, undeniable grade-one thinker and breathless aspirant. With true greatness, Professor Einstein realized that any contact was better than none. He pointed to a trout wavering in midstream. He spoke: "Fisch." My brain reeled. Here I was, mingling with the great, and yet helpless as the veriest grade-three thinker. Desperately I sought for some sign by which I might convey that I, too, revered pure reason. I nodded vehemently. In a brilliant flash I used up half of my German vocabulary. "Fisch. Ja. Ja." For perhaps another five minutes we stood side by side. Then Professor Einstein, his whole figure still conveying good will and amiability, drifted away out of sight.[20]

Lindemann arranged for comfortable accommodation for Einstein in Christ Church looking out on to the majestic Tom Quad but, like Schrödinger, he did not take to college dinners and described them as "bizarre and boring".[19] Einstein was also awarded an Honorary Degree in a ceremony in Christopher Wren's Sheldonian Theatre with an oration in Latin on his great works. There was a misunderstanding over the publication of his lectures by the authorities at Rhodes House but, after returning to Berlin, Einstein did write a letter of thanks to Lindemann emphasising "the wonderful weeks he had spent in Oxford".[19] However, in 1931, the political situation in Germany was changing rapidly and Einstein told Lindemann: "The situation here is horrible. All money values have disappeared, and the people are disturbed and embittered against the Government. The future lies threatening and dark."[19]

Encouraged by the success of the visit, the huge publicity it received, and the emerging problems in Germany, Lindemann managed to persuade his colleagues in Christ Church to offer Einstein a Studentship (the Christ Church version of a Fellowship) with a stipend of £400 per annum, a dining allowance and a set of rooms for five years. The condition associated with this generous offer was that Einstein would spend a month each year in Oxford. Einstein was also having discussions with the California Institute of Technology on a similar visiting position but Oxford was much closer to his home at that time. In 1936 Schrödinger was to manoeuvre an analogous arrangement for himself with Magdalen College when he departed to Austria.

Einstein then returned again to Oxford in May 1932 and gave the Rouse Ball Lecture on how to include electromagnetism in his general theory of relativity, and this time he did speak a little English in his lecture.[17] However, while in Oxford, Einstein had a visit from Abraham Flexner, the Director of the newly created Institute for Advanced Study (IAS) in Princeton, New Jersey, USA. Flexner raised the possibility of Einstein moving permanently to his Institute with a very high salary.[19] By 1933, with the situation getting much more difficult back in Berlin with even his life under threat, Einstein gladly accepted this offer.

Therefore, by the time Einstein had his third visit to Oxford in June 1933 his mind was elsewhere. Nevertheless he did lecture on "The Method of Theoretical Physics" and also on "The History of Atomic Theories". A sketch in chalk by the artist F. Rizzi was completed and signed by Einstein, and this still hangs in the Senior Common Room of Christ Church.

Einstein's introduction to Oxford in 1931 with all its publicity and wide interest was very different to that of Schrödinger in 1933. Despite the remarkable coincidence that he heard he won the Nobel Prize in Oxford just hours after he had been admitted as a Fellow at Magdalen College there was minimal local and national publicity on Schrödinger's new appointment. There was not even an introductory lecture. Schrödinger was not then a household name and the significant advances from quantum mechanics had not grasped the imagination of the public. Sir William McCrea was a distinguished astronomer and mathematician who in 1936 became a Professor at the Queen's University Belfast. From there he was to engage with Schrödinger when he moved to Dublin during the Second World War. McCrea wrote on Schrödinger's time at Oxford:

> He seems at the time to have made little impact here in England, at any rate outside of Oxford. During those years I was at Imperial College (making occasional visits to Oxford) yet I did not even know that Schrödinger was in the country. It seems true to say that during this interval he published no major original work in physics. His thought was evidently turning rather toward the philosophy, in a broad sense, of physical sciences.[21]

Over the years the appreciation of Schrödinger has changed. With the arrival of the quantum age in the 21st century, and with the continuing public interest in his imaginary cat, the name of Schrödinger is much better known to the academic community and the public in 2021 than it was in 1933.

Once Einstein had started at Princeton with his exceptional salary of $16,000 a year and an atmosphere of safety compared to the developing turmoil in Europe he wrote to Lindemann saying he would be unable to continue to visit Christ Church and asked if the stipend could be used for other German refugees in need.[19] However, Lindemann would not give up and after learning that Schrödinger was to visit Princeton in the spring of 1934 he asked him to meet with Einstein and see if he could persuade his close friend to continue his visits to Oxford.

Princeton

Just after the announcement of his Nobel Prize, Schrödinger had received the telegram from Hermann Weyl, who had also been recruited to the Institute for Advanced Study at Princeton, sending congratulations to him and his very close friend Anny and asking: "What about the Princeton invitation?"[22] With the dramatic advances in theoretical physics in Europe, and particularly in quantum mechanics, the leaders of American universities were very keen to make appointments in this field of research. In the 1930s American universities were expanding at a very fast rate and wanted urgently to become internationally competitive in the sciences. It was realised that with the mass exodus of the leading theoretical physicists from Germany a significant opportunity was arising. In due course, this opportunity was to revolutionise American science.

By 1933 there were already a small number of American scientists who had been appointed to US universities and had spent time in one of the leading European centres for quantum mechanics. John C. Slater had worked with Bohr and Kramers in Copenhagen and had applied the old quantum theory to the interaction of radiation with matter. With the emergence of Schrödinger's wave mechanics he developed several practical applications of the theory including his Slater Determinant. This is a concise and general mathematical method which ensures a wave function of a multi-electron system satisfies the Pauli principle of no two electrons having the same quantum numbers. Slater also showed how the mathematical form of the orbitals for the electrons in the hydrogen atom can be modified to be used in calculations on larger atoms. He became the head of the MIT department of physics in 1930 and subsequently worked largely on solid state wave mechanics. He was nominated several times and was unlucky not to win the Nobel Prize.

Close by at Harvard, John Van Vleck had just been appointed to a Chair. He had hosted Schrödinger on his visit to Wisconsin in 1927 and was an expert on the quantum theory of magnetism. He eventually won the Nobel Prize in 1977. Another Nobel Prize winner already mentioned in this book who came to Zurich to work with Schrödinger was Linus Pauling. By 1934, he was a Professor at the California Institute of

Technology in Pasadena and was applying wave mechanics to molecules, research that would lead to his Nobel Prize in Chemistry in 1954. Also advancing this field at the University of Chicago was Robert Mulliken. He had worked with Born and Friedrich Hund in Göttingen just at the time Schrödinger's papers were published. With Hund, Mulliken developed Molecular Orbital Theory which won him the Nobel Prize for Chemistry in 1966.

A new appointment in the USA was the brash and brilliant J. Robert Oppenheimer. He had gone to Europe, first working with Patrick Blackett at Cambridge with whom he fell out in a serious way. However, he redeemed himself in Göttingen for his doctoral research with Max Born where they showed how the motion of electrons and nuclei can be separated in wave mechanics, an approximation which underlies much of the theory of molecular quantum mechanics to the present day. He had described Schrödinger's wave mechanics theory as "perhaps one of the most perfect, most accurate, and most lovely man has discovered".[23] On returning to the USA, Oppenheimer was given an unusual joint position between the University of California, Berkeley and the California Institute of Technology.

Oppenheimer became world famous for his leadership of the Los Alamos Laboratory in the Second World War which produced the first nuclear weapons. Due to his previous work in Cambridge and Göttingen he had got to know many of the leading scientists in Europe at that time whom he recruited to the Los Alamos project. This included Wigner, Bethe, Franck, Szilard, Teller, Peierls, Frisch, Chadwick, Cockcroft, von Neumann, Goeppert Mayer, Fuchs, Fermi and even Niels Bohr.

After the war Oppenheimer became Director of the Institute for Advanced Study at Princeton. However, in 1934 the University of Princeton felt it had to keep up with its Ivy League competitors and appoint a famous expert in quantum mechanics. In typical American style they decided to go straight to the top and Luther Eisenhart, the Dean of the Graduate School, had named Heisenberg and Schrödinger as the leading candidates. He was friendly with both Weyl and Einstein and it is clear from their letters that they were encouraging an approach to Schrödinger.

Accordingly, Schrödinger was invited by Dean Eisenhart to visit Princeton to give some lectures for a month and he left England on 8 March 1934 on the ship President Harding. In 1928 he had not been so impressed by the cities on the US East Coast but he was to look again at Princeton. At Princeton University was Rudolf Ladenburg who was one of Schrödinger's close former colleagues in Germany. They had overlapped in Breslau and Berlin, and Ladenburg had done important experimental work on atomic spectroscopy which had confirmed some predictions of wave mechanics. Also at Princeton University as a refugee was Schrödinger's former colleague from Berlin, Eugene Wigner, who then had a visiting position.

On 29 March 1934 Schrödinger wrote from Princeton to Lindemann in English on the possibility of Einstein returning again to Oxford:

> I suppose you are waiting for news from me concerning our friend A.E. and his coming to Oxford. I did not want to write you before I had the feeling his answer was really definitive – and of course I did not wish to urge him, because I feared that might still reduce the likelihood of a positive answer. But now I asked him once more adding that it would of course be desirable for you to know as soon as possible. Well I am sorry to say he asked me to write you a definite no. I really wanted to cable but unfortunately I said so and he had objections to it.
>
> The reason for his decision is really that he is frightened of all the ado and the fuss and the consequent duties that would be laid upon him if he came to Europe at all. He considers the only way of escaping is to stay in America this summer. I also told him the idea of so to speak transferring the grant that was at disposition for this purpose to some other refugees who need it was not feasible. He understands this, of course, though he regrets it. That is all that has to be said in this matter.[24]

Schrödinger added some news that would have been of some concern to Lindemann of an offer from Princeton University and also some comments about his uncertain position in Oxford:

> Other things have come up here on which we shall have to have confidential talks after my return. There is not much use of discussing them

in a letter yet since – in spite of all the confidentiality – rumours spread out almost in contradiction to the speed of light, I should like to avoid you hearing of them in this way.

Yesterday a permanent chair at this university was offered to me. Strange situation! It is only in three weeks from now that my wife will be busy to move us at last into definite and permanent quarters in Oxford. And precisely during these weeks I am forced to take a preliminary consideration of a new aspect, under which the state reached in Oxford would turn out to be not so very "permanent". I am not frightfully happy at that, yet I am aware that I have to consider things seriously.

For after all and in spite of all so called scientific reputation I am without what a man of my age and metier considers a Lebenstellung (job for life). And if *e.g.* I were drowned in the passage, I am afraid that my wife would neither live upon the German pension nor the "Schrödingergleichung" (Schrödinger equation).

Well this thing will take its due time and of course everybody here understands that my utterance of yes or no, not even of probably or improbably, can be expected from me before personal discussion with (1) the Oxford people and (2) with my wife. I also pointed out at the first talk my great indebtedness to Oxford and that for this reason alone (if not for many others) the decision would be a very difficult one.

Well, we'll see. I really did not want to enter in any details at the moment, the whole idea is still too new. I just wanted to tell you about it and in order that you should not hear it from somebody else. Please take the thing as confidential, in that way you don't speak about it except to those whom it might be in some way or other a concern. I don't write to anybody else in Oxford, except to you, about it.[24]

At Princeton, Abraham Flexner was Director of the Institute for Advanced Study (IAS). This was a separate institution to the University of Princeton but did share some of the same buildings known as Fine Hall. IAS had been set up by a significant donation just four years before with a mission, in some ways similar to All Souls College in Oxford, to undertake top-level research at the postdoctoral and senior level. The IAS had already recruited both Einstein and Weyl. In making a report on Schrödinger's visit, Flexner wrote:

> His (Schrödinger) opinion of the situation in Princeton in mathematical physics...coincides with that I have received from other sources. While he is technically a member of the Princeton University faculty he has had quite as much contact with our own faculty as with the Princeton faculty. Professor Einstein and all our other members of staff attended his regular seminars. One morning I had to see Professor Einstein on an urgent matter... When I entered I found Professor Einstein and Professor Schrödinger in their shirt sleeves at the blackboard engaged in an animated discussion... This is a general illustration of the kind of cooperation which is taking place at Fine Hall. Though the time during which we have been in operation is brief, there has not been a ripple to disturb the smooth surface of intellectual and spiritual cooperation and activity.[25]

This statement, at first seeming favourable to Schrödinger, at once raised a problem associated with his visit. In Princeton, the IAS had very quickly become the dominant intellectual institution over the University in theoretical physics. When Schrödinger had been in Zurich he was at the University there and not the more prestigious ETH. Since then he had taken up a most distinguished Chair in Berlin, replacing Max Planck, and after that a Fellowship at an Oxford College. He did not want to move again to a position in an institution ranked inferior to another more prestigious one very close by. A Professorship at IAS would have been ideal for him with his close friends, Weyl and Einstein, established there. Flexner, however, had recently taken a leading mathematician James Alexander from the University to the Institute and both Einstein and Weyl had also been transferred this way. Flexner was firm he did not want to embarrass the University again, and do the same with Schrödinger.[25]

After his return to Oxford, Schrödinger had a visit from Hermann Weyl who had made a trip back to Europe. On 19 June 1934 Weyl wrote in German to Flexner from Switzerland:

> I visited Schrödinger in Oxford. Of course we talked a lot about Princeton. On the essential point that Fine Hall was a paradise, I didn't need to persuade him anymore. Other concerns, too, which he had

earlier harboured about America in general, were also dwindling. Still, it is clear to me he is more inclined to decline the call to the university than to accept it. I find such a decision in his position bold. It is obviously an independent hope that it would certainly contribute a lot to his decision if he did not allow any future appointment to the Institute to be obstructed. He said you told him it is likely the Institute will be expanded in theoretical physics.

What I heard about Heisenberg in Europe seems worse and worse with the conditions in Germany but he is indispensable there, and there is a very small prospect to win him over to Princeton. Of the leading theoretical physicists: Bohr, Heisenberg, Schrödinger, Dirac (and Fermi?) it would probably only be Schrödinger and after that Dirac who may come to Princeton. From the latter – I briefly visited him in Cambridge – I have no aversion as to how he himself would react to an appeal from Cambridge. From Schrödinger, however, I have the definite impression that if he has a call from the Institute he would accept safely and smoothly. He finds the university offer inadequate, especially as regards the pension conditions, *etc.*, but next to it also the much more appealing, freer spirit at the Institute.[25(t)]

Schrödinger then wrote in English to Flexner on 25 June 1934, from Oxford:

This is to tell you that I have just written two letters to Dean Eisenhart and President Dodds telling them I could not accept the offer of the Professorship offered to me by the University of Princeton. The very kind and amiable interest you let me feel in this affair encouraged me to make some further remarks, but which are so unusual that before I venture to do so I must apologise in detail, telling you the single arguments that encourage me. The first as I said was your very kind interest and concern in the question. The second, that you told me, you very probably would continue the building up of your Institute in the direction of theoretical physics.

The third is that as long as the offer from the side of the University was in suspense, your loyalty towards the latter would absolutely prevent you from hinting at any other possibility. The fourth is that, hearing I had refused, you might easily be induced to infer that I do not like

Princeton after all and that I have put off the idea of going there all together. Now I should like to be allowed to tell you that this is not so.

My enchantment at the beautiful scientific milieu of Fine Hall is genuine and undiminished. Moreover my initial apprehensions with respect to a transplantation from Europe to the USA. I have been overcome to a considerable extent by all what my friends Ladenburg, Einstein, Weyl, Veblen spoke to me and wrote me in very kind letters. It was after this that I carefully reconsidered the financial side, especially also the conditions of pension, security for my wife, *etc.* And these considerations formed one of the most important points why I finally could not make up my mind to accept the University Professorship.[25]

Flexner then at once wrote to Eisenhart at the University saying:

I have no recollection of speaking with Schrödinger regarding the possibility of the Institute expanding on the side of theoretical physics though I may have mentioned it in talking of Dirac's visit. It is evident however, I think, that his declination of the offer is based on the hope of an offer from the Institute... under no circumstances will the Institute compete with the University for anyone... what I hope is that he will write to you in such a way as to make it possible to reopen the matter.[25]

Flexner then wrote back with a negative response to Schrödinger on July 4, 1934:

I need to proceed with great caution because of the financial situation in the country, and the anxiety of the (IAS) Trustees that in the short time I have left I would start other schools either in the humanities or economics or both. I am delighted that you like Princeton and hope that under the circumstances you will be inclined to reopen the Princeton offer with Dean Eisenhart and you should have no hesitation in doing so. Unfortunately, I am unable to advise as I do not know the terms upon which either Oxford or the Princeton Professorship is held.[25]

In fact, the position Schrödinger had at Oxford had unclear financial security and no tenure. It was renewed annually on an uncertain basis and had no prospects of a pension. The disappointed Schrödinger was still hoping Flexner would reconsider and make an offer from the IAS. On 9 July 1934 he followed up with another letter to Flexner, this time written in German:

> I beg you to be convinced that I answered the call from Princeton from myself alone without consideration of future possibilities. Decisive were circumstances which I had discussed with Dean Eisenhart in the first days after receiving the invitation. It seems to be mainly a question of the pension for myself and my wife. She would only receive something like $800 a year. The distance between that and what I would like seems too great to be bridged.[25(t)]

Einstein was not very happy with the decision and wrote a year later on 4 September 1935 to Flexner saying that he had been in correspondence with Schrödinger who he "held in high esteem and he would in my opinion be a real achievement for our Institute".[25] Flexner wrote back saying in frank terms:

> The Schrödinger matter is a delicate one, which I cannot completely settle without talking to you as well as the Princeton people... Schrödinger made a blunder that embarrassed both me and the Institute but I shall handle the matter with the utmost possible discretion and with every desire to do the best for him as well as for you.[25]

In subsequent years, Einstein and Weyl were involved more than once in trying to make an appointment for Schrödinger at the Institute. However, when the next opportunity came up in 1940, Schrödinger's case was considered again but he lost out to Wolfgang Pauli who had been forced to leave Zurich. Then, in 1947 after Pauli returned to Zurich, there was a major search conducted by Weyl and he reported to the Trustees of the Institute that Bohr, Schrödinger, Heisenberg, Dirac and Fermi were all in the "top rank".[25] However, Robert Oppenheimer had just stepped down from the very high-profile post of Director of the Los Alamos Laboratory. He was appointed to the Institute and also became the Director. Schrödinger, however, clearly liked the format of an Institute for Advanced Study as one was quite soon to be created for him in Dublin.

It should be remarked that it is sometimes mentioned that the Institute at Princeton did not want to make an offer due to gossip on the unusual personal position of Schrödinger back in Oxford.[14] However, his illegitimate daughter was born only after his visit to Princeton. Furthermore, there is no reference to this in the extensive set of papers in the archives at Princeton referring to Schrödinger's visit and correspondence, some of which were marked as confidential and are of an informal and personal nature.[25]

Shortly after Schrödinger's visit, Paul Dirac was invited for a whole year to the Institute for Advanced Study at Princeton with a generous stipend. There was the hope that he would be able to interact closely with the other genius of theoretical physics, Albert Einstein. However, their characters and ways of working did not mix well. Dirac spent the time writing his definitive book *The Principles of Quantum Mechanics* and also met Manci, the sister of Eugene Wigner. She had been married before and had two children. She at once fell for Dirac and they were soon to be married and settled down at 7 Cavendish Avenue in Cambridge. When the couple subsequently met other scientists Dirac would frequently introduce his wife as "Wigner's sister".[26]

In 1938 Wigner himself was appointed to the Jones Professorship at Princeton University which Schrödinger had declined. He would go on to win the Nobel Prize in Physics in 1963 for "contributions to the theory of the atomic nucleus and the elementary particles, particularly through the discovery and application of fundamental symmetry principles". Schrödinger, however, had some reservations about Wigner's work on symmetry. He told Wigner in 1934: "This may be the first method to derive the root of spectroscopy. But surely no one will still be doing it this way in five years."[27]

At the end of May 1934, Max von Laue came to Oxford from Berlin and stayed with the Schrödingers. He was not Jewish but, like Schrödinger, was anti-Nazi and was concerned with the developments in Germany. After his visit, von Laue wrote:

> Schrödinger feels himself to be not at all happy, although he is the only one of the German scholars who has had the honour to be made a Fellow of a college. He speaks of leaving and alternates in his mind between the most distant lands of the earth.[14]

It seems there was some discussion about von Laue taking over Einstein's visiting position at Christ Church.[19] However, nothing came out of the negotiations and von Laue was to spend the Second World War in Berlin.

In the summer of 1934 Magdalen College published a summary of the main events of the previous year for its members and Fellows. Alongside several other achievements, it is clear that the College was proud that Schrödinger had been admitted as a Fellow before the announcement of his Nobel Prize. The account included:

> H.R.H. the Prince of Wales was appointed by H.M. The King to be Honorary Colonel of the Oxford University Officers' Training Corps.
>
> The President was elected Professor of Poetry.
>
> The Archbishop of Canterbury was appointed High Almoner to H.M. The King.
>
> Mr Geoffrey Dawson and Sir Henry Miers received the Hon. Degree of D.C.L. at the Chancellor's Encaenia.
>
> Sir Charles Sherrington was awarded the Harben Gold Medal.
>
> Professor Robinson was elected a Foreign Associate of the National Academy of Sciences, Washington.
>
> Mr Kenneth Clark was appointed Director of the National Gallery and also, more recently, Surveyor of the King's Pictures.
>
> Dr Erwin Schrödinger was elected to a Supernumerary Fellowship, and was subsequently awarded, jointly with Professor Dirac of Cambridge, the Nobel Prize in Physics.[28]

Spain

Not long after he had returned from Princeton Schrödinger made a visit to Spain in August 1934 to give a course of lectures spoken in French at Santander.[29,30] A new Spanish Republic had been established in 1931 and there were many new reforms and initiatives. This included the creation of an international summer school in Santander where foreign experts could be invited to give lectures to Spanish students. The government had gone as far as converting the king's summer residence, the Palacio de la Magdalena, for this purpose. By 1934 the conversion was complete and it was very appropriate that Schrödinger, a Magdalen Fellow, could enjoy the new accommodation with almost the same name. At one time Weyl had also intended to take part but this did not prove possible.

The invitation to Schrödinger had come from the poet Pedro Salinas. The Dean of the Summer School was the physicist Nicolás Cabrera. He was an experimental expert on crystal growth and had attended the 7th Solvay meeting in 1933 where he had the opportunity to discuss a visit to Spain with Schrödinger. When the Civil War erupted in Spain in 1936, Cabrera himself became a scientific refugee and moved to the University of Virginia in the USA. He returned to Spain in 1971 and the Nicolás-Cabrera Institute at the Autonomous University of Madrid bears his name.

The Director of the University of Santander was a well-known philosopher José Ortega y Gasset. It is likely that the invitation had arisen from the Spanish philosopher of science Xavier Zubiri who had studied in Berlin in the early 1930s where he had met Einstein, Planck and Schrödinger. Zubiri put the six lectures that Schrödinger delivered into a book. They describe in non-mathematical terms the developments of wave mechanics and the relationship to the work of Heisenberg and Dirac. Schrödinger wrote to Zubiri with great enthusiasm about his visit:

> Yours is a happy country. A little thanks to the external gifts that God has given her, but above all thanks to the internal divine gift of a temperament that is receptive to happiness, one which seeks joy and creates it. Four weeks are not enough to find one's feet there, when one has come from coarse Europe. Only later did I realize, and felt a bit ashamed, that I did such a bad job of adapting to it. There will be a next time. I am so glad that Spain exists, that it exists for me, now. Not only because I will come back. It is the knowing that a people exist who see the world in such a way, a people who shape and change, support and improve the very concept of the world.[29]

Schrödinger enjoyed this first visit to Spain so much he decided to return for the spring of 1935 to motor around the country with Anny.[14] He stopped off in the Academy of Sciences to give a talk on "The principle of indeterminacy and its influence on the concepts of World Geometry". He also spoke at the Sociedad Española de Física y Química (Spanish Society of Physics and Chemistry) and at the Academia de Ciencias Exactas, Físicas y Naturales (Academy of Exact, Physical and Chemical Sciences).[30] He was appointed a Foreign Member of the Spanish Academy of Sciences and he also published a short article in Spanish on "Are the true equations of the electromagnetic field linear?".

Einstein had visited Spain in 1923 to great acclaim and had met King Alfonso. He had won the Nobel Prize just two years before and was the star of the Spanish social life during his visit. Einstein was even offered a position in Spain as was the case with Hermann Weyl in the next year. On 17 May 1935 Schrödinger wrote with much enthusiasm to Einstein about his own visits to Spain:

> If I think about it that you had the opportunity to go there and probably so much you want to live outside Madrid. It is difficult for me, to your illustrious and beloved person, to hold back grossly disparaging verbal incursions. So the trip was wonderful. I was with my wife in our little car. We drove a figure of eight through the Spanish peninsula visiting Madrid, Valencia, Gibraltar, Cadiz, Salamanca and Altamira.[30]

Cabrera wrote some years later on Schrödinger's visits. He said:

> Perhaps the visit I remember most for several reasons is that of Erwin Schrödinger. During the year 1935 he visited Madrid, where he gave a seminar in Spanish on quantum mechanics which greatly benefited all the young people who were at that time playing with physics.[30]

However, with the Civil War erupting in 1936, the commentators on science in Spain do not consider that Schrödinger's two visits had an opportunity to influence significantly the development of physics in Spain in the 1930s. With the Princeton matter resolved and the first visit to Spain a fine memory he was able to divert his attention back to matters in Oxford.

Don

In the 1930s newly admitted Fellows at Magdalen College were placed on probation for one year after being appointed. On 17 October 1934 Schrödinger was admitted to what was called "actuality" and he signed the Book of Admission as E. Schrödinger. The previous signature in the book was "Johannes Carew Eccles" which was the Latin for John Eccles. Other signatures on the same page from previous years included "Robertus Robinson", from 1929, and Clive Staples Lewis from 1926.[31] Schrödinger was now an official Oxford don.

Schrödinger could now attend Governing Body meetings and contribute to the decisions of Magdalen College. He was appointed to the Fellowship by Examination Committee. An interview was carried out on 10 October 1934 of candidates to make an appointment providing three years' funding for a Research Fellowship. Two candidates had been shortlisted: D.A. Bell in Engineering and J.H.E. Griffiths in Physics. The majority recommendation from the Fellowship Committee was for Mr Bell. However, Schrödinger was in the minority and wrote:

> I wish to state that of the three examiners it was I whose vote was in favour, I may say decidedly in favour, of Mr J.H.E. Griffiths' claim and contrary to that of Mr D.A. Bell. Trying my best to form an adequate opinion of the standard required for a Fellowship of this college I do not think that Mr D.A. Bell comes up to the standard at the present moment.[32]

The Governing Body of the College was impressed with Schrödinger's report, overturned the majority recommendation of the Committee and elected Mr. Griffiths to the Fellowship instead. Some 34 years later, James Griffiths was elected President by the Fellows of Magdalen College and it is very unlikely that this would have happened if he had not got his first foot on the Magdalen Fellowship ladder due to Schrödinger's intervention. Griffiths was the first student of physics to be elected to the headship of an Oxford college.

Griffiths did research on radar valves in the Second World War, and shortly after that made the first measurement of ferromagnetic resonance which had been predicted by Lev Landau. This is an important effect in solid state physics which has subsequently been exploited in many electronic devices. For this research advance, Griffiths was considered for Fellowship of the Royal Society (FRS) and was unlucky not to be elected. In 1955 Lindemann was asked by the Royal Society to give his view on several physicists being considered for FRS.[33] He recommended Nicholas Kurti and Griffiths, but only Kurti was elected. Kurti had come to Oxford in 1933 as one of the refugee scientists with Francis Simon.

Another physicist on the Fellowship of Magdalen College in the 1930s was Patrick Johnson who gave the tutorials to undergraduates. Like Griffiths he was an experimentalist. He was the only Fellow of Magdalen to ever get a "blue" rowing for Oxford University against Cambridge – but this is something that would not have interested Schrödinger. Johnson and Griffiths could often be found in the Lamb and Flag public house near the Clarendon Laboratory of Physics. In Zurich and Berlin, Schrödinger was known to enjoy a glass of wine with the students but it is unclear if this interest extended to the English pub. Schrödinger had been asked to give some lectures and Johnson remarked:

> In their approach to physics Schrödinger and most of the faculty were poles apart, but we were keen to learn from him. He conducted a series of lectures for staff for our benefit, but they were devoted to the most recent developments of his theories when we badly needed something more basic.
>
> At one of these lectures he was developing an analysis based on hyperspherical harmonics (my term not his) when Lindemann interrupted quite irrelevantly to say that what he thought was important was the transfer of action. After some to and fro of remarks Schrödinger said: "I don't know what you are talking about" and went on with the lecture. Lindemann looked "daggers" and did not turn up subsequently.[34]

Another member of the Clarendon Physics laboratory was Carl Collie who also taught physics at Christ Church. He said:

> The refugees were so aware that theoretical physics was in the most exciting period since Newton that they could not understand how anyone of ability could go on working on the same old problems. Schrödinger thought the Oxford mathematicians were mad. The idea that the old

> syllabus had to be followed, students taught and examinations passed and these matters could not be quickly changed was something quite outside their understanding.[34]

Collie was more positive about Schrödinger than some commentators. It was reported that "Collie regretted the impression historians have formed that Schrödinger had no part in Oxford Physics, having himself been much influenced by him, and having treasured some notes the great man wrote in answer to a query he raised after one of his lectures."[35]

Some of Schrödinger's notes in perfect English for his lectures in Oxford are still available.[22] They are in a form that would be suitable and understandable today for final year undergraduates specialising in quantum mechanics. However, as we have seen from the above remarks, the students and physics faculty in Oxford in the 1930s did not have the background to make best use of the lectures. They start off with some simple basic principles of quantum mechanics and then go on to consider examples. Schrödinger intentionally avoided long-winded mathematics in his lectures, using in most cases the operator formalism developed by Dirac. His examples included the treatment of commuting operators, coupled oscillators, degenerate states and magnetic quantum numbers.

It is interesting that Schrödinger avoided the detailed treatment of the hydrogen atom in his Oxford lectures, which was the work in his first wave mechanics paper which made him famous. He was well aware, for example, that the details of associated Laguerre polynomials would not be welcomed by the students. In one of his lectures he stated with a good sense of humour: "We might proceed to treat as a further example the hydrogen atom. This would fill the blackboard with lengthy formulas which you can find in every textbook (except for those which use the excuse, that you find it in every textbook, for not repeating it!)."[22]

Schrödinger also was sympathetic to typical needs of students. For his first lecture at the start of 1935 he stated: "This lecture is a continuation of that delivered last term, but I'll take into account of the fact that not all of you attended on the last Saturday of last term."[22] Then, as now, Saturdays are not the most popular days for students to attend lectures on quantum mechanics.

It was clear, however, that Schrödinger was becoming frustrated with his position at Oxford. He wrote on 17 March 1935 to Einstein with concern that he had to wait for a Chair to become available:

> There is no value at all for me here. And when I think further, I say to myself: I sit here and wait for the death or the complete decrepitude of one very dear old gentleman (Love) and on the fact that they might make me his follower. I don't mean to pretend that this hurts my feelings, but it hurts my sense of honour.[36(t)]

This letter implies that the Chair at Oxford held by "dear old gentleman" Love was under consideration for Schrödinger. The prestigious Sedleian Chair of Natural Philosophy at Oxford was founded in 1621 and had been occupied since 1899 by Professor Augustus Love. He worked on the mathematical theory of elasticity and also on wave propagation – a subject clearly linked to Schrödinger. The L-wave in seismology is named after Love. A winner of the Royal Medal of the Royal Society, he was 72 years old in 1935 but there was then no retirement age at Oxford and he was still firmly in post. Professor Love died in 1940 and it was only after the war that he was replaced with the applied mathematician Sydney Chapman who developed some of the fundamental equations of stochastic processes and also discovered the key role of ozone in the earth's atmosphere.

Another possibility for Schrödinger might have been the Chair of Sir John Townsend who was the Wykeham Professor of Physics. He had been appointed to this Chair in 1900 at the very young age of 32 but by the 1930s was very much out of date in physics terms. The University of Oxford was not able in those financially restricted days, even under pressure from Lindemann, to create a brand-new Chair and it was necessary for one to become available through retirement or death. Townsend still had some years to go and Lindemann himself held the other Chair in Physics.

In due course, the University managed to retire Townsend in 1941 when he refused to instruct Royal Air Force cadets. However, this was all too late for Schrödinger. The Wykeham Chair was converted eventually to the subject of Theoretical Physics in 1946 and the first holder was Maurice Pryce who had obtained his PhD with Max Born in Cambridge on the wave mechanics of the photon and had also married Born's youngest daughter Gritli. Like many others, Pryce did not find it easy to get on with Lindemann and left after eight years to take up a Chair at Bristol.

Pryce was replaced by Willis Lamb in 1956. He had just won the Nobel Prize for measuring the fine structure of the electron in the hydrogen atom, known as the Lamb Shift. This was quite a daring appointment for Oxford as he came from Stanford University and had no previous UK experience. Lamb did both experiments and theory and had been a student of J. Robert Oppenheimer. His work was crucial for confirming the new quantum electrodynamic theories for the interaction of electromagnetic radiation with charged particles which were developed by Richard Feynman, Julian Schwinger and others. They were a new group of quantum theoreticians, mainly trained in the USA, who had interacted closely with the refugee scientists from Europe in World War Two scientific projects.

Lamb only stayed in Oxford for six years before being replaced in the Wykeham Chair by Rudolf Peierls in 1963, a German refugee scientist well known to Schrödinger. He had worked with Sommerfeld, Heisenberg and Pauli and did novel research on applying wave mechanics to metals, solids and nuclear problems from the late 1920s right through to the 1960s. With the rise of the Nazis, and with the help of the Academic Assistance Council, Peierls had first moved to Cambridge and then Manchester. He was elected to a Chair in Applied Mathematics at the University of Birmingham in 1937 and subsequently did very important scientific work related to the war effort. Of all the many refugee physicists assisted by the Academic Assistance Council only three, Peierls, Born and Simon, were able to find permanent positions in Britain before 1940. All were elected as Fellows of the Royal Society and both Peierls and Simon were knighted. Like Peierls, Simon was also elected, in due course, to a prestigious statutory Chair in Oxford. Both Born and Peierls were awarded the prestigious Max Planck Medal after the Second World War, but Born was the only one of the three to win the Nobel Prize.

Schrödinger did find some like-minded academics in Oxford. Henry Whitehead was a Tutor in Mathematics at Balliol College and was an expert in algebraic topology. He held joint seminars with Schrödinger that were attended by undergraduates. Brebis Bleaney, who attended one such seminar and would 20 years later take Lindemann's Chair, said "a space of infinite dimensions meant little to me".[37] Two years later, after he had escaped from Austria and returned to Oxford, Schrödinger stayed with Whitehead who also helped other refugees from Europe. Whitehead worked in operations research in the Second World War in Patrick Blackett's team and then became the Waynflete Professor of Mathematics at Magdalen College in 1947. He died in 1960 while on sabbatical at the Institute for Advanced Study in Princeton, a place which has already been discussed in the Schrödinger context.

Schrödinger was well-read in the classical languages and he had a strong interaction with Gilbert Murray who was the Regius Professor of Greek at Oxford. Anny recalled:

> Sometimes if he was not working very hard on a certain subject he read a lot of books. He was immensely interested in Bertrand Russell and in Gilbert Murray. This was a beautiful time in Oxford when we were together with Gilbert Murray because my husband was very, very fond of antique languages – Latin and Greek. He was a very good scholar; he was always first in his class during the whole time he studied. Gilbert Murray translated the Greek plays and we often were there and they spoke together. I found one letter in which Gilbert Murray said: "I'm so pleased that you think so highly of my translation." It was very nice, indeed.[38]

Hansi Bauer-Böhm also said:

> Schrödinger may not have been popular with his neighbours but he also had friends and fans: the famous and charming Gilbert Murray and Lady Mary whose beautiful home he was always welcome at; Whitehead the philosopher and several other professors whose company he enjoyed and who were smitten by his theories and his delicate features – his facial expressions that could quickly display the full range of emotions

> and his many different interests – the mountain guide, the poet and the philosopher – his Erwin Schrödinger amused smile, flashing eyes full of agreement or anger, his forgiving tolerance towards the ignorant.
>
> His mimics also gave away the frailty of his reactions – those of a finely tuned and overly sensitive instrument that registers everything in a slightly exaggerated way, be it mental or physical: sweater on, sweater off, glasses on, glasses off, a regiment of equipment to meet his many inclinations, whether in Portmerion in Wales or in the Vienna Woods.[4]

The indefatigable President Gordon was continuing to write many letters to help secure the annual funding for Schrödinger. The initial arrangement was a salary of £1,000 per annum for three years from ICI with a £100 supplement from the Rockefeller Foundation. A research grant of £150 had also been awarded from Magdalen College and it seems that Schrödinger used this sum largely for travel expenses which he had got used to with his position in Berlin. On 5 October 1934, Gordon wrote to Professor Lauder Jones at the Rockefeller Foundation:

> Your Foundation, at a meeting held in Paris on November 23, 1933, was so generous as to vote a sum of £100 in supplementation of the salary of Dr. Erwin Schrödinger for the year 1934... As the year is now drawing to a close I think it proper to report on the financial disposition of that grant, and on the other financial arrangements made on Dr Schrödinger's behalf, and to express also the hope that the Foundation may see its way to repeat the grant for another year.
>
> The whole £100 has now been paid through the College, in your name, to Dr. Schrödinger and I take it as certain that it will have been expended by the end of the year. The basic £400 from outside sources which I mentioned in correspondence with you last year has been paid, and is guaranteed for 1935. Further, this College on its part, besides extending to Dr. Schrödinger the status of a Fellow of the College has, in joint supplementation with you of the basic £400, paid him £150 from its own funds for 1934. This payment we propose to continue for 1935, and should be happy to think that it might be possible for your Foundation to continue its grant for that year.
>
> I should add that you no doubt have observed that Dr. Schrödinger was awarded, jointly with another scientist, the Nobel Prize for Physics

last November, the divided value of the Prize was, I believe, £3,500, but we considered here that it would be ungenerous to regard that sum as other than a very necessary windfall to Dr. Schrödinger – something which an exile might well be allowed to regard rather as a security for future contingencies than as a present case.

Dr Schrödinger's year in Oxford has, I may say, been a complete success, and everyone who has met him or worked with him here hopes that nothing may occur to cut short his stay. He has taken part in the regular advanced teaching of the University, and has already had a most stimulating effect not only on the best of our younger physicists but also on a number of his scientific colleagues.[32]

However, the Rockefeller Foundation was by now supporting many other refugees and agreed with George Gordon's well-argued request but only for one more year. On 21 December 1934, Gordon wrote to Lindemann:

I have just had notice from the Paris office of the Rockefeller Foundation that they will renew for 1935 the £100 grant which they made for 1934 in supplementation of Schrödinger's stipend here. They at the same time give me notice that their fund for émigré scientists is now exhausted, and that they will not be able to renew this grant for 1936 or following years. I should be glad some time if we might have a conversation about Schrödinger's real financial position and prospects. The more we see of him here, the clearer it is that Oxford ought not to lose him.[24]

Under the terms of his agreement with ICI Schrödinger had to write a report on the research he had undertaken during the previous year. On 9 January 1935 he wrote:

The general trend of research in which I have been engaged during this period was the same as in the preceding seven years. It is closely connected with the revolutionary change in physical theory that took place in 1925 and 1926. I endeavour to outline and develop the new line of thought by (1) clearing its foundations, (2) reconciling the various new points of view with one another and with those of the older ones which must be maintained in spite of the change, (3) trying to push on the progress of ideas into those domains of physical thought which more or less seen to oppose themselves to their intrusion; more especially into the theories of relativity, electrodynamics and gravitation.

Aims of this general kind automatically lead one to study and re-study alternatingly the various presentations of the various theories involved and the new papers that are published in connection with them. Definite progress is seldom reached as the intended result of a unidirectional effort, continued for months and years, it arises rather from lucky contingency, to which the preceding studies bear only the relation of giving it a certain chance to appear.

In the period under consideration there was, I think, one definite case of such "good luck", which has occurred quite recently (December 1934) and is connected with Professor Max Born's unified theory of matter and electromagnetism. By using certain complex imaginary combinations of the electromagnetic vectors to express the important functions of the field, as *e.g.*, the densities of energy and momentum, one finds, in terms of the new variables, extremely simple expressions for these quantities, *viz.* rational and homogeneous functions, hardly more complicated than in Maxwell's theory (which is very astonishing, since Born's theory actually is much more complicated). It is hoped that this form of the theory will lend itself best to introduce in it the concepts of quantum mechanics.

A small success was reached in finding the so-called "fundamental solution" (German: Grundlösung) of Dirac's equations. The solution is equivalent to the Dirac-operator, continued over a finite interval of time. On the other hand it is closely connected with Professor Milne's cosmological theory. I learned later about that at about the same time when I gave a lecture on these questions in Princeton. Dirac himself had made use of the same type of solution in developing his theory of the positron. A more intimate synopsis of these different aspects has not yet been reached.[1]

Schrödinger then gave a brief summary of his visits to Princeton and Spain, and also his interest in new discoveries in cosmic rays and artificial nuclear physics:

During the elapsed period I had frequent opportunity of meeting my scientific colleagues and friends from all parts of the world, in some cases with ample time for exchanging not merely "views" but knowledge. In the Easter vacation I spent one month in Princeton (New Jersey, U.S.A.) gaining insight by extended discussions into the present aims of O. Veblen, H. Weyl, A. Einstein, J. v. Neumann and others (mainly with respect to the work on unified field theory of electrodynamics and gravitation). The month of August I spent in Spain, where I first assisted to the Santiago meeting of the Spanish Society for the progress of science and later delivered four lectures at the "Universidad de Verano" (summer university) in Santander. In the early days of October I attended the physics meeting in London and Cambridge. This meeting may be called a historical event in physical science, both by its numerous and distinguished attendance and by the exciting new evidence of cosmic rays and artificial disintegration, which was presented and discussed.[1]

Schrödinger was starting to make himself known in the UK outside of Oxford. The *Nottingham Evening Post* on 5 February 1935 reported, with some errors, on a "Noted Professor's Visit to Nottingham":

> A distinguished figure respecting research into the composition of the atom, and one whose discoveries rank next to those of Einstein, paid a visit to Nottingham yesterday in the person of Professor Ernst Schrödinger. Prof Schrödinger delivered a lecture at University College, Highfields, under the auspices of the Physical Chemical Society, upon the subject of "Quantum Mechanics, Geometry and Relativity".
>
> He is the originator of the theory of the wave mechanism, and is now following up his studies in this direction at Magdalen College, Oxford, of which he is a Fellow. His discovery of wave mechanics, upon which he has published several books, ranks as one of the most important scientific achievements of the day, and compares in importance with Einstein's theory of Relativity. In fact, he shared with Prof. Dirac, of Cambridge, the Nobel Prize for the most distinguished work in physics.[18]

In May 1935 the *British Broadcasting Corporation* (BBC) made an announcement in the *Times* that there would be a new radio series on "Uncensored Talks on Freedom":

"Freedom" is the title of the principle series of talks given by the BBC from next month to June. The British conception to freedom and how it applies to everyday life is to be discussed by 12 speakers.

The BBC has been at pains to prevent this from becoming a political series in a party sense. But the talks will be uncensored. The Church, the law, letters, industry, employer and employee will be represented as well as politics.

Among those taking part are Mr Bernard Shaw, Mr G.K. Chesterton, the Bishop of Durham, Sir Ernest Benn, Mr Herbert Morrison, Mr J.A. Spender, Lord Eustace Percy, Sir Thomas Barlow, Prof Erwin Schrödinger, Mr Wyndham Lewis and a representative working man. The talks will be given on Tuesday evenings at 10 pm.[18]

Schrödinger was in extraordinary company for this series (including the "representative working man") and was the only scientist and the only refugee from Germany. His 20-minute talk on "Equality and Relativity of Freedom" was listened to by a large audience and Schrödinger had never given a talk like this before. Anny commented on how much he enjoyed this project.[9] The *Radio Times* for May 28, 1935 had a rather flattering photograph of a tanned Schrödinger with a bowtie that needed straightening and said:

Tonight's broadcast is unique in the fact that it is the only one to be given in the series by a foreigner. Professor Erwin Schrödinger had the distinction of winning the Nobel Prize for Physics in 1933. He is at present over in England, and was especially invited by the University of Oxford to conduct his researches there. He knows England very well, and is well-known for his tolerant and witty comments upon affairs in general.[18]

Schrödinger started his talk by saying: "My acquaintance in politics in this or any other country is as feeble as my interest in them." He also spoke with approval of how the individual is protected in Britain and the vital independence of justice. He emphasised that freedom depends on a balance of demands by different members of society. He said: "There have been communities who felt aggravated at persons who did not think and believe the same as everybody was supposed to think." He was careful not to refer directly to Nazi Germany in such a public forum but the implications were clear.

Cat

While in Oxford, Schrödinger published four papers which have been very influential for physics research and the public understanding of science to the present day.[39-41] These papers extended the debate on the appropriate interpretation of quantum mechanics, a topic that Schrödinger had started with Bohr nearly ten years before and had been a major point of discussion at the 1927 Solvay Conference. The first of these papers was submitted to the *Mathematical Proceedings of the Cambridge Philosophical Society* on 14 August 1935 and communicated by Max Born.[39] In those days, to ensure the integrity of a paper submitted to a journal of a learned society it needed to be communicated by a member of the society. That system is still used by a small number of distinguished societies today.

The paper, written in English, had the title "Discussion of Probability Relations between Separated Systems". The first paragraph stated:

> When two systems, of which we know the states by their respective representatives, enter into temporary physical interaction due to known forces between them, and when after a time of mutual influence the systems separate again, then they can no longer be described in the same way as before, *viz.* by endowing each of them with a representative of its own. I would not call that *one* but rather *the* characteristic trait of quantum mechanics, the one that enforces its entire departure from classical lines of thought. By the interaction the two representatives (or ψ-functions) have become entangled. To disentangle them we must gather further information by experiment, although we knew as much as anybody could possibly know about all that happened.[39]

This problem had become quite well known in quantum chemistry when two electronic wave functions (orbitals) based on different atoms in a molecule, which are placed quite far apart, can still be mixed together if there is an interaction between them. The mixing becomes stronger if the two original states have similar energies, even if they are quite far apart in distance. This paper by Schrödinger has become highly cited in modern times as it is an early work discussing some of the principles behind the contemporary areas of research known as quantum information processing and quantum computation. It also introduced the word "entangled" into quantum mechanics. The problem expressed in the final sentence above, with regards to disentangling the states in a practical experiment, remains one of the key challenges in quantum computation.

The paper referenced one published by Einstein, Podolsky and Rosen in which it was argued that quantum mechanics provides an incomplete description of physical reality.[42] That paper had considered two interacting particles and stated that if the position of the first particle was measured, under the principles of quantum mechanics the result of measuring the position of the second particle could be predicted. However, they claimed this was unsatisfactory as no action taken on the first particle could instantaneously influence the other particle as information would then be transmitted faster than light. The phrase "spooky action at a distance" is often associated with Einstein in connection with this paper although it is not stated explicitly there but can be inferred from a letter Einstein wrote to Born on 3 March 1947.[43]

Schrödinger then followed up his first paper on entanglement with a second one to the same journal entitled "Probability relations between separated systems".[40] In this case, the paper was communicated to the Cambridge Philosophical Society by Paul Dirac. This paper develops the concept of mixtures of states that had been introduced by John von Neumann. The paper concludes by stating on mixed states that: "These conclusions, unavoidable within the present theory but repugnant to some physicists including the author, are caused by applying non-relativistic quantum mechanics beyond its legitimate range." As an acknowledgement, these papers stated: "My sincerest gratitude is due to Imperial Chemical Industries, Limited, to whose generosity I owe the leisure for carrying out these studies."

The third paper from Schrödinger in Oxford was published, this time in German, on 29 November 1935 in *Die Naturwissenschaften*,[41] the popular science journal whose editor had been Schrödinger's close friend Arnold Berliner. In the paper, to attempt to illustrate the interpretation of quantum mechanics in an absurd thought experiment, Schrödinger stated:

> One can even set up quite ridiculous cases. A cat is penned in a steel chamber, along with the following device (which must be secured against direct interference by the cat): in a Geiger counter there is a tiny bit of radioactive substance, so small that perhaps in the course of the hour one of the atoms decays, but also, with equal probability, perhaps none; if it happens, the counter tube discharges and through a relay releases a hammer which shatters a small flask of hydrocyanic acid. If one has left this entire system to itself for an hour, one would say that the cat still lives if meanwhile no atom has decayed. The ψ-function of the entire system would express this by having in it the living and dead cat (pardon the expression) mixed or smeared out in equal parts.[41(t)]

The paper had the (translated) title "On the Current Situation in Quantum Mechanics". It was stated in the heading that the paper was by Von E. Schrödinger, Oxford. There was no acknowledgement this time to the Imperial Chemical Industries. There were two other papers on this topic from Schrödinger in the same issue.[41]

Because the great physicist Schrödinger used the example of the life or death of a cat in the paper it has caught the imagination of the public more than any other of his works to the present day. Accordingly, "Schrödinger's Cat states" is also a phrase often used by physicists in connection with highly entangled quantum states or by a microscopic quantum effect having an influence on a macroscopic phenomenon. There have been even more citations to this paper than his greatest Nobel Prize-winning work, the 1926 paper on the Schrödinger equation for the hydrogen atom. However, his equation is taken so much for granted in modern science that very few authors feel the need to cite it. The Schrödinger Cat paper highlighted the possibility of quantum mechanical effects in biology which has also become quite an active field in recent years.

DIE NATURWISSENSCHAFTEN

23. Jahrgang | 29. November 1935 | Heft 48

Die gegenwärtige Situation in der Quantenmechanik.

Von E. Schrödinger, Oxford.

Inhaltsübersicht.

§ 1. Die Physik der Modelle.

In der zweiten Hälfte des vorigen Jahrhunderts war aus den großen Erfolgen der kinetischen Gastheorie und der mechanischen Theorie der Wärme ein Ideal der exakten Naturbeschreibung hervorgewachsen, das als Krönung jahrhundertelangen Forschens und Erfüllung jahrtausendealter Hoffnung einen Höhepunkt bildet und das klassische heißt. Dieses sind seine Züge.

Von den Naturobjekten, deren beobachtetes Verhalten man erfassen möchte, bildet man, gestützt auf die experimentellen Daten, die man besitzt, aber ohne der intuitiven Imagination zu wehren, eine Vorstellung, die in allen Details genau ausgearbeitet ist, *viel* genauer als irgendwelche Erfahrung in Ansehung ihres begrenzten Umfangs je verbürgen kann. Die Vorstellung in ihrer absoluten Bestimmtheit gleicht einem mathematischen Gebilde oder einer geometrischen Figur, welche aus einer Anzahl von *Bestimmungsstücken* ganz und gar berechnet werden kann; wie z. B. an einem Dreieck eine Seite und die zwei ihr anliegenden Winkel, als Bestimmungsstücke, den dritten Winkel, die anderen zwei Seiten, die drei Höhen, den Radius des eingeschriebenen Kreises usw. mit bestimmen. Von einer geometrischen Figur unterscheidet sich die Vorstellung ihrem Wesen nach bloß durch den wichtigen Umstand, daß sie auch noch in der *Zeit* als vierter Dimension ebenso klar bestimmt ist wie jene in den drei Dimensionen des Raumes. Das heißt es handelt sich (was ja selbstverständlich ist) stets um ein Gebilde, das sich mit der Zeit verändert, das verschiedene *Zustände* annehmen kann; und wenn ein Zustand durch die nötige Zahl von Bestimmungsstücken bekannt gemacht ist, so sind nicht nur alle anderen Stücke in diesem Augenblick mit gegeben (wie oben am Dreieck erläutert), sondern ganz ebenso alle Stücke, der genaue Zustand, zu jeder bestimmten späteren Zeit; ähnlich wie die Beschaffenheit eines Dreiecks an der Basis seine Beschaffenheit an der Spitze bestimmt. Es gehört mit zum inneren Gesetz des Gebildes, sich in bestimmter Weise zu verändern, das heißt, wenn es in einem bestimmten Anfangszustand sich selbst überlassen wird, eine bestimmte Folge von Zuständen kontinuierlich zu durchlaufen, deren jeden es zu ganz bestimmter Zeit erreicht. Das ist seine Natur, das ist die Hypothese, die man, wie ich oben sagte, auf Grund intuitiver Imagination setzt.

Natürlich ist man nicht so einfältig zu denken, daß solchermaßen zu erraten sei, wie es auf der Welt wirklich zugeht. Um anzudeuten, daß man das nicht denkt, nennt man den präzisen Denkbehelf, den man sich geschaffen hat, gern ein *Bild* oder ein *Modell*. Mit seiner nachsichtslosen Klarheit, die ohne Willkür nicht herbeizuführen ist, hat man es lediglich darauf abgesehen, daß eine ganz bestimmte Hypothese in ihren Folgen geprüft werden kann, ohne neuer Willkür Raum zu geben während der langwierigen Rechnungen, durch die man Folgerungen ableitet. Da hat man gebundene Marschroute und errechnet eigentlich nur, was ein kluger Hans aus den Daten direkt herauslesen würde! Man weiß dann wenigstens, wo die Willkür steckt und wo man zu bessern hat, wenn's mit der Erfahrung nicht stimmt: in der Ausgangshypothese, im Modell. Dazu muß man stets bereit sein. Wenn bei vielen verschiedenartigen Experimenten das Naturobjekt sich wirklich so benimmt wie das Modell, so freut man sich und denkt, daß unser Bild in den wesentlichen Zügen der Wirklichkeit gemäß ist. Stimmt es bei einem neuartigen Experiment oder bei Verfeinerung der Meßtechnik nicht mehr, so ist nicht gesagt, daß man sich *nicht* freut. Denn im Grunde ist das die Art, wie allmählich eine immer bessere Anpassung des Bildes, das heißt unserer Gedanken, an die Tatsachen gelingen kann.

Die klassische Methode des präzisen Modells hat den Hauptzweck, die unvermeidliche Willkür in den Annahmen sauber isoliert zu halten, ich möchte fast sagen wie der Körper das Keimplasma, für den historischen Anpassungsprozeß an die fortschreitende Erfahrung. Vielleicht liegt der

The first page of the “Schrödinger Cat” paper.[41]

Schrödinger had submitted some of his papers previously to *Die Naturwissenschaften* and he wanted to support the editor Arnold Berliner who he knew was in difficulty in Germany as he was Jewish. Just three days after Schrödinger had written to Berliner stating that both Born and London had read a draft of his new paper, Berliner replied on 14 August 1935 to say he had been removed from the editorship of his journal:

> Your dear letter with the friendly trust that you have placed in my judgment has genuinely pleased me, like every letter that comes from you. But this time with a large percentage of emotion because it will be the last letter you will send me as editor of *Naturwissenschaften*. I was removed from my position at noon yesterday. You know the reasons, but I won't go into it further without saying a word...
>
> Can you write to Planck to make the Kaiser Wilhelm Society aware? They need to do something to keep *Naturwissenschaften* in its present state.[44(t)]

The dismissal was even reported in *Nature*:

> We much regret to learn that on August 13 Dr. Arnold Berliner was removed from the editorship of *Die Naturwissenschaften*, obviously in consequence of non-Aryan policy. This well-known scientific weekly, which in its aims and features has much in common with *Nature*, was founded twenty-three years ago by Dr. Berliner, who has been the editor ever since and has devoted his whole activities to the journal, which has a high standard and under his guidance has become the recognised organ for expounding to German scientific readers subjects of interest and importance... Albert Einstein said, 'His journal cannot be imagined as absent from the scientific life of our time.'[45]

Both Schrödinger and Einstein attempted to find a position in the USA to enable Berliner to leave Germany but they were not successful. Berliner had been a very close friend of the great Austrian composer and conductor Gustav Mahler, whose work was banned in the Nazi era. Berliner remained in Berlin right up until 1942 when, under house arrest and at the age of 80, he tragically took his own life shortly before he was to be taken to a concentration camp. A small number of scientists mentioned in this book and still in Berlin attended his cremation including von Laue, Rosbaud and Hahn.

Schrödinger spent just three years in Oxford and his time there was clearly unsettled from several points of view. However, he did write these papers that continue to have much influence on modern science and are of interest to the general public right to the present day. He spent twice this amount of time in Max Planck's Chair in Berlin but did not produce any such influential works then. In science, it is the quality and impact of publications that counts the most and there is a strong argument that, at least from the point of view of publication of influential research papers, Schrödinger's short period of time spent in Oxford was a success.

London

Since September 1933, Schrödinger's former assistant Fritz London had been working in Oxford on the quantum theory of low temperature systems which was being studied experimentally by Simon and his collaborators in the Clarendon laboratory. It is of interest to consider London's movements over the next two years, which continued in some ways to be linked with Schrödinger and had several parallels to Schrödinger's own subsequent moves.

London's younger brother Heinz also emigrated from Germany to work in Simon's team in Oxford in 1934 and, together with Fritz, he collaborated on applying quantum theory to superconductivity. When working in the group of Simon in Breslau on experiments for detecting superconductivity, Heinz London found he was better suited to developing a theory for the problem. He was one of the last Jewish scientists to receive their doctorate in Germany in the 1930s. He stayed with his brother Fritz and his wife Edith at Hill Top Road in Oxford and their two-year collaboration there was very productive.

At the invitation of Douglas Hartree, a pioneer in the approximate solution of the Schrödinger equation for many-electron systems, Fritz London visited Manchester in 1934 and discussed his latest ideas on superconductivity with Rudolf Peierls and Hans Bethe. They were both

Jewish refugees from Germany who had been brought to Manchester by W.L. Bragg. Their critical interest encouraged London, and shortly afterward he and his brother Heinz published their new theory of the quantum electrodynamics of a superconductor.[46]

During his visit to Schrödinger in 1934 in Oxford, Max von Laue discussed the new work with the London brothers. He had some criticisms and they continued a long and sometimes fraught correspondence over several months. However, the work of the London brothers was a major component of a meeting of the Royal Society held in 1935 on superconductivity and other low temperature phenomena. It was perhaps a pity that the Londons could not get Schrödinger involved in this exciting field but he was never a collaborator and always did research on his own.

Fritz London had been paid £600 by ICI for each of two research years in Oxford and in April 1935 he was very concerned to be informed by ICI that his funding would come to an end in August 1936. He consulted with Lindemann about further funding but he was not helpful. London felt that Lindemann wanted to reserve the ICI support to keep Schrödinger and Simon in Oxford and the funds would not extend further.[47]

Michael Polanyi put Fritz London in contact with Chaim Weizmann who suggested he could move to the Hebrew University of Jerusalem. London also wrote to Wigner and said that there was no prospect of finding a position in Oxford while Schrödinger intended to stay and "everything points to the USA".[47] However, by this time many scientists had left Germany and the competition for positions was great, even for someone of London's calibre. The stress on London was becoming serious and his wife Edith had a stillbirth. He asked Hartree and Bragg about the possibility of a position at Manchester but, after some initial encouraging signs, they could not help. Then he visited Paris and finally got an offer of a permanent position at the Institut Henri Poincaré. Reluctantly, Simon encouraged him to accept the post.

Hearing this, colleagues in England worked even harder and Bragg and Hartree did then come up with a position in Manchester but London had already made up his mind to go to Paris. Leslie Sutton had been elected a Tutorial Fellow in Chemistry at Magdalen College alongside Schrödinger. He had worked with Debye in Leipzig and was eventually elected a Fellow of the Royal Society. He was aware of the very significant contribution London had made to understanding chemical bonding and intermolecular forces using wave mechanics. He wrote to London to say that it was "a disgrace Oxford had not been able to find a position and after his departure theoretical physics would become insignificant again".[47]

London's brother Heinz moved to Bristol University and Fritz continued his brilliant work in Paris over the next two years on low temperature phenomena by developing a theory of superfluidity. However, several of his friends in Paris were involved with the anti-Franco movement in Spain and he was well aware of the likelihood of an impending war with Germany. With the political situation becoming very difficult in Europe, he nearly accepted the post at the Hebrew University but was then offered a secure Professorship at Duke University in the USA. He and his wife Edith boarded the Ille de France from Southampton to New York on 1 September 1939, the day Germany invaded Poland.[47] The timing was fortunate as he still held a German passport and war was declared just two days later.

Schrödinger had written to ICI to get clarification about the funding of his salary. They responded on 24 September 1935:

> We have to refer to your letter of 1st September 1935 and note that you accept the arrangement whereby payment of your fee will cease on 14th October 1938.
>
> With regard to your comments in connection with the possibility of being offered other academic positions, we quite understand what you say and you may take it the Company will not make any difficulties in releasing you at an earlier date should it be considered desirable.[1]

The reference to a fee confirms the impression of Fritz London that ICI was giving priority in funding Schrödinger at least until 1938. Many of the scientific refugees had hoped their exile would only be temporary and the Hitler regime would not last so they could return to their previous positions in Germany. However, by the end of 1934 the grip of the Nazis had become even tighter and there were disturbing stories emerging from those who remained in Germany like von Laue and Berliner. Accordingly, Schrödinger had written to the University of Berlin to resign from his Chair there.

Hitler

The appointment to a University Chair in Germany had to be approved at the top level by the Chancellor of Germany himself. The same was true for a resignation. Accordingly, Schrödinger received the letter which is shown here. It was dated 20 June 1935 from Berlin and signed by Der Führer und Reichskanzler, Adolf Hitler.[1] Headed "Im Flamen des Reichs" (In the Name of the Empire), the English translation of this Entpflichtungsurkunde (certificate of discharge) states:

> According to your request, I relieve you, with effect from the end of March 1935, of your official obligations in the Philosophy Faculty of the University of Berlin. I express my thanks to you for your academic achievements and for the services you have performed for the State.

One month later, Max Born also received a similar letter, this time sent from Hitler's vacation resort of Berchtesgaden.[48] Like Schrödinger, Born's position in Cambridge was not a permanent one and he was receiving tempting approaches from elsewhere.

Born had kept up a regular correspondence with Max Planck hoping there may be a possibility to return to Germany if the political situation improved. However, Planck's letters became more and more negative and the influence of the Nazi scientists such as Lenard and Stark had become even more significant. After the death of Fritz Haber in 1934, just a month after spending Boxing Day with the Borns in Cambridge, von Laue had written two obituaries and attended a commemoration event in Berlin, and had been reprimanded by the authorities. Then the Nuremberg Laws were passed which defined citizens of the Reich as those who had "German or kindred blood".

Hedi Born was finding it difficult to settle in Cambridge and after a nasty bout of flu went to a sanatorium for three months in the Harz

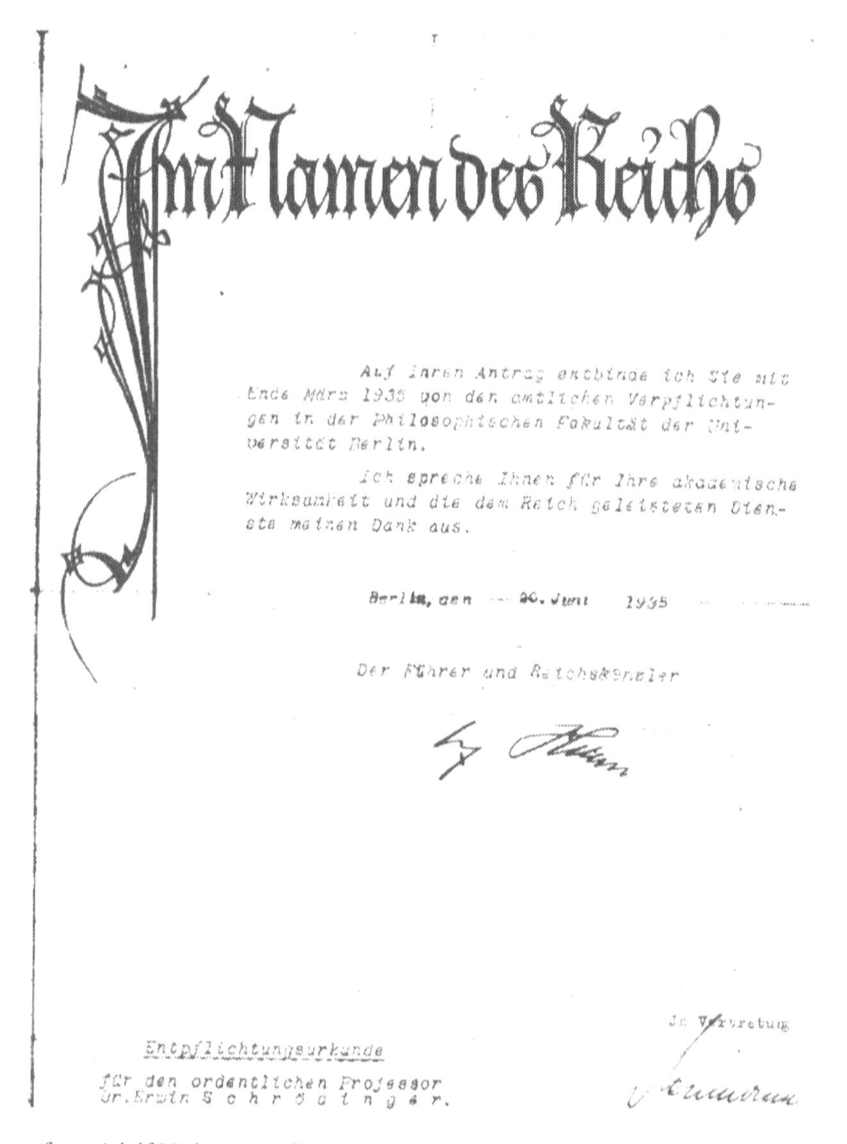

Im Namen des Reichs

Auf Ihren Antrag entbinde ich Sie mit Ende März 1935 von den amtlichen Verpflichtungen in der Philosophischen Fakultät der Universität Berlin.

Ich spreche Ihnen für Ihre akademische Wirksamkeit und die dem Reich geleisteten Dienste meinen Dank aus.

Berlin, den 20. Juni 1935

Der Führer und Reichskanzler

In Vertretung

Entpflichtungsurkunde

für den ordentlichen Professor
Dr. Erwin Schrödinger.

Letter from Adolf Hitler to Professor Dr Erwin Schrödinger, 20 June 1935. Photograph taken by Heather Clary at the Schrödinger archive in Alpbach.

Mountains in Germany. Their daughters had by now grown up but their son Gustav was just 12. Anny Schrödinger, described by Max Born as "the perfect and German housekeeper", very nobly came over to help the family.[48] Heisenberg also came to lecture in Cambridge. He informed Born that the German Physical Society could offer him a research position to return to Germany and this had been approved by the National Socialist government. However, the offer did not include the possibility for Hedi and the children to return permanently.[48] There were not many good stories about Heisenberg after the Nazis came to power.

Max Born was closely involved in trying to help the many German academics who had been forced to leave Germany. He arranged for physicists in the USA, including Franck and Ladenburg, to send money to assist the refugees. Born had received a surprising offer from India. C.V. Raman had received the Nobel Prize for Physics in 1930 for his work on the Raman effect on the scattering of light and invited Born to come to Bangalore so that the latter's theoretical work could complement his own experimental research. At that time India was still part of the British Empire and was a long way from the looming trouble in Europe. So once Hedi had returned feeling better from her German visit they took a steamer for the two-week sea journey to India, leaving the children with friends in Cambridge.[48]

In Bangalore, the Borns had a comfortable bungalow and were charmed by India. However, the poverty in India was upsetting and they were worried about Gustav's education. Born also found himself in the middle of a dispute between Raman and the Senate of his institute which put out the statement: "Mathematical Physics has little contact with industry, and in this respect cannot compete with Chemistry as a subject likely to be of service to India."[48] Accordingly, a permanent position could not be offered and the Borns returned to Britain. Just a few years later Schrödinger would also be tempted by an offer of a position in India.

The academic politics at Cambridge University was similar to Oxford and it was very difficult to make a senior permanent appointment unless a Professor in post died or retired. There had been talk from the theoretical physicist Ralph Fowler of finding a Readership for Born. This was a position lower than the status of a Professor but above that of a Lecturer. However, nothing had emerged.

The position of Schrödinger also remained uncertain. On 17 December 1935, George Gordon, who was convalescing in Spain, wrote once more to Lindemann to say:

George Gordon, President of Magdalen College from 1928–42.

The position of Schrödinger demands attention. The Rockefeller people have no more money for this purpose and have given adequate warning that this would be so. All payments from this source cease at the end of the present year.

There is still I.C.I. but Schrödinger himself grows, not unnaturally, a little restless, and would like to get settled more securely.

I am prepared to take the matter up at Magdalen when I return (from Spain) and when I know the position fully. But I thought it wise to get in touch with you now.

I don't think Schrödinger should be approached or worried at present. The other and quite impersonal side is better to be tackled first.[24]

Austria

Schrödinger had kept up a very regular and friendly written correspondence with several other theoretical physicists, especially Einstein and Born, and had discussed possible appointments he might be able to take up. On 17 May 1935 he wrote to Einstein:

> Letters from friends tell me that I should soon be asked if I want to come to Graz to replace Michael Radaković. I think I would assume to go, if I was, extraterritoriality, allowed my Swedish securities and freedom from taxation for any foreign lecture fees, *i.e.*, travel expense allowances. It is not that I will not be able to stay anywhere in the long term. Up to now I have liked everywhere I have been, except in North Germany. It is also not that people here (in Oxford) are not very nice and friendly towards me. But it still increases the feeling that you have no office and that you are living on the generosity of others.[36(t)]

Michael Radaković had been appointed to the Chair of Theoretical Physics in Graz, Austria in 1906 and had recently died leaving his Chair available for a new appointment. Schrödinger would have been an obvious choice. It should also be noted that Schrödinger had, most fortunately, left his Nobel Prize award in a Swedish bank account which meant the authorities were not able to confiscate it when he later left Austria.

Another regular writer to Schrödinger was his old friend from student days Hans Thirring who by now was Professor and Head of Theoretical Physics at the University of Vienna, a Chair previously held by Boltzmann. Thirring was a key person in keeping Schrödinger linked back to Austria over a period of some 30 years. On 30 May 1935, Thirring had written to Hanfried Ludloff in Breslau about a position for his assistant Theodor Sexl. Ludloff was a Jewish theoretical physicist still trapped in Germany and Thirring was trying to assist him. In this letter, Thirring also mentions a possible move of Schrödinger to the Chair in Graz and other related appointments:

> How the situation will develop later depends, among other things, on the occupation of the currently vacant Chair for Theoretical Physics in

> Graz. There is a certain hope that Schrödinger can be won over. On the other hand, however, this probably has the consequence that Arthur March, who is currently with Schrödinger in Oxford, would return to Innsbruck for his post, which brought my assistant Sexl, who supplements March in Innsbruck. In this case, the prospects for theoretical assistants in Vienna are negligible. If, on the other hand, Schrödinger refuses, Sexl might come to Graz and the osmotic overpressure would decrease.[22(t)]

Einstein, meanwhile, was still hoping that an appointment could be made for Schrödinger at the IAS in Princeton. In his letter of 4 September 1935 to the IAS director Flexner, Einstein said:

> Lately I've had an in-depth scientific correspondence with Schrödinger, which again has shown his importance as a researcher. He would be in my opinion a real achievement for our institute. He wrote me of the intention to accept a call to Graz (Austria), and that he had only been called to Oxford, so to speak, from courtesy, without a real need for a

Hans Thirring at the University of Vienna.

position for a teacher of theoretical physics. I believe the senior men in Princeton University after an open discussion of the situation feel there is nothing wrong with Schrödinger being employed at our institute despite his rejection of the appointment there. After all, such an outstanding researcher cannot be held back if he strives for a position that allows him to use his scientific strength as completely as possible.[25]

Clearly there was a lot of discussion going on elsewhere about a possible move of Schrödinger to Austria, but his supporters in Oxford seemed to be unaware of these approaches. Thirring wrote again to Schrödinger on 30 October 1935 about arrangements for a visit and lectures in Vienna:

We have just learned from Ehrenhaft that you have already received a letter from him in which Viennese guest lectures are in prospect. Now that the negotiations have progressed so far and a new reorganization has only caused you concern, we both believe that it is best to leave things with the arrangements made. The different suggestions that Mark made to you in his last letter are therefore outdated and irrelevant. We both know through comments from you that you do not like the usual major fuss of the guest lectures, and we will therefore try to ensure that he imposes a certain reserve on himself in this regard. In any case, we look forward to seeing you again soon in Vienna and we hope that we will then have the opportunity to do ski tours with you.[22(t)]

Schrödinger visited Austria in December 1935 to give his guest lectures in Vienna. He also visited Graz where his old friend Kohlrausch was based and the possibility of a post there was discussed further. After returning to Oxford he wrote to Hans Thirring on 16 February 1936:

In the last four weeks (and more) I have let the Austrian affair go through my head and I regretted that after the visit to Graz and the second conversation in the Ministry I could no longer talk to you in a frank manner. My conversation did not refer to further detailed requests, but had the general tenor, which I would like to express privately between us as follows: in the Graz Physical Institute, everything is missing. Nobody can really learn physics there, the apparatus is Victorian (as they would

> say in this country). Why are they going to buy those expensive theorists? That's a waste of money – it doesn't make the herb fat.
>
> From this, in connection with my reluctance to go into exile there, the following dream has now developed for me: couldn't you get people around to turn the professorship in Graz into something that plays between Graz and Vienna? I don't want to officially accept a full teaching commitment in Graz, but would like to do a small amount also in Vienna.
>
> I haven't told the Grazers (who might not be very delighted) about this yet, because I consider it more important to consult you about it first ... I am no less grateful to you for your kind, friendly attitude and active support in this matter from the very first moment. It's a long and slow process. But maybe something sensible will come out of it.
>
> I would like to go back. I felt so warm and at home in Austria during the weeks I was there... I am told that out of prestige and especially for political reasons that my acquisition is important. Well, one will come up with something. When there is a will there is a way.[22(t)]

On 20 February 1936, Thirring then wrote to Schrödinger in Oxford about the possibility of giving up his own position in Vienna for him:

> When I arrived, I wanted to telegraph you in the first impulse, "Idiot, of course you are welcome in Vienna!" But it might be better if I print it out in more detail. So: your question as to whether you should have a position between Graz and Vienna I would agree with one hundred percent.
>
> I go even further to say that I have always considered you to be the only worthy representative of the Boltzmann chair and that in the current situation it is one of the fairest solutions if I go to Graz and you came to Vienna. If my family ties and the thought of the future of my children did not tie me to Vienna very much, I would try to find this solution myself on my own initiative – although the question remains whether Graz would agree to this. I would of course prefer it if we could make arrangements according to your own suggestion so that we are both in parallel. My joy about having you here in Vienna would be so great that I wouldn't give a damn what other people were saying about "being virtually pushed aside".

> Of course, I'm not doing anything for the time being, because first you have to get in touch with your Graz colleagues. Otherwise, Benndorf could rightly reproach me with the fact that we did not give the Grazers the best and wanted to catch the "valuable bird" ourselves. So I won't do anything for the time being, but will wait for your further messages. I just want to add that for a long time I haven't been as happy about anything as your letter and about the thought that we might be able to win you over to Vienna, at least partially.
>
> I would like to thank your dear wife for the friendly card from Gurgl. I was very happy that you had done so well back then and that Anny was satisfied with my friendship. If you come to Austria again at Easter, perhaps we could go somewhere up into the high mountains together.[22(t)]

Hans Benndorf is mentioned here. He was Professor of Physics at the University of Graz and had been Rector of the University. Like Schrödinger, he had been a student of Professor Exner in Vienna, although at a previous time. He had made several contributions to the fields of seismology and atmospheric electricity. He interacted with the Nobel Laureate Victor Hess in Graz on cosmic ray research.

Schrödinger responded to Thirring at once on 22 February 1936 about the possibility of a joint appointment between Graz and Vienna, and Thirring's suggestion of giving up his Chair for Schrödinger:

> What you say about Graz and Vienna, I hardly dare to listen, it makes me blush. No, no dear friend, our relationship is really not like that… we shouldn't think of such a flagrant violation of ancient academic principles. As for the people of Graz, I actually didn't want to put the matter up for discussion there first, but to the Ministry. That will kindly take care of Victor Hess, with whom I had a meeting in the Ötztal station (near Innsbruck) in January. As a "state culture council" the Ministry has quite an influence and has been incredibly nice and friendly towards me.
>
> You know the institute I was introduced to in Graz – you have no idea! And that is of course due to Benndorf's sleepiness. Why does it look so European to Kohlrausch? Perhaps you will tell Herman Mark (in Vienna) about my plan. I think he's quite safe not to promulgate it prematurely.

> I am terribly happy in the prospect of what will perhaps be achieved. The two of us together again in the old Millihaferl (Exner's word), that would be a genius arrangement. It would be like turning back time with everything in between a bad dream.[22(t)]

Then, to complicate matters further, on 27 February 1936, Schrödinger had a letter from Nevil Sidgwick, the theoretical chemist at Lincoln College, Oxford. Sidgwick had been one of the Oxford academics who had written to congratulate Schrödinger on his Nobel Prize. He was well aware of how wave mechanics was revolutionising the understanding of chemical bonding. He said:

> The Chair of Natural Philosophy (*i.e.*, Applied Mathematics) at Edinburgh is vacant, as you probably know, through the appointment of Charles Darwin to the Mastership of Christ's Cambridge. Some of the electors have asked me to find out whether if it were offered to you, you would be willing to accept, or at any rate consider it.
>
> The position, I understand, is approximately this. The salary would probably be £1,200 and it might rise later to £1,300 when the University is more prosperous. The duties are not heavy, consisting of usually six lectures a week, but often three: at present the position is particularly easy, because of the departmental lecturer Schlapp who is very helpful, experienced and obliging. The prospects of a research school are reasonably good: at any rate Whittaker has been pretty successful in pure mathematics.
>
> Of course you will understand that this inquiry is purely unofficial. I don't like to do anything that may lead to your leaving Oxford but I felt bound to put the matter before you.[49]

Schrödinger at once wrote back to Sidgwick on 1 March 1936:

> I thank you very much for your letter of February 27th. I had to think the matter over for a day and then, yesterday, I could not find sufficient leisure to write it down properly.
>
> The answer cannot be doubted. If the chair were offered to me, I should consider the offer very seriously and, I think, there would be a high probability of my accepting it. In correspondence with your very

> kind remark that you "don't like to do anything that may lead to my leaving Oxford", I should say that certainly I, in my turn, would regret that very much. But you surely know that I, like many of my co-nationals, live at present on the munificent generosity (in order not to say charity) of this country, without being able to return any true service. To reject any possibility of relaxing this burden would hardly be justifiable.
>
> But this argument is so clear that I would not have had to think it over for one minute. I think I ought to tell you what was the only point of deliberation at all, in order that you should not believe I was pondering whether a chair at the famous Scotch University was "good enough" for me. Well, the only reason for having to think about it at all was that I am a man of decidedly southward orientation, who feels less happy the further he recedes from the Mediterranean.
>
> But as I say, that would not be allowed to come into consideration under the present circumstances.[49]

The University of Edinburgh, in the capital of Scotland, had a distinguished academic history going back to 1585. Like Oxford at that time it had an excellent reputation in the humanities and also a fine medical school. By 1936, the University had one Nobel Prize winner, Charles Barkla who had won the Prize in Physics in 1917 for his work on the characteristic X-rays of elements. Charles Darwin, the holder of the Tait Chair of Natural Philosophy, was the grandson of the very famous biologist with the same name. He had used Dirac's theory to do a calculation of the fine structure of the hydrogen atom. Edmund Whittaker was a mathematician who was expert on special functions and had written several well-known advanced mathematics textbooks. Before coming to Edinburgh he had been Professor at Trinity College Dublin, a link which became crucial for Schrödinger just a few years later.

Schrödinger visited Edinburgh and talked to Darwin and Whittaker about the Tait Chair. However, by this time, Schrödinger's negotiations with the Universities of Graz and Vienna had been nearly settled. The pull of his home country was too great, even if the political situation there was getting complicated, and he declined the approach from Edinburgh. The University then offered the Tait Chair in July 1936 to Max Born who accepted at once.

Born had also just been offered a Chair in Pyotr Kapitsa's new institute in Moscow. He had been to Russia before in 1928 but sensed that the situation there was becoming nearly as dangerous as in Germany. The case of Hans Hellmann is a tragic example. Hellmann had worked as the assistant to Erwin Fues at the Technical University of Hanover. Fues himself had been an assistant to Schrödinger in Zurich at the time his great papers were published in 1926. Hellmann made several novel contributions to quantum chemistry and wrote one of the earliest books on the subject. His wife was Jewish and was born in Ukraine. In 1933 they came under the new laws in Germany and so, in 1934, they moved to the Karpov Institute of Physical Chemistry in Moscow. There Hellmann saw Paul Dirac lecture during one of his several visits to see Kapitsa. Hellmann derived an important theorem on molecular quantum mechanics which is today known as the Hellmann-Feynman theorem and allows for electrostatic forces to be obtained from Schrödinger's wave-functions. However, because of his nationality, Hellmann became under suspicion in Moscow and less successful colleagues were envious of his academic success. In 1938 he was taken away by Stalin's secret police and, most tragically, was never seen again. Over 50 years later his son tracked down documents which showed Hellmann was forced to sign a false confession and was shot on 29 May 1938.[50]

Max Born settled down to the safe life of a chaired professor in Edinburgh but found the students were not up to the very high standard he had seen in Göttingen or in Cambridge. At Edinburgh, he published mainly single-author papers and his new work on the dynamics of crystal lattices was influential. With a refugee from Germany, Klaus Fuchs, he also published three joint papers on the fluctuations of electromagnetic radiation. Born was a pacifist and, unlike a large number of his students and scientific collaborators, declined to take part in science projects related to the Second World War. Fuchs joined the Manhattan nuclear bomb project in the USA and, after returning to Britain after the war, became a most infamous scientific name when he was arrested for giving key nuclear secrets to the Russians.

By the spring of 1936, Schrödinger was having detailed negotiations with the federal authorities about a position back in Austria. On 18 March

and the Russian emigrant zoologist Vladimir Tchernavin to the *Times* to express gratitude to the Academic Assistance Council which had helped hundreds of scholars who had been forced out of their home countries and which was now to be replaced by the Society for the Protection of Science and Learning. They wrote: "The Academic Assistance Council is coming to an end in its emergency form, but we and our friends will endeavour to remain it unforgotten. May we hope that the continuation of our scientific work – helped in no small measure by its activities – will be an expression of our gratitude?"[18]

16 of the scholars helped by the Academic Assistance Council won Nobel Prizes. This included not only some of the greatest physical scientists of the 20th century such as Einstein, Schrödinger and Born but also several Nobel Laureates in the life sciences. For instance, there was Ernst Chain, who with Howard Florey discovered the clinical action of penicillin, Bernard Katz, who worked on transmitters in nerve terminals, and Hans Krebs, who established the chemical processes in the citric acid cycle. It was a significant loss for the other countries, but a great gain for Britain.

It should be noted that one of the signatories of the letter, Vladimir Tchernavin, had achieved a remarkable escape from the dark depths of the Soviet Gulag and then went on to undertake research at the National History Museum in London. With his signature, the letter emphasised that the good work extended to helping scholars from other countries beyond Germany, and in just two years, many more were to be assisted from Austria as well. The newly created Society for the Protection of Science and Learning was shortly to help Schrödinger again when he got into difficulties with the Nazis in Austria. The work of this organisation has continued to the present day and is now called the Council for At-Risk Academics. Very recently, it has even assisted refugee scholars from countries such as Syria and Iraq.

It is likely that the letter to the *Times*, written in collaboration with Einstein who was a major public enemy of the Nazis, would have been read carefully by the German Ambassador to Britain Joachim von Ribbentrop. This would cause Schrödinger some difficulties when he started to make moves to leave Austria some two years later and

Ribbentrop himself was consulted on the matter by the British Foreign Secretary Lord Halifax.

By the summer of 1936, Schrödinger had settled the arrangements for a Chair at the University of Graz and an Honorary Professorship at the University of Vienna. He wrote on 22 September 1936 to his employers at the Imperial Chemical Industries:

> As I mentioned several times in the course of the last year, I have been negotiating with the Austrian government about a position offered to me in Graz. These negotiations have now been successfully terminated and last week, on my way back to this country, a letter from the Austrian Ministry of Education reached me in Zürich, telling me that the President of the Austrian Federation had agreed to the proposal of the Minister of Education and signed by appointment as a Professor at the States-University of Graz.
>
> Although I hope that I shall have the opportunity of keeping up relations with the University of Oxford, returning here for a few weeks every year, I am afraid that my new appointment will prove to be irreconcilable with the terms of my employment in your company (stating that I should normally reside in Oxford) and will therefore compel me, to my deepest regret, to resign from this employment, leaving, for obvious reasons, the precise date when you wish to end it to your discretion.
>
> As was arranged with Dr. Rintoul, this will involve my giving up the tenancy of 24 Northmoor Road. Professor Lindemann suggests that you should write to him, if you think he could be helpful in connection with this matter.[1]

This development had been anticipated by the Academic Assistance Council and Walter Adams, the General Secretary, who was aware of the attempts to find the finances to keep Fritz London in Oxford. On 16 June 1936 he wrote to Lindemann to say:

> I hear confidentially that Professor Schrödinger has decided to accept a position in Austria. Do you think that there is a possible chance of securing from the I.C.I. the balance of the funds which would have been used for the further support of one of the other German scholars, in particular of Dr Fritz London?

> As you know the I.C.I. made us a general donation of £2,500 to assist the German scientists, so that we could help London for a few months, but it would be far more satisfactory if a grant for a period of years could be obtained.[52]

However, this request for transfer of ICI funds to Fritz London came too late. By this time, London had already agreed to leave Oxford and accept an appointment in Paris. Several of the other refugee physicists who had moved from Germany to Oxford, however, had been able to settle and remained in Oxford for the rest of their careers. Francis Simon and Nicholas Kurti had developed a magnetic method to undertake studies of materials at very low temperatures and had used it to discover the superconductivity in metals including cadmium, zirconium and hafnium. Their work had been published in both *Nature* and the *Proceedings of the Royal Society*. However, Simon was a senior physicist and, like Schrödinger, he was living on short-term funding from ICI. He applied for a professorial position at the University of Birmingham but that went to Mark Oliphant who had been a member of Rutherford's team in Cambridge.

Simon had spent a sabbatical at the University of California, Berkeley in the early 1930s and there had got to know Cyril Bailey, a Fellow in Classics from Balliol College Oxford who was also on sabbatical. After his arrival in Oxford, Bailey then arranged for Simon to be a member of the Senior Common Room at Balliol where he got to know the Fellows of the College. At Balliol, the Reader in Thermodynamics was Alfred Egerton who had worked with Nernst in Berlin just before the First World War. Egerton was appointed to the Chair of Chemical Technology at Imperial College London in 1936 and Simon was then elected to his Readership, an appointment with a status between a Lecturer and a Professor. This was a rare and permanent appointment for a German refugee and released some ICI funds for the other refugee physicists to enable them also to stay at Oxford after 1936. Simon was given an annual salary of £500 from his University post and this was supplemented by another £500 from ICI.[35]

Accordingly, Kurti was now able to keep working with Simon, Kurt Mendelssohn could continue his own productive experiments on superconductivity and Heinrich Kuhn could extend his work with Derek Jackson on high-resolution spectroscopy. As is described in Chapter 5, all of these refugees from Germany then did vital work for the government in the Second World War. Afterwards they all obtained permanent positions at the Clarendon Physics Laboratory in Oxford, were elected to College Fellowships at Oxford and even to Fellowships of the Royal Society. Kurti, a colourful raconteur, gave a summary of his first day at Oxford, when on the pillion of Mendelssohn's motorbike: "We came over Magdalen Bridge. The sun was shining. It was like a fairyland. Why should I ever leave? And of course I never did!"[53]

In Berlin, just before he left to come to Oxford, Kurti had been examined for his doctoral thesis by Nernst and Schrödinger, and had also taken several of Schrödinger's lecture courses. The accounts of the early years of Kurti and his other experimental colleagues in Oxford are extensive, but no direct links with Schrödinger are mentioned.[54] There was no shortage of stimulating visitors to the Clarendon Laboratory in 1934 and 1935 but none appear to have had interactions with Schrödinger. Arthur Compton was the visiting Eastman Professor. He was one of the leading experimental physicists in the USA who had won the Nobel Prize for Physics in 1927 for demonstrating the particle properties of electromagnetic radiation and had attended some of the Solvay Conferences. Also supported by ICI for a short period at Oxford was Leo Szilard, who had known Schrödinger in Berlin and had collaborated with Einstein. He had help set up the Academic Assistance Council and had just secretly patented, with Lindemann's assistance, the first idea of a nuclear chain reaction. He was working on nuclear physics in the Clarendon Laboratory with Collie and Griffiths.

In the 1920s and early 1930s the leading research group for high magnetic field and low temperature physics in Britain was that of the Russian physicist Pyotr Kapitsa in Cambridge, whom Schrödinger had visited in 1928. Kapitsa had returned to Russia in 1934 to see his parents but the authorities there would not then let him return to Britain. However, he managed to continue his research and eventually was awarded the Nobel Prize in 1978 "for basic inventions and discoveries in the area of low temperature physics". His departure from Cambridge, however, was an excellent opportunity for the Oxford group to become the leaders in the UK on magnetism and low temperature physics.

Lindemann, originally an experimentalist with Nernst on low temperature physics, would have been well aware of this opportunity and gave the group led by Francis Simon full backing. The research of the group was so impressive that Lindemann skilfully exploited their need for extra space to persuade the Oxford University authorities to provide the funds to build an extension of the Clarendon Laboratory. This was opened in 1939.[54]

It seems that Schrödinger was thinking of other things when much exciting experimental work on new quantum mechanical effects was going on at the Clarendon Laboratory. Despite his early experimental training in the group of Exner in Vienna, Schrödinger had by now become more interested in the philosophical aspects of science. However, the young refugee physicists in the Clarendon Physics Laboratory in Oxford, such as Kurt Mendelssohn, still looked favourably on Schrödinger as he had left Germany after Hitler came to power and did not remain in the same way as Planck or Heisenberg.[55]

On 8 July 1936 Schrödinger wrote in German to Lindemann from Innsbruck on his new appointments in Austria:

> I want to finally report to you myself on the current situation. The dealings with Vienna are now clearly concluded and with human foresight I will be settled at the University of Graz before this year. We are still waiting for a contract, above all the specified amount for the retirement years and the minister's signature, through which everything is actually fixed.
>
> My stay as a Fellow of Magdalen has been extended and this is very welcome. I'm very happy about that. Perhaps this is due to Baron Frankenstein, or possibly to your friendly intervention. To President Gordon I have already written and I hope that the matter goes smoothly with a lack of financial difficulties with grants.
>
> Unfortunately, I won't be able to reside in Michaelmas term in Oxford. I only had a small seminar with Whitehead – the same one that has been held informally for two semesters. Since London is also departing, the spirit of the Whitehead seminar may be lost.
>
> That leaves Imperial Chemicals, who have not yet written again to me. I sent a communication of the matter to Dr Rintoul already in

> August of last year and received full approval. He responded: "We do not wish to prevent anybody from obtaining a more stable position." When the grant was continued for another three years, I let them know there was nothing wrong with my appointment as a Professor...
>
> In any case, I will return to Oxford in September and can be contacted by the mailing address: c/o Prof March, Innsbruck, Pradlerstr.18. The small Zurich congress from which I have just come was very nice. Besides the young Italians and Frenchmen were Bothe and Geiger, who gave a fine Viennese award-winning lecture.
>
> I wish you, dear friend, nice holidays and good relaxation. Do not be too burdened with politics, it is harmful to your health
>
> All the best wishes for your sincere devotion. E. Schrödinger.[24(t)]

Therefore, despite his impending move from Oxford to Austria, Schrödinger had been successful in his clever plan to keep his Fellowship at Magdalen College. He also secured the arrangement, somewhat similar to the one Einstein had received from Christ Church, that he would return each summer to Magdalen and give lectures in the University of Oxford. There remained the matter of whether he would receive a research grant and this would continue to occupy the Grants Committee, President, Bursar and Governing Body at Magdalen. The fact that Schrödinger wanted to keep his Fellowship at Magdalen and come back every summer does imply that he had some affection for the College. As for Lindemann, he did not take Schrödinger's advice to "not be too burdened with politics, it is harmful to your health". He was to become the Chief Scientific Advisor to Prime Minister Winston Churchill in the Second World War.

With Schrödinger returning to Austria, Arthur March, together with his wife Hilde and two-year-old Ruth, left Oxford for Innsbruck. He was appointed there to a full professorship and the period spent at Oxford had not done his career any harm.

Schrödinger's final letter sent from Oxford to Thirring, his old friend and soon-to-be colleague, was on 24 September 1936 from Northmoor Road:

> My appointment to Graz has been completed and the Viennese one probably too. I should like to tell you right away that it has been already

reported in a newspaper and I was ashamed... I'm afraid I may be overwhelmed if I start with both appointments at the same time, especially since the treatment of the successor to Benndorf, which unfortunately has become necessary, must certainly not cause any annoyance, and will require many sessions.

Moving is no fun but it is a means to an enjoyable end. Four latitudes to the south (instead of five to the north as would have been the case with Edinburgh!)...

I was sorry not to see you in the summer, it was due to my indolence, which let the days run by in rainy idleness, always expecting that the next day would be really good. There was only one hot week in Vienna and later in September almost as hot in Graz, when I was looking for an apartment. The wonderful day on the Danube, which I owe to you, remained a high point of this summer. Hopefully again![22(t)]

In 1933, Schrödinger had been appointed a Fellow of Magdalen College for five years, a term which would finish in 1938. President Gordon was to continue to work very hard to facilitate his return to Oxford after his departure to Austria in 1936. He wrote in his usual friendly terms to Schrödinger on 16 October 1936:

The question on Wednesday's College Meeting agenda, which concerned you, I had put there mainly in order to inform the College about the nature of your appointment, and the possibility of your being able to return here for the summer terms. I believe this latter piece of news was received with much satisfaction by the assembled Fellows, but, as no question immediately arose (your Fellowship and present grant being unaffected by your Graz appointment), we merely resolved that the Grants Committee should report on the proposed summer term arrangement to the January Meeting. That date was chosen because, by that time, our now famous economy committee will have reported, and we shall be better able to see what sums will be available for that and other purposes.

Privately, I have little doubt that, when the time comes, we shall accept the summer scheme with great satisfaction, and repeat your present grant as often as it is wanted.

I hope that Mrs. Schrödinger and you had a pleasant journey, and that you will soon be comfortably established in your new quarters.[22]

Portrait of George Gordon by Sir William Coldstream, Magdalen College, Oxford.

One senses from this letter that President Gordon was almost relieved that he would not have to be hunting around again for funds to support Schrödinger and the "summer term arrangement" kept his link with the College. Gordon had just had his portrait painted (as shown above) and he looks reasonably contented. It should be noted, however, that not all the Fellows at Magdalen were so enthusiastic about the arrangements being made for Schrödinger. On 17 June 1936 at the College Meeting the Bursar Arnold Forster had formally noted his dissent on the recommendation to reserve a research grant of £150 for Schrödinger.[32]

This was becoming a tense time for Magdalen College as its most prestigious member, King Edward VIII, who had been a student there from 1912–14, had started a close relationship with a twice-divorced American lady, Mrs Wallis Simpson, and the story was starting to leak out in the international press. It would lead to King Edward's abdication on 11 December 1936 in one of the most dramatic announcements of the

In the late 1920s the Christian Socialist Party was dominant in Austria and did not favour an Anschluss or association with the Nazis. However, after Hitler was made Chancellor of Germany in 1933 there was considerable unrest in Austria from the National Socialist groups there. Engelbert Dollfuss had been appointed Chancellor of Austria in 1932 and from 1933 ruled by decree. There was civil war in 1934 and all political parties, except the Christian Socialists, were banned. Dollfuss was assassinated by Nazis in an attempted coup in 1934. He was replaced by Kurt Schuschnigg who was anti-Hitler but close to the fascist regime in Italy of Benito Mussolini.

Unlike other leading physicists such as Nernst, Planck, Einstein and Lindemann, Schrödinger did not have close interactions with politicians and at times seemed unaware of the problems emerging around him. His letters often showed his irritation with officials and ministers who had to approve his appointments and he did not seem to be aware that the funding for science depended on a good relationship with administrators and politicians. By the summer of 1936, it seemed obvious to many observers that the political situation in Austria was rapidly deteriorating but the naïve Schrödinger failed to realise this. It seems his friend Hans Thirring, whom he corresponded with many times during his stay in Oxford and was largely responsible for bringing him back to Austria, was not fully aware of this either.

As has often been the case, his wife Anny was able to give a concise summary of the situation:

> In 1936 my husband got a call to Edinburgh and at the same time another one to Graz. It was rather difficult to choose, the decision fell on Graz. We gave up our nice little house in Oxford and moved to our house in the country. It never occurred to us that this step might turn out to be rather foolish, even dangerous. Many of our friends shook their heads and soon we understood their attitudes.[1]

Graz is the second largest city in Austria lying 95 miles southwest of Vienna. Like Oxford, Cambridge, Leiden and Heidelberg it is a city known for its University. The Karl Franzens University of Graz was founded in 1585 and has a distinguished history. Ludwig Boltzmann had done some of his greatest work while holding Professorships there in mathematical physics from 1869–73 and in physics from 1876–90. His students in Graz included the great Nobel Laureates in Chemistry, Arrhenius and Nernst. Anny and Erwin found a house at 20 Merangasse in Graz and Schrödinger gave his introductory lecture at the University on October 1, 1936.

Victor Hess, who had interacted quite closely with Schrödinger when he was a student in Vienna, had himself gone to school in Graz and was to very shortly win the Nobel Prize in Physics while a Professor in Innsbruck for his discovery of cosmic rays. He would soon be returning to Graz and, like Schrödinger, would run into serious difficulties with the Nazi regime. In addition, Otto Loewi had been a Professor at Graz since 1903. He was from a Jewish family and won the Nobel Prize in Physiology or Medicine in 1936, jointly with Sir Henry Dale, for his research on neurotransmitters. He would also be treated badly by the Nazis.

A close colleague for Schrödinger in Graz was Fritz Kohlrausch whom he had known since the Exner days in Vienna some 25 years previously. He had been Schrödinger's best man at his wedding to Anny in 1920. From a family of distinguished scientists, Kohlrausch's father Friedrich had done very important work on electrical conductivity in solutions. Fritz Kohlrausch had held the Chair of Physics at the University of Graz since 1920 and had also been the Rector there. On 14 January 1936 he had written to Schrödinger in Oxford to summarise the administrative and financial challenges for doing research in Graz and give Schrödinger some wise advice:

> If my institute has survived these years moderately well, there are special reasons for this:
>
> A second physics chair at Technology was closed and its inventory was merged with mine, and in the previous year also the chair at the Leoben University, and at least a part of their inventory, was handed over to me. I have had the Rockefeller support for three years and I had the nerve to request 50,000 schillings, and I think accidentally received the money. The requested amount was paid out to me in three years.

> We must not forget how we all struggled to put an end to these intolerable conditions; you do not know how many entries, approvals, requests and threats have been sent to the Internal Ministry for teaching; you don't know how tired and bitter we are in this... The power is held by the political party and the finance ministry; the latter is entirely anti-scientific, and so is the party, because they need the money for their own purposes. And if you should perhaps see nice speeches in the newspaper about the immense Austrian culture, counter them with these concerns.[2(t)]

These cautious comments did not prevent Schrödinger from accepting the Professorship in Graz. He was given an annual salary of 20,000 schillings by the University of Graz which was then equivalent to about £750. He was also given a supplement of 10,000 schillings for his post at the University of Vienna which was an Honorary Professorship. In total, this was a superior sum to what he had been receiving in Oxford.

Scientists have always complained about the problems with bureaucracy and funding but perhaps Schrödinger should have read some of the warnings from Kohlrausch more carefully. On the very day of the announcement of his Nobel Prize in 1933 there was a violent celebration in Graz on the tenth anniversary of Hitler's Beer Hall Putsch and over 200 people were arrested. Graz was a centre for the Nazis in Austria just as Munich had been in Germany.

In the first month of his appointment at Graz Schrödinger was elected a foundation member of the Pontificia Academia Scientiarum (the Pontifical Academy of Sciences in Rome). This Academy had been originally founded in 1603 with the title Linceorum Academia. It was refounded in 1887 with the title Pontificia Academia dei Nuovi Lincei and then finally established again in 1936. The Academy had a remarkably broad and enlightened mission for:

> Promoting the progress of the mathematical, physical and natural sciences, and the study of related epistemological questions and issues.
>
> Recognising excellence in science.
>
> Stimulating an interdisciplinary approach to scientific knowledge.
>
> Encouraging international interaction.

> Furthering participation in the benefits of science and technology by the greatest number of people and peoples.
>
> Promoting education and the public's understanding of science.
>
> Ensuring that science works to advance the human and moral dimension of man.
>
> Achieving a role for science which involves the promotion of justice, development, solidarity, peace, and the resolution of conflict.
>
> Fostering interaction between faith and reason and encouraging dialogue between science and spiritual, cultural, philosophical and religious values.[3]

As a founding member with broad interests and influence Schrödinger essentially ticked all these boxes. Even though the Academy is based in the Vatican, membership of the Catholic Church or any religion is not required. Like many scientists, Schrödinger was not a follower of a religion but had spoken, such as in his BBC address and in several publications, on wider philosophical issues. While in Oxford in 1935, he had published *Science and the Human Temperament*, a collection of his popular German essays and lectures which he had delayed publishing due to the political situation emerging in Germany.[4] In Schrödinger's words the essays show: "The old links between philosophy and physical science...are being more closely renewed. The further physical science progresses the less can it dispense with philosophical criticism."[4]

Other distinguished physicists elected as founding members of the Academy included the Nobel Prize winners Bohr, Debye, Millikan, Planck, Rutherford and Zeeman. In addition, Schrödinger's colleague on the Fellowship at Magdalen College, Sir Charles Sherrington, was elected as was Thomas Morgan, the American who was awarded the Nobel Prize in Physiology or Medicine in the same year as Schrödinger's award. Also elected was Sir Edmund Whittaker, the mathematician from the University of Edinburgh who had discussed a possible position there with Schrödinger. Whittaker was going to prove helpful for Schrödinger in 1938 and 1939 when he made the move to leave Austria and eventually took up a position in Dublin.

Up until the reformation of the Pontifical Academy in 1936, the Catholic Church had not had a very distinguished link with science. The

Copernicus picture of planet earth orbiting round the sun was not accepted and led to Galileo being forced to recant this fact by inquisitors presided over by Pope Paul V in 1616. It was only in 1979, after Pope John Paul set up a Committee to look into the matter, that a declaration was made before the Pontifical Academy of Sciences that Copernicus and Galileo were right after all.

The first meeting of the new academicians was held in the Vatican on June 1, 1937.[5] Not many of the new members could attend but Schrödinger and Debye were present, together with some cardinals and many Vatican officials. Due to a serious illness, Pope Pius XI was not able to be present and he lived on for just two more years. The cardinals sat in the front row for the lectures which were given by Italian academics working mainly in mathematics or astronomy. The opportunity was not taken to invite the distinguished new members of the Academy to give lectures.

Following this interesting event, Schrödinger found himself busy giving lectures in Graz and also one lecture a week at the University of Vienna. He seemed to be enjoying his move back to Austria and said:

> I am happy to have moved here. Not that I did not like Oxford, but it was not meant to be a permanent position. I am glad to have found a quiet place back home.[6]

In Vienna, he would often meet with his old friends Hans Thirring and Herman Mark. Mark was a physical chemist who was born in Vienna and had worked with Michael Polanyi in Berlin. He had met Schrödinger on military service towards the end of the First World War. He was an expert on X-ray diffraction and was one of the first scientists to apply this technique to deduce the macromolecular structure of polymers. Linus Pauling had learned the techniques of X-ray diffraction under Mark's supervision that he was to apply so effectively to biological molecules.

Mark moved on to work for the German company I.G. Farben in Ludwigshafen where he was involved in some of the first commercialisation of polystyrene, PVC and synthetic rubbers. Together with the breakthroughs in high-pressure catalytic chemistry of Haber and Bosch in producing ammonia, this helped I.G. Farben to become one of the largest companies in Europe. At that time, I.G. Farben had links with the German People's Party but after the Nazis took power the company became almost part of the State. One of its subsidiaries, Degesch, produced the Zyklon B cyanide compound that was used for murder in Auschwitz and other extermination camps. Another part of the company developed the first chemical nerve agents tabun and sarin, although these were not used in the Second World War.

Mark was from a Jewish family and he was very concerned about the rise of the National Socialist party in Germany. Accordingly, in 1932 he left I.G. Farben and moved from Germany to the University of Vienna as a Professor of Physical Chemistry. In the First World War he had served alongside Dollfuss, the previous Chancellor of Germany who was assassinated by the Nazis. He was therefore considered by the Nazis in Austria as an opponent.

With his attendance at the meeting of the Pontifical Academy of Sciences in Rome, and other activities in Graz and Vienna, Schrödinger found himself too busy to return to Magdalen College for the summer of 1937 as he had previously arranged. He had written to inform President Gordon who replied on 13 February 1937:

> Many thanks for your letter. I am sorry to hear that you are so closely held to your work at Graz. In view of what you tell me we passed an order, which you will see in due course, generalising the situation. It reads as follows:
>
> "That so long as Dr Schrödinger shall continue to hold his Fellowship, a grant of £100 will be made to him in any year in which, during one of the three academic terms, he resides and takes part in the teaching of the University."
>
> I hope you feel that this meets the situation. We shall be delighted to see you this summer, but if that should prove impossible, then we should expect you in 1938, and need no further order about it. Please tell me some time, at your leisure, how this strikes you and how your prospects are shaping.[2]

Schrödinger replied in positive terms and Gordon also followed up again on 2 March 1937:

> I was delighted to have your letter of 21st February, and to know that you so warmly approved our action. I am sorry that your work at Graz should look like depriving us of your company this next Summer Term, but we must endure that and look forward to 1938.
>
> It is busy here, as usual, at our old game of committee meetings; the college flourishes, and your friends here keep enquiring about you and asking to be remembered.[2]

In the summer of 1937 the scientific refugees Simon, Kurti and Mendelssohn, who were working in Oxford on low temperature physics, were delighted to welcome from Berlin their previous research supervisor Walther Nernst who was receiving an Honorary Degree from the University of Oxford. He had won the Nobel Prize for Chemistry in 1920 for his work on the third law of thermodynamics which states that the entropy of a system approaches a constant value as its temperature approaches absolute zero. He had overlapped closely with Schrödinger in Berlin. One of the great men of German science, Nernst had resigned his post of Director of the Physical Chemistry Institute of Berlin and membership of the Board of the Kaiser Wilhelm Institute in 1933 when he was asked to fill out a form on his racial origins. Lindemann himself had studied under Nernst in Berlin.[7] He was instrumental in arranging the Honorary Degree. Nernst was delighted to see that his work on low temperature physics and chemistry was being continued so well by his protégés in Oxford.

The Max Planck Medal had emerged as the most prestigious award after the Nobel Prize to be won by theoretical physicists. It was awarded annually by the German Physical Society, the Deutsche Physikalische Gesellschaft (DPG), "to a scholar for achievements in theoretical physics, specifically for those extending Max Planck's work".[8] Together with Einstein, von Laue and Sommerfeld, Schrödinger had been involved in establishing this gold medal and German industry contributed significantly to its endowment. Planck was the first awardee and the

Committee making the award was all previous medallists. Before 1937 the awardees had been Planck, Einstein, Bohr, Sommerfeld, von Laue and Hèisenberg.

In 1934 the two candidates who had emerged were Born and Schrödinger. However, the DPG was nervous about making an award that would not be approved of by the National Socialists. With Born from a Jewish family and now working in Cambridge, and Schrödinger in Oxford, it was decided to postpone the award. However, Planck urged the DPG to make the award in subsequent years and the matter was considered again in 1937. Born, by now in Edinburgh, was still considered inappropriate but with Schrödinger returning to Austria he was now a feasible candidate. It seems that he was still listed as a member of the University of Berlin despite his dismissal letter to Schrödinger from Adolf Hitler.[8] The minutes of the DPG Board Meeting on 10 March 1937 stated:

> Mr Zenneck notes that the Privy Councillor Planck upholds his proposals: Born and Schrödinger for the award of the Planck Medal and moves that the award go to Mr Schrödinger, providing that the Society's fall convention takes place in Salzburg. Should the convention take place at another venue then the award should be postponed. At the proposal of Messrs. Geiger and Mey it is resolved: provided the convention takes place in Salzburg, an inquiry should be directed at the responsible authorities in advance about whether any reservations exist against an award of the medal to Mr Schrödinger.[9]

However, due to financial problems with foreign exchange, a meeting in Salzburg was not feasible and the annual DPG meeting was arranged in Germany in Bad Kreuznach. Planck then made clear that he would propose the elimination of the award of the medal if it could not be made without the restriction of being presented in a certain place. The Board agreed to his demand providing the Reich Ministry for Science, Education and Culture had no objection to the Medal been awarded to Schrödinger. With Schrödinger now established at the University of Graz, the Ministry responded that they had no objection. The award was then made to

Schrödinger *in absentia* at the DPG meeting in Bad Kreuznach.[8] After the Nobel Prize, this was the award to Schrödinger that he valued the most.

The Nazi physicist Johannes Stark was not happy with the decision. In an anonymous article in *Das Schwarze Korps* he stated that "Professor Schrödinger demonstratively placed himself politically in opposition to Germany; a more sensitive sign of national debasement cannot be imagined."[8] However, as the DPG had already obtained approval from the Reich Ministry, Stark's comments did not prevail. This debate showed clearly that there was strong opposition in some quarters in Germany to Schrödinger despite his appointment back in Austria. The award subsequently made for 1938 was to Louis de Broglie and this was considered to be more politically neutral. It was to be presented at a symposium in Berlin celebrating the 80th birthday of Max Planck in April 1938. Schrödinger was determined to attend this symposium as he rated both Planck and de Broglie so highly, and also because he was very grateful for the award of the Max Planck Medal.

By 1937 Schrödinger was aged 50 and was enjoying being invited to the more celebratory events arising from being a great man of science. In October a major conference was organised in Bologna, Italy on the 200th anniversary of the birth of the Italian scientist Luigi Galvani who discovered animal electricity while working in Bologna. Founded in 1088, Bologna is thought to be the oldest university in the world and this was a special occasion. Many of the Nobel Prize winners in Physics had agreed to attend including Raman, Heisenberg, Bohr, Richardson, de Broglie, Siegbahn, Debye, Aston and Hess. None of these, however, were scientific refugees at the time. Einstein was invited but was keeping to his rule of not returning to Europe.

Many of the leading universities in Europe, including Oxford, were invited to send a representative to the Bologna meeting. Accordingly, Schrödinger wrote to President Gordon suggesting he might be able to represent the University of Oxford as he was still a Fellow at Magdalen College. He also mentioned to Gordon that he intended to return to Oxford in 1938 and give a course of lectures. Gordon replied on 20 September 1937 in his usual very polite way to say:

It was very pleasant to hear from you and to know there is some prospect of our seeing you here next summer. I have inquired into the question of the Galvani celebrations in Bologna and find that last term, unfortunately, the University informed the authorities at Bologna that they would be unable to send a representative, but were forwarding an official address of congratulation. Had we known here that there was any chance of you being able to represent the University of Oxford at these ceremonies, I need hardly say we should have been delighted to invite you to do so. It seems, however, too late to alter an official reply sent long ago. I may say I went round and saw the University Registrar about this matter and found that, although he shared my regret, he took this view.

There should be no difficulty in arranging the announcement later on of any course of lectures you would be prepared to give here when you come in the Summer Term. A course on Atomic Physics or on Eddington's Cosmology would, I am sure, be most welcome. May I make enquiries? It will be no trouble for me to consult the physics people and get all that put through to you. I will send you a note, once term has begun, making all this a little more precise.

I hope that you and Mrs. Schrödinger are well, and that you find it possible to get some peace and tranquillity for your own private work.[2]

Schrödinger at once, on 22 September 1937, replied in a slightly sarcastic tone:

My dear President,

Thank you so much for your kind letter of Monday. I am awfully sorry for having given you the trouble of these inquiries. I, of course, fully understand that an alteration of the first answer to the authorities at Bologna is impossible, because that would clearly say: "Well, of course it was not worthwhile to send somebody, but since Mr. so and so happens to be there anyhow, he may as well." I am terribly sorry that the honour of, possibly, representing the University of Oxford has escaped me. I would have been very proud of it. I ought to have told you earlier. Now I apologize for giving further trouble, but I hope not much.

If you do not disapprove of it, I would like to wear my Oxford gown after all at that occasion. I obviously must ask your consent. If you approve, would you kindly have the enclosed address slip handed to the porter, asking him to use it for dispatching my cap and gown (which are in the common room) to Bologna. If you disapprove would you just have him or your secretary or anybody drop a post-card to me saying "No" or "The President disapproves".

I am awfully glad you consider my idea of a course of lectures fit and shall be very thankful if you will kindly make inquiries about it at the time you think fit. I am increasingly enthusiastic about Eddington's cosmology and would prefer this subject matter. I have written an article in *Nature* and I am going to lecture on it in Bologna and God knows where else. One ought to have but one sweet-heart at a time, ought one not? Please remember me and my wife to Mrs Gordon and believe me to be.[2]

It is rather ironic that Schrödinger himself should mention to the President of Magdalen that "one ought to have but one sweet-heart at a time". Schrödinger, still feeling perhaps rather annoyed, then sent a bizarre letter to Gordon on 14 October 1937:

My dear President,

Please answer me the following question. I am awfully sorry for troubling you, but I cannot help it. The question is: Could the theoretical fact, that every Fellow is a joint owner of the College, lead to a liability of the College to pay a Fellow's income-tax if, for some reason, he does not pay it? If the answer is no, I am satisfied. If it is yes, I should, in a definite practical case, have to alter my behaviour, in the way of putting up with an insolence. But at least I would then know a sound reason for putting up with it.

Only in order not to leave you before a conundrum, let me explain briefly. When I entered the house in which I lived in Oxford, I had to purchase the fixtures for £30. Imperial Chemical Industries Ltd. have sold them in November 1936 without my consent. This I was ready to give "behindehand", provided that they paid me the £20 they had received from the purchaser. They refused to do this, telling me (in August 1937) that they believed to have counter-claims. That is what, above, I called an insolence.

Second act. The collector of taxes claims £29 odd from me. I wrote him the affaire, and that I had reckoned to pay at least the main part of this debt by what I would get for those goods; that the I.C.I. would probably not refuse to pay him the £20, because otherwise he could put execution on the goods, qua mine; that I was in the position to pay the rest. He says he cannot do that. Tired out and vexed by the whole affair, I had just resolved myself to "wait and see" — because I need my brain for other purposes. But in that moment the idea stroke me, which led to the question in the beginning of this letter.

I am just starting for Bologna, that is why this letter is pretty hurried. Thank you very much for your last letter. I think all that about the lectures will be quite all right. I ought to say, though: I do not think I should have to say much about Milne's theory. It is a rather special thing, much out of the way of the general trend of thought. More politely, I might justify my silence officially by saying that I refrain from carrying owls to Athenes.[10]

The perplexed President Gordon, who was always very patient with requests from Schrödinger, then wrote on 18 October 1937 to Lindemann, with some irritation, to say:

Can you make anything of this? I mean the I.C.I. part? I should like to help the poor man (whose English has receded strangely since he got to Graz).[10]

British science barons often go to the top of an organisation to sort out a problem, even if it is trivial. Lindemann consulted with Lord Melchett, a Director of ICI, who replied on 5 November 1937:

I have now had an opportunity of looking into this matter, and I must say I think Professor Schrödinger has been very fairly treated. However, in order to get the matter cleared up and so that he shall not bear any feelings of ill will towards the Company, our claim will not be pressed, and in addition the Company will pay Schrödinger the sum of £20 which they have received from the sale of the fittings. A cheque for this amount will be sent to him in due course.[10]

Francis Simon back in Oxford was most concerned by this interchange. The funding for his research was crucially dependent on good relations between Oxford and ICI. He would refer the matter again to Schrödinger when he returned to Oxford in 1938.

The congress in Bologna, meanwhile, was opened by the King and Queen of Italy. The journal *Nature* made a report on the meeting in which it stated:

> Bologna was made a public holiday. The streets were lined with troops; girls from the villages paraded in traditional costume, the Balilla were out in force, and the sober black of the party uniform contrasted vividly with the splendour of historic costumes and academic robes.[11]

The Balilla was an Italian fascist youth organisation and it was clear that the conference celebrations had political connotations. Perhaps it was fortunate that Schrödinger did not, after all, represent the University of Oxford in such a procession.

A highlight of the meeting was the talk by the Italian physicist Enrico Fermi on the discovery of new elements from neutron bombardment.[11] Fermi was to win the Nobel Prize for Physics just one year later for this work. His wife was from a Jewish family and new racial laws had just been passed in Italy. Therefore, straightaway after collecting the Prize in Stockholm in December 1938, Fermi emigrated to the USA where he initially settled at Columbia University in New York. He then contributed significantly to developing nuclear fission, reactor physics and the Manhattan nuclear bomb project. Fermi was also to assist Schrödinger after he escaped from Austria to Italy in 1938.

Most of the physicists who attended the Bologna conference spoke about their own recent research, including de Broglie, Debye, Heisenberg, Pauli and Sommerfeld, but Schrödinger gave an overview in French of Eddington's cosmological theory which attempted to bring together quantum mechanics and relativity. When he was in Oxford, he had visited Eddington in Cambridge and was enthused by his ambitious idea of linking wave mechanics with relativity to describe the whole universe. Schrödinger was not happy, however, with a formula that Eddington had derived which related the energy of an elementary particle to the square root of the total number of particles in the universe and they had a lengthy correspondence on the topic. Like Einstein, Schrödinger was to continue a deep interest in a unified theory between quantum mechanics and relativity for the rest of his career.

On returning from Bologna, Schrödinger apologised to George Gordon about his letter on the ICI problems. Gordon, in his usual considerate way, responded on 20 November 1937:

> You are far too kind in your appreciation of my small effort to clear things up for you with I.C.I. It really took very little trouble and only demanded a little firmness on my part and a little attention on theirs. I should add that the last thing they desired was that any unpleasantness should survive your relations with them, and this undoubtedly did much to secure the equitable arrangement now concluded. I am delighted to think that it should have taken a load off your mind.[2]

Schrödinger was still keen to follow up his suggestion of giving some lectures at Oxford on his ideas on Eddington's theory. The kind President Gordon had already contacted colleagues to set things in motion. On 21 December 1937 he wrote to the Vice-President of Magdalen, Henry Parker:

> I think you can tell Schrödinger that it will be in order for him to come into residence about 24th April, which is, in fact, the first day of Full Trinity Term. You might advise him at the same time to get into touch with Dr Griffith, 37 Banbury Road, who is Chairman of the Board of the Faculty of Physical Sciences. I have already told Griffith that Schrödinger is willing to do some University lecturing and teaching next Summer Term; Griffith welcomed this warmly and made some suggestions, which I passed on to Schrödinger some time ago. Schrödinger had better now deal direct with Griffith about dates, hours and subjects.
>
> Schrödinger doesn't raise the question of where he is to live. It would be appropriate, if possible, to put him up in College during his stay. You might add that we should naturally like his stay to be as long as his other engagements allow.[2]

Schrödinger was treating his lectures to be given in Oxford very seriously and had even written some of them out in detail. He clearly wanted to keep up his close link with Oxford through his Fellowship at

Magdalen. He was even prepared to live in the College. He followed up with another letter to Vice-President Parker on 28 January 1938:

> In a letter which the President wrote to Weldon on Dec. 21st and Weldon kindly forwarded to me the next day, the President suggests that it would be appropriate for me to live in College during my intended stay in Oxford, which is to be for about four weeks (at any rate not less than that) in the first part of Trinity term, beginning from April 24th. The President's suggestion meets exactly with my own wishes — so much so, that in the first moment I thought I had already written about it (but this seems not to be the case, following the President's wording).
>
> At any rate, I shall be very glad and thankful indeed, if that proves possible, and I intended to ask for this favour spontaneously, for it seems to me the best and surest way for my really taking up my relations again with all of you and living with you at least "intensely" in order to compensate for the fact that it cannot be for very long. Besides it is, of course, extremely convenient for me, if I have not to bother about the little necessities of life and can therefore devote all my veracity to lecturing or to other tasks, which I should welcome if they came.
>
> It was Weldon who advised me to write to you about accommodation and to "let you know, what I should like". But as regards the latter, I don't think there is much to be said. Anything will do, of course. One room with a bed and a table to write on and the possibility to put up half a yard of books (at most) would be quite sufficient.[2]

The Magdalen Fellow Harry Weldon had got to know Schrödinger quite well when he was in Oxford and corresponded with him when he moved to Austria. Parker then wrote to Schrödinger in friendly terms on 2 February 1938:

> There is no difficulty in finding you a set of rooms in College while you are in residence in Trinity Term. Hunt, our new Fellow by Examination, is going to be in Greece in the spring and early summer and he says he will be very glad to put his rooms at your disposal, they are in cloisters. It will be very pleasant to have you with us again.[2]

Einstein had delivered some very well-publicised public lectures when he was briefly in Oxford but Schrödinger was not given this opportunity in his first period there in 1933–6. He was clearly keen to make up for this and was making detailed arrangements. He wrote on 28 January 1938 in some detail to Dr Griffith in the Faculty of Physics to say he fully intended to begin his lectures in Oxford on 24 April:

> Dear Doctor Griffith,
>
> I had been informed by the President of Magdalen, already in September, that you think the Board of the Faculty would welcome a course of four private lectures in Atomic Physics and a public lecture on Eddington's Cosmology, to be delivered by me during a four-week residence in Magdalen in the first half of Trinity term. The President has now asked me to get in direct touch with you about it.
>
> My stay in Oxford is intended to begin on April 24 and to last four weeks or maybe a few days longer. Any hours to be fixed for the lectures will of course suit me. As regards the subject matter, I should like to suggest, if you think it fit, that the private lectures should specialise on the same subject as the one mentioned for the public one which, as you know, really is a subject of Atomic Physics, though a very new and undeveloped one. What I actually should wish to do is to use these private lectures for telling about some work of my own (incomplete as it is, in the present moment) which was elicited by and is in intimate connection with Arthur Eddington's studies on the connection between Cosmology and Atomicity.
>
> The appropriate title for this course would perhaps be: Wave Mechanical Aspects of the Cosmological Problem. As regards the public lecture, I shall almost certainly be in the position to make it attractive by showing very beautiful slides of celestial objects, from photographs taken at the Mount Wilson Observatory. I had a cable from my friend Richard Tolman a few days ago, saying that he has, very kindly, despatched the slides to me from Pasadena. It is true that these photographs have already been published; yet even if one or the other of the audience happens to have seen the paper-prints, even he will enjoy, I think, to see the originals on the screen.
>
> Under these circumstances my public lecture will, to a great part, be devoted to showing these pictures and explaining them, and to a small part to explaining Eddington's theory. But I think there is no harm in that, for to explain that theory to a large audience in any detail within an hour is an impossibility anyhow. I also think there is no harm in keeping

> to the title "On Eddington's Cosmology". Just perhaps one might add in brackets: "with slides from the Wilson Observatory".
>
> The last item to discuss is the present degree of certainty of my coming at all. The only reason for touching it (except for the well-known brick on my head, in the meantime) is that I have not yet the written grant of leave in my hand but I don't suspect any difficulty from this side. I mentioned my plans when I called in the Ministry before Christmas. They cannot really refuse me the leave for Oxford, which was made all but a necessary condition when I accepted the call to Graz.
>
> I am greatly looking forward to my visit. Please give my compliments to Professor Lindemann when you see him.[2]

Schrödinger was taking his proposed lecture course in Oxford very seriously. This was also the first communication from Schrödinger sent to Oxford which gave reference to possible problems with the authorities back in Graz. He had been negotiating with the Ministry to take leave to come to Oxford and now some doubts were arising.

Anschluss

Just three days before he wrote this letter, the headquarters of the Austrian National Socialist Party in Vienna was raided by the Austrian police who discovered a huge cache of arms and plans for a putsch. Things were moving very quickly. On 12 February Schuschnigg, the Chancellor of Austria, had met with Hitler in his residence at Berchtesgaden. Hitler proposed that Nazi sympathisers be appointed to positions in the Austrian Government. This included making Arthur Seyss-Inquart the Minister of Public Security, which could control the police. In return, Hitler said he would affirm his support for Austrian sovereignty.

With the gathering storm in Austria, Schrödinger gave a lecture on 18 February 1938 at the University of Vienna with the title "World building on a large and small scale".[6] His notes suggest he was planning to give the same talk as a public lecture in Oxford.[2] The lecture was reported in the press and acted as a something of a premonition as to what would very soon happen to Austria. Schrödinger described how the science of small

The audience at the University of Vienna before Schrödinger's lecture on 18 February 1938. Schrödinger is in the front row at the left.

particles could have much larger implications and his lecture ended with the words: "Can anyone who deals with these things still be of the opinion that any of the small nations that live on our earth owns the privilege of being God's Children and is infinitely better than any other?" The photograph shown above, taken before the lecture, captures the apprehensive atmosphere in Vienna at that time.

Then, on 27 February 1938, Schrödinger wrote in English, from the house of Anny's mother in Vienna at Ferrogasse 7, to his close friend Patrick Blackett who by now had been appointed to the Langworthy Chair at Manchester University, previously filled by both William Lawrence Bragg and Rutherford. Schrödinger often wrote letters of a more personal nature to Blackett.

In his letter, Schrödinger first mentioned the meeting of the two chancellors and then discussed the recent work by his former colleague Arthur March on quantum electrodynamics. Following this he summarised his thoughts on Arthur Eddington's ideas:

> I am myself now engaged in studies, the relevance of which is denied by many of our most important theorists (including Bohr, Heisenberg, but not including Dirac, I believe). It is Eddington's idea of explaining atomicity by the finiteness of space. Since I first grasped it, I am so thoroughly surprised by its overwhelming importance that I can hardly spend any interest in anything that is not connected with it. Of course, in the long run everything must be connected with it unless the idea is misleading. I am most sorry that the really good mathematicians among us (like Pauli) don't think much of the thing. The development would be ever so much quicker if they could find it attractive.[12]

Here Schrödinger mentions that Dirac was also developing his own cosmological theory. In a paper published in *Nature*, Dirac had discussed, with little mathematics, the numerical aspects that describe the universe on its largest scale.[13] This included the number of protons in the universe and the distance across the observable universe. His ideas, however, did not receive favour from astrophysicists. Herbert Dingle, for example, wrote in *Nature* that Dirac "was a victim of the great universe mania".[14] Dirac was discouraged by the reaction to his ideas and reverted back to his research on pure quantum mechanics. Schrödinger, however, did not take this direction and continued to work on cosmological theory for many years.

Finally in his long letter to Blackett, and unusually for Schrödinger, he describes the political situation developing in Austria:

> I am sure, you read much about us in Austria in these days. My ironical fate has brought me into the place of the maximum Nazi concentration within the frontiers of this small country. You can imagine how I feel at home. I think I shall have to ask the ministers of public education presently to be allowed to leave this charming city and stay in Vienna where things are going on quite all right as they do in most parts of Austria.
>
> I wonder what will come out of all of that. The present situation is, I think, that a pronounced, though not overwhelming majority in this country, including the government, is willing to grant peace to everyone, whereas a strong minority is prepared to bully us all by the help of their foreign friends. The ultimate decision on our fate will be with the

great powers. For the moment our chancellor is doing his best, I believe, to secure peace both outside and inside. If he does not succeed, I do hope he will not give way, but let the Germans march into Vienna. If that happens without the Western powers stirring (like they did not with Mandchuko and Danzig and Abyssinia and Spain) then we can at least be sure that Western civilisation is before the dogs — and can go to live a little while in some quiet place and know that nothing is to come after us.

Maybe that fascism and bolshevism destroy each other to an extent, similar to the biblic flood; leaving only a few couples here and there in a desert, who are unable, or at first unwilling, to manufacture poison gas and bombs; maybe that then, in the course of centuries, a new culture shall rise on our planet.

If anyone had told us ten years ago that Great Britain would see her flag sink on the high seas about every fortnight and get into the habit of answering: Ah, you naughty boy, I'm sure you won't do that again, will you...

Why did you not use your victory over us barbarians in a more enduring way! Or was it no victory? Was it just a chance that we happened to collapse a second earlier? Were you after all too exhausted to prevent that devil of a German to rise on his feet? Or were you too stupid for it, you too?

All my love to all of you, my dear friend. Let us hope... E. Schrödinger.[12]

Less than two weeks after this letter was written Chancellor Schuschnigg called a referendum on 9 March proposing to keep Austria independent of Germany. Hitler was furious and insisted that the referendum must be cancelled. Under much pressure of an invasion, Schuschnigg agreed on the cancellation.

Lord Halifax, the newly appointed British Foreign Secretary, stated: "His Majesty's Government cannot take the responsibility of advising the Chancellor to take any course of action which might expose his country to dangers against which HM Government are unable to guarantee protection."[15] The tone of this statement clearly confirmed Schrödinger's prediction that "Great Britain would see her flag sink on the high seas".

Hitler then demanded Schuschnigg's resignation and replacement by Seyss-Inquart. The President of Austria, Miklas, reluctantly agreed to this demand.

Then, on 12 March 1938, German troops entered Austria and were greeted by jubilant crowds with Nazi salutes and flags. Seyss-Inquart, now the Chancellor, signed an act making Austria a province of Germany. Hitler himself then entered Austria and went first to his birthplace of Braunau am Inn and then on to Linz, the town of his boyhood. After this he went to Vienna and in front of a jubilant crowd of 200,000 in the Heldenplatz said: "The oldest eastern province of the German people shall be, from this point on, the newest bastion of the German Reich." The Anschluss was essentially complete.

Heinrich Himmler and Reinhard Heydrich entered Austria with members of the SS and arrested several prominent politicians, political dissenters and Jewish citizens. The unification of Austria with Germany was then put to a vote and three weeks later Hitler returned again to Austria. He visited Graz on April 4 and, on the same day, the *New York Times* reported on his visit: "This Styrian Nazi stronghold (Graz) gave Chancellor Adolf Hitler a reception today that surpassed even his triumphal entry into Vienna a fortnight ago, and the hurricane of "Heils" that greeted his arrival was the final outlet of the Styrian longing for unification with Germany."[16] Films of the visit to Graz show Hitler being mobbed by enthusiastic crowds of all ages giving the Nazi salute and shouting Sieg Heil.

Josef Dobretsberger, the Rector of the University of Graz who had been close to the Social Democrats, was forced to resign. He was made to write a letter to the Senate of the University saying: "As a civil servant I have made this decision with the utmost loyalty to the new government."[17] He was replaced by Hans Reichelt who was a known anti-Semite. He was given sole decision-making authority and reported directly to the Reich Minister of Science.

Schrödinger had written to President Gordon on 16 March to say he would not be able to come to Oxford in April after all. Gordon replied, in his usual very polite tones, on 23 March 1938:

I was glad to have your letter of 16 March. There will be great regret here if you should find it impossible, after all, to carry out your plan to visit Oxford next term, we very much hope it will still be possible for you to come. We had been looking forward to having you living in College here, and your old colleagues in the university are equally anxious to hear your lectures and profit by any informal instruction that you might feel able to give during your stay.

It is most polite of you to hand over to the Senior Common Room your Manchester Guardian subscription. I am telling the Vice-President who will doubtless wish to thank you himself on behalf of the Room.[2]

However, on 31 March 1938, Gordon also wrote to Lindemann in a much more concerned tone:

I hear you rang me up about Schrödinger on Tuesday last when I was in London. I had a letter from him on 16 March very cautiously worded in which he said that it now seemed "not quite certain" whether it would be possible for him to visit Oxford next Term, as he had planned.

He went on to add: "the great and important events we are living through in these days might make my presence necessary, so that I could not get away". This all sounds like custody of some sort. I have been in touch with Sir William Bragg who informs me that he has been told, though he cannot confirm it, that Schrödinger is in a concentration camp. I have written, on his suggestion, to his informant for any evidence that there may be of this and have also enquired of Professor C.S. Gibson, Secretary of the Society for the Protection of Science and Learning. Have you any information?

I propose, if only I can be made more secure about the facts, to apply to Halifax in his double capacity of Foreign Minister and Chancellor, to do something for a man who is still a Fellow of an Oxford College. The danger, of course, is doing more harm than good by one's efforts.[10]

The rumours had also reached the British press. On 29 March 1938, the *Daily Mail* wrote:

Professor Otto Loewi of Graz University, winner of the Nobel Prize for Medicine and Physiology in 1936 with Sir Henry Dale of London, is in

> "protective custody". His colleague, Professor Erwin Schrödinger, who shared the Nobel Prize for Physics in 1933, has been dismissed from the university, where he is a Professor of Physics.[16]

Reflecting the sign of the times in Austria, the *Daily Mail* in the same issue reported on the former Chancellor Schuschnigg:

> The trial of Dr Schuschnigg, announced by Field-Marshal Göring of Saturday will, it is understood, come before the Supreme Court at Leipzig, where Dimitrov was tried and acquitted for burning the Reichstag. He will be accused of having intended to falsify the plebiscite he had called.[16]

The worrying rumours on Schrödinger had also reached Tess Simpson and Walter Adams at the Society for the Protection of Science and Learning based in London. On 5 April 1938 Michael Chance from Hadlaub Strasse 52, Zurich wrote to Tess Simpson:

> I saw Dr Franz V. Brücke in Vienna and he told me the following: Schrödinger, Hess and Loewi are in prison in Graz. Of these, Loewi's condition is much the worst. He has been made to take hard physical exercise and this has led to his contracting a severe cold. He is now transferred to hospital. His position is serious.
>
> The reason why all three are in prison is that officially they are members of the National Jewish Organisation which is reputed to have harboured a secret broadcasting station and to have hidden arms. Actually it is more likely that it is because they are all still prominent in the life of Graz. I had no hope of writing them as Hitler was there when I might have gone to Graz.[18]

It was exceptional for any university in Europe at that time to have three Nobel Prize winners as working professors. Schrödinger, Hess and Loewi were, therefore, particularly prominent in the Austrian public eye and especially in Graz. Franz Theodor von Brücke, who is mentioned in the above letter, was a pharmacologist who had worked with Loewi and his co-Nobel Prize winner Henry Dale in London. In the Second World

War he was conscripted into the German armed forces and eventually became a prisoner of war in England. When he returned to Austria after the war he was appointed to the Chair of Pharmacology at the University of Vienna.

A concerned Henry Dale had written on 9 April 1938 to Tess Simpson from the National Institute for Medical Research in London:

> I hear also that Professor Victor Hess and Professor Schrödinger have been imprisoned. This morning however, I have a letter signed only by an initial, but certainly from Frau Loewi herself, which tells me that Hess has been in prison but is now released. Perhaps the same may be true of Schrödinger.[18]

Walter Adams had also written to Guy Keeling at Queen Mary College on 7 April 1938:

> Among the Austrians who are either known to have been displaced or are certain to be displaced are:
>
> 1. Professor Erwin Schrödinger, Nobel Prize winner in Physics.
> 2. Professor Victor Hess, Nobel Prize winner in Physics.
> 3. Professor Karl Przibram, a physicist of international reputation at the Radium Institute in Vienna.
> 4. Professor Hans Thirring, Director of the Institute of Theoretical Physics, Vienna.
>
> Any of the above Austrians would be a great capture for any English or American academic institution. Each has considerable publicity value and, what is more relevant, great scientific importance.
>
> Two other names which I may mention are Professor Ludwig Hopf, the mathematician dismissed from Germany whose work has brought him near to the subject of Aerodynamics. And Professor Hermann Müntz, who went from Germany to Russia in 1929 but has now lost his position there.
>
> A couple of years ago we had confidential enquiries from Dr Riabouchinsky, the very well-known Russian refugee specialist in Aeronautics who was feeling that his position in France was insecure.

> I think it possible that he could be tempted to leave France if a more secure position were available here. He is one of the great names in the history of Aeronautics.
>
> I enclose a summary of the particulars of these individuals. I shall be pleased to send further particulars if wanted.[18]

Karl Przibram had overlapped with Schrödinger at the Radium Institute in Vienna during his student days. He came from a Jewish family and managed to escape from Vienna to Brussels. In the Second World War he joined the resistance movement in Belgium. He returned to his Professorship in Vienna after the war. His brother Hans Leo founded the first station for experimental biology in Vienna and was an internationally known zoologist. He escaped to Amsterdam where he continued research on the chemistry of the pineal gland. In 1943 he was arrested and deported to the Theresienstadt concentration camp where he died.

Ludwig Hopf was a German theoretical physicist whose work was well known to Schrödinger as he had published papers with Einstein on the classical statistical theory of radiation. From a Jewish family, he lost his position in Aachen in 1934 but remained in Germany until 1939. He then managed to escape and took up a position for a short period in Cambridge before being appointed to a Professorship in Mathematics at Trinity College in Dublin. He died on 21 December 1939 of thyroid failure. Schrödinger, who had only been in Dublin for three months, gave an oration at his graveside saying: "a friend of the greatest geniuses of his time, indeed, he was one of them".[19]

Herman Müntz was a mathematician who worked with Einstein in 1927 after Schrödinger's arrival in Berlin. He took up a Professorship at Leningrad University but was expelled from Russia in 1937 and managed to move to Sweden.

On 25 April 1938 J. Kendrick at the British Passport Office in Vienna wrote to Walter Adams about the Austrian professors:

> Professor Hess apparently has not been arrested and still retains his professorship at Graz University. Professor Mark (the gentleman mentioned 5th on your list) informed me today that he understands Professor Hess

has not been interfered with and he intends to continue in his post. In view of this information, which I think is reliable, I would be glad to know whether you desire me to take any further action in his case.

Professor Loewi is, as you state, under arrest and has been deprived of his Professorship at Graz University. I will let you know later the results of certain action which is being taken to secure his release and permit to leave the country.

Professor Mark has been dismissed from the University. He very greatly appreciates the extreme kindness of your Society, but considers it his duty, however, to remain here in order to give German industrial interest the opportunity of acquiring his services. This decision, I imagine, is influenced by a desire to avoid giving the impression of flight out of fear (which conceivably is also a form of nerves). Professor Mark intimated, however, that should his life be made impossible, he would gratefully accept the kind invitation extended to him to go to the UK. He stated it would distress him if he thought that by going to the UK it would deprive somebody of a post to which that person might have more claim than he had, and he would be prepared to devote himself to practical industrial chemistry under a commercial body. He will keep me informed of developments.

Professor Schrödinger has not been arrested and is still at Graz University. Professor Mark had a letter from him on April 22 in which he wrote that he was leaving Graz on that date to deliver a lecture in Berlin on April 23. Professor Mark believed that Professor Schrödinger has no intention of leaving the country. Should you desire any further action to be taken in this case I would be glad to do anything I can.[18]

Walter Adams then wrote on 29 April 1938 to President Gordon at Magdalen College in reassuring terms:

In case the news about Professor Schrödinger has not reached you from any other source, I write to inform you that I have now heard from a completely reliable authority that Professor Schrödinger has not been arrested and is still at Graz University. I understand that Professor Schrödinger has expressed no intention of leaving the country, was intending to give a lecture in Berlin on April 23rd and seems therefore not to be in political difficulties.[18]

Patrick Blackett had also been aware of the concern about the welfare of Schrödinger. He wrote to Walter Adams on 2 May 1938 to say:

> Many thanks for your letter about the Austrian professors. I had, as a matter of fact, heard from other sources that Schrödinger was all right. It may be that rather unnecessary alarm was felt by us about the matter, but of course it is quite possible that the various steps that have been taken to obtain information have in fact had a practical and good effect.
>
> Do I hear aright that you are going to the London School of Economics? If so, may I congratulate the London School of Economics and you too.[18]

Walter Adams was clearly a most attentive General Secretary of the Society for the Protection of Science and Learning. He was moving on to be the Secretary of the London School of Economics (LSE). He eventually became Director of the LSE in 1967 and was knighted in 1970.

Lindemann in the meantime had taken up the suggestion of George Gordon and wrote to Lord Halifax. As well as being Foreign Secretary of Great Britain, Halifax was also Chancellor of the University of Oxford, a role which put him in regular and personal contact with both Gordon and Lindemann. Halifax responded to Lindemann on 9 May 1938 from the Foreign Office:

> With reference to your telephone message of 6 April I am writing to let you know that our Ambassador in Berlin has now had a reply from the German Minister for Foreign Affairs to the enquiries he made at my request about Professor Schrödinger.
>
> I am sorry to have to tell you that Herr von Ribbentrop's reply is unfavourable. According to the German authorities Professor Schrödinger left Germany for political reasons and, after a brief stay in Oxford, settled in Graz where he proceeded to busy himself as a fanatical opponent of the new Germany and National Socialism. A recent examination of his case had shown that Professor Schrödinger had, up till very lately, remained in constant communication with German emigres abroad and it was therefore feared that permission for him to visit Oxford would merely offer him a further opportunity to resume his anti-German activities.

> In these circumstances Herr von Ribbentrop fears that a further course of lectures by Professor Schrödinger at Oxford could only harm Anglo-German relations and is therefore obliged to refuse my request.
>
> I much regret that my intervention on behalf of Professor Schrödinger should not have been more successful but I fear that, in view of the attitude adopted by the German authorities, there is nothing more that I can do.
>
> Yours sincerely. Halifax.[10]

In the context of this letter it should be noted that, before he was the German Minister for Foreign Affairs, Ribbentrop was the Ambassador to Great Britain and in that role had several personal interactions with Lord Halifax. In addition, Halifax had visited Germany in November 1937 and had met with Hitler at Berchtesgaden, Göring at his estate in Karinhall, and Goebbels. Göring told him "that the incorporation of Austria and the Sudetenland and a settlement of the problems of Danzig and the Polish Corridor were essential to Germany".[20] Halifax had replied that England wanted these questions resolved peaceably. Göring declared that the issues had to be settled, but said, smiling at Halifax, "We will never use force."[20] Both Ribbentrop and Göring would be condemned to death at the Nuremberg Trials in 1946 but Göring took cyanide to avoid his execution.

The discussions with the German leaders influenced Halifax and the British Government significantly in their subsequent appeasement policy as the Nazis marched into Austria, Czechoslovakia and Poland. Subsequently, after the German invasion of Norway in April 1940, the British Prime Minister Neville Chamberlain resigned. It was expected that Halifax would take his place as Prime Minister but the King chose Winston Churchill.

Halifax refers to the comment from Ribbentrop that Schrödinger "settled in Graz where he proceeded to busy himself as a fanatical opponent of the new Germany and National Socialism". Schrödinger's interaction with friends in Vienna such as Herman Mark and Hans Thirring, who were known to be close to the previous Austrian government, may have been a factor here. In addition, Ribbentrop's comment that Schrödinger had "up till very lately, remained in constant communication with German emigres abroad" may refer to his letters to friends such as Max Born, to whom Schrödinger was still writing in January 1938. However, there is no evidence that Schrödinger was corresponding with Born and other German emigres in the months straight after the Anschluss but it is clear his letters abroad to people such as George Gordon were using careful language.

There is evidence that the Nazis had been keeping a close eye on Schrödinger. As he had taken German citizenship when he took up the Chair in Berlin the German Consul in Graz had called for his immediate expatriation on 24 January 1938 after he had voted in favour of a German Professor who had been in prison for insulting Hitler and had been proposed for employment at Graz University. This information was passed on to the Gestapo but no action was taken.[21]

Lindemann replied to Halifax on 12 May 1938:

> Many thanks for your letter about Schrödinger and for all the trouble you have taken. I am extremely sorry that nothing can be done to help him, more especially I think it most unlikely that he should have taken part in any political activities after his return to Graz. Here in Oxford he certainly was most reticent, a fact I think all the Fellows of Magdalen would confirm.
>
> I do not know whether in the ordinary course you will refer again to this matter in your correspondence with the Wilhelmstrasse, but if you do it might be helpful and would certainly be accurate, if you mentioned this fact.
>
> Once more with my warmest thanks for your having consented to intervene.[10]

By a coincidence, the Magdalen Fellow in Politics William Mackenzie was travelling in Austria just after the Anschluss. At the suggestion of President Gordon he managed to arrange a meeting with Schrödinger and Anny. Mackenzie wrote to Lindemann on 21 April 1938 with reassuring news:

> The President has suggested to me that it might be of use to you to have some details as to my meeting with Professor Schrödinger last week, as

there seems to have been a good deal of conflicting rumour here as to his position.

I was staying at the Seekarhaus in the Radstädter Tauern, and took the chance of sending a postcard in very non-committal form to Mrs. Schrödinger at Graz; I enclose the card which I got in return. The Vienna address may be useful if it is absolutely essential to get in touch with him. This led me to expect that I should see them in Vienna, and I was rather surprised to get a phone call last Tuesday, the 12th, to say that they had just arrived at the nearest hotel, a couple of miles or so from the Seekarhaus. They both came up to lunch next day, the 15th, and were with me most of the afternoon; they intended to stay where they were for a short holiday, but unluckily I had to leave for Vienna the next day, and may not see them again.

We had a very long talk about things in general, and I was charged to give best wishes to all friends in England, but there is so little we can do that I haven't any very specific message, or suggestion for assistance. The following points may be of interest:

(1) Schrödinger himself is in good health, has not at any time been under arrest, and has not been personally molested; Mrs. Schrödinger is still convalescent after an operation but seems to be getting on well. Their correspondence, however, is probably being tempered with, though there didn't seem to be any direct evidence of this except some curious dealings with their copy of the *Manchester Guardian Weekly*; so that no letter from him is to be taken at its face value, and anything sent to him should be as colourless as possible.

(2) The University had been shut since the disturbances in February, and after the Nazis took over several of the professors were arrested. S. mentioned that Hess had been in custody and later released, and that Loewi was still in gaol; no specific charge, and no signs of any ill-treatment beyond the natural extreme discomfort. The University reopens this week, and as far as he knows S. will proceed as usual with lectures in Graz and once weekly in Vienna.

(3) His personal position is one of complete uncertainty, for better or for worse; he summed it up by saying that they ought logically either

promote him to a better job or put him in a concentration camp, but he had no idea which it would be. He doesn't know who is responsible for deciding his fate; nor what his attitude to the regime is officially supposed to be; nor whether he is debarred completely from leaving the country. But for the present he certainly cannot come to Oxford to lecture without specific authorisation from some high power, and doesn't even know whether he needs permission to travel to Berlin for some celebration there (I think in honour of Planck). It is even possible that later in the term he may get definite leave to come to Oxford, but he has no means of hurrying a decision.

(4) In regard to future plans; he has clearly no feeling that it is necessary for him to get out of the country at all costs as soon as possible, and would on the whole like to make his peace with the regime if they will let him. So that his policy is to establish a reputation as a non-political person, with reasonably patriotic principles; I don't think Schuschnigg's fate troubles him much, as a matter of principle the followers of Schuschnigg are only getting as good as they gave; but his dislike of the anti-Jewish policy is as strong in him as ever, and may perhaps cause some difficulty.

(5) There is practically nothing we can do to help him directly, unless things get much worse than they are; letters to him would be rather dangerous, but he said he would be very glad to see anyone from Oxford who was in Austria — he apparently thought it would be quite safe. When I saw him, I hadn't heard of the various demarches that had been made through the Vice-Chancellor, and hadn't discussed them with him; but I think in principle that anything which stresses his international importance may be useful, and may in the long run filter through to the relevant department — provided always that it contains no hint whatever of political importance or associations.

We had four or five hours of talk, and naturally there was a great deal more than this said, but I think these are the main points.[10]

Gordon was much relieved to hear the encouraging news. He wrote on 1 May 1938 to thank Mackenzie:

> I am very much obliged to you for your note about Prof. Schrödinger. It confirms what had reached us from other sources lately and is, so far, very comforting to us. I fear that we shall not see him, however, in Oxford this summer to give his advertised course of lectures.[10]

Mackenzie was an unusual teacher of politics in that he always seemed to be where the action was. In a subsequent summary of his papers he wrote:

> One of the pleasantest duties was to get around in Europe, taking good care not to get damaged. I managed to be in Vienna just after Dollfuss was shot in 1934, in Paris just after the Popular Front strikes of 1936, in Geneva when Samuel Hoare made his appeal for sanctions against Italy over the Ethiopian war (I heard him), in the Austrian mountains just after the Nazis walked in, in Prague when Runciman was there, in Russia just before Ribbentrop arrived at the end of August 1939.[22]

Lindemann, always with his ears to rumours, wrote back to thank Mackenzie on 22 April 1938 and had an additional worrying query about a letter from Schrödinger published in the German papers:

> Very many thanks for your letter and all the information about Schrödinger from which I am relieved to hear that he has not been having any trouble. I cannot understand how these very definite rumours got about.
>
> I wonder whether Schrödinger said anything about a letter which I was told was published in the German papers, alleged to have been written by him, recanting anything he may have said against the Nazis. I have not seen it myself, but its appearance rather lent colour to the fact that he was under pressure.[10]

Here, Lindemann was referring to a letter that had been published under Schrödinger's name on 30 March 1938. The local newspaper, the *Grazer Tagespost*, gave the letter the headline "Acclamation of the Führer: A Leading Scientist offers himself for the Service of Nation and Fatherland":

There is in German Austria naturally also a number of scientists who have lived for their special work alone, and who have troubled little about contemporary political questions. The Dollfuss-Schuschnigg system managed to enrol just such unpolitical people for its propaganda. In recent days when an unprecedented storm of enthusiasm swept over the whole country the scales fell from our eyes. They realised that just the contrary of that which the former holders of power told them, was in fact the truth. The call of blood led also those men back to their people and thus to Adolf Hitler.

The Senate of the University of Graz has forwarded to us the following declaration of Professor Schrödinger because of his request for its publication:

"In the midst of the exultant joy which is pervading our country, there also stand today those who indeed partake fully of this joy, but not without deep shame, because until the end they had not understood the right course. Thankfully we hear the true German word of peace: the hand to everyone willing, you wish to gladly clasp the generously outstretched hand while you pledge that you will be very happy, if in true cooperation and in accord with the will of the Führer you may be allowed to support the decision of his now united people with all your strength.

It really goes without saying, that for an old Austrian who loves his homeland, no other standpoint can come into question; that — to express it quite crudely — every "no" in the ballot box is equivalent to a national suicide.

There ought no longer — we ask all to agree — to be as before in this land victors and vanquished, but a united people, that puts forth its entire undivided strength for the common goal of all Germans.

Well-meaning friends who overestimate my importance consider it right that the repentant confession that I made to them should be made public. I too belong to those who seize the outstretched hand of peace, because, sitting at my writing desk, I have misjudged up to the last the real will and the true destiny of my land. I make this confession readily and joyfully. I believe it is spoken out of the hearts of many, and I hope in doing this to serve my country. E. Schrödinger."[23(t)]

Back in England several others had been alarmed to read the statement. Gertrud Bing, a German art historian who had studied under

Ernst Cassirer, a correspondent with Schrödinger, worked at the Warburg Institute of Cultural History. This research institution, which had been founded by the Jewish Warburg family in Hamburg, had managed to relocate to London in 1933 as the Nazis came to power. She wrote to Walter Adams on 29 April 1938:

> I am sending you the photographic copy of a letter signed by Schrödinger which has appeared in the Grazer Tagespost. I thought you would like to see it. I must confess I do not know what to think of it. At any rate, it must be assumed that the letter was written under considerable pressure, but even so it seems so preposterous that I should prefer to doubt whether it was written by him at all. There seems to be no doubt that at some time or other Schrödinger was arrested and there are rumours that he is not in Austria at all.[18]

The journal *Nature* also drew the letter to the attention of many scientists in a News and Views article:

> Since the recent incorporation of Austria into Germany, little precise knowledge has been available as to the result of the change upon the position of some distinguished Austrian men of science. Upon inquiry we are informed that Prof. E. Schrödinger will continue to occupy the chair of theoretical physics in the University of Graz. The *Tagespost*, Graz, of March 30, published a letter from Prof. Schrödinger to the Senate of the University, in which he explains that he has not hitherto taken the active part expected of him in the National Socialist movement but is now glad to be reconciled to it.[24]

The journal also quoted the last paragraph of Schrödinger's statement starting "Well-meaning friends...". His scientific colleagues in Europe and America felt particularly let down by the statement. Some of his former colleagues and associates, including Lindemann, Simon and Peierls, never forgave him. He also had to do a lot of explaining with close friends, including Einstein, Born and Dirac, when he met or corresponded with them personally in the years to come. A year later, he wrote to Einstein on 19 July 1939 with an explanation: "I hope you did not seriously denounce my subsequent, certainly quite cowardly, behaviour.

I wanted to stay free — and couldn't do it without gross hypocrisy."[25] On 19 January 1943, he also wrote to Born to complain about the *Nature* article:

> I fail to find that *Nature* has any case against me; but the reverse is almost true. They once brought a rather ugly notice about me, hardly short of denouncing me as a Nazi. For this notice they had used my own information that I gave them on their request — and indeed misused it. For they exchanged my statement, that I was holding office in Graz into the future tense that I would be going to hold office...which was a mistake.[26]

Dirac discussed Schrödinger's statement with him later in 1938 when he visited Cambridge and, in the obituary he wrote for *Nature* in 1961 after Schrödinger's death, he gave his own view:

> With the German annexation of Austria in 1938 Schrödinger was immediately in difficulty, because his leaving Germany in 1933 was taken to be an unfriendly act. He was forced to express his approval of the Nazi regime, and he did this in as ambiguous a way as he could.[27]

It remains unclear whether Schrödinger signed the letter in a moment of great duress or if he truly meant it at the time. It was clear that President Gordon back at Magdalen College was becoming increasingly concerned with his mental condition during these months from the language in the communications he had been receiving from Schrödinger. The letter was written in a celebratory language that was being used at that time by the Nazis and did not use phrases and words normally used by Schrödinger (of which there are many examples in this book).

William Mackenzie had met with Schrödinger two weeks after the statement was published and reported: "He would on the whole like to make his peace with the regime if they will let him. So that his policy is to establish a reputation as a non-political person, with reasonably patriotic principles." This implies that Schrödinger's state of mind at the time was to go along with the new regime in Austria.

The German Consulate in Graz had also seen Schrödinger's public statement seeming to support Hitler and had forwarded it to the

Department of Foreign Affairs in Berlin. The response came that: "The question of withdrawing his Professorship might still be addressed by the local administration."[21]

There were other well-known Austrians who were forced to write in somewhat similar ways. The world-famous psychoanalyst Sigmund Freud was born in Vienna and was from a Jewish family. He was one of the most prominent authors whose books had been burned by the Nazis. After the Anschluss he was an obvious target and he was visited by the Gestapo several times. His daughter Anna was arrested by them.

There was considerable international interest in Freud's case including from President Roosevelt in the USA. Eventually Princess Marie Bonaparte managed to negotiate his release but only after the confiscation of much of his property and bank accounts.[28] On his departure from Vienna for Paris and then London he was forced to sign a statement which enabled him to obtain an exit visa:

> I hereby confirm that after the Anschluss of Austria to the German Reich I have been treated by the German authorities and particularly by the Gestapo with the respect and consideration due to my scientific reputation, that I could live and work in full freedom, that I could continue to pursue my activities in every way I desired, that I found full support from all concerned in this respect, and that I have not the slightest reasons for any complaint.[29]

Back in Germany, the Nobel Prize winner and Nazi, Johannes Stark, had lectured at Berlin University on "Dogmatism and Experience in Atomic Research" in which he stated that the models of the atom of Rutherford and Bohr could not be relied on as they were based on the theories developed by Jewish scientists. This inspired the *Sunday Times* on 24 January 1937 to respond with great sarcasm and the headline "The Militant Atom":

> Nowadays we paint not the rose but the atom; and we paint it with national colours. The President of the German Physical Institute complains that Bohr's atom had a spherical form in "company with the ideas of the Jewish physicist, Abraham". The excellent professor has changed

> all that by substituting a German atom of "several rings of etheric perturbations, all set in different planes, with the nucleus of the atom between the planes." Really high class atoms, presumably, are in the form of a swastika with "Heil Hitler!" set between the arms. Every little boy and girl, we used to be told is born "Either a little Liberal or else a little Conservative". Today every atom is either a little brown-shirt or else a little Ikey Mo. Brave New World indeed! In the next war gallant scientists will let off the mustard gas with cries of Atta-Atom.[16]

Even the German patriot Heisenberg, who had famously met up with Schrödinger in Stockholm in December 1933, was having problems with Stark and Lenard at this time. Heisenberg's great ambition was to replace Sommerfeld at the University of Munich where he had started his research career. When Sommerfeld retired in 1935 Heisenberg, then at Leipzig, had been delighted to receive the call to the Chair. However, the official journal of the SS, the *Das Schwarze Korps*, then published an article from Stark that called Heisenberg a "white Jew". Heisenberg's mother Annie raised her concern on the matter with her friend Anna Maria Himmler who was the mother of the infamous Head of the SS Heinrich Himmler.[30]

The SS attacks on Heisenberg were then called off but he was still not appointed to the Munich Chair. This went to Wilhelm Müller who had never published a research paper in theoretical physics and was a member of the SA, the paramilitary wing of the SS. Even Ludwig Prandtl, a famous pioneer in fluid mechanics at Göttingen, who in the Second World War became the Chairman of Research for the Reich Air Ministry, complained to Himmler and subsequently to Hermann Göring about the appointment of Müller.[31] His long letter to Göring of 28 April 1941, in referring explicitly to the quantum theory, stated:

> But, for the supporters of Lenard, this was also considered to be "Jewish physics", perhaps according to the principle which I occasionally have heard, "That which I do not understand I consider to be Jewish." There were naturally also contributions from Jews to this part of scientific development, but the greatest contribution was made by men such as Planck and Sommerfeld, Heisenberg and Schrödinger.[32]

The implications of this letter from Prandtl, although well-meaning in criticising Lenard, were sadly typical of some of the German scientists who remained in Germany after 1933 and worked for the Nazis. The letter diminished to Göring the significant contributions to quantum theory of great Jewish theoretical physicists such as Pauli and Born, Prandtl's former Göttingen colleague.

Like Schrödinger, Heisenberg had changed direction in his research in the mid-1930s. He wrote several papers on cosmic rays, which had become very topical due to the experimental work of Victor Hess and Blackett. His attention was also drawn to nuclear physics following the work of Chadwick, Fermi and others. In his correspondence he was well aware of the loss to German science from the forced removal of many physicists.

As was mentioned in the letter from William Mackenzie, and also in letters from others, Schrödinger had been invited to Berlin on 23 April 1938 for the 80th birthday celebration of Max Planck held in the Harnack-Haus. The publication in the *Annalen der Physik* associated with the event had a paper from Schrödinger on the ambiguity of the wave function in which he could not resist mentioning the work of Eddington. Other papers were also given by Bohr, Millikan, Nernst, Debye, Sommerfeld and Heisenberg who wrote on the universal length — the problem being addressed by Arthur March. There were no papers from the refugee physicists such as Einstein and Born.

It is possible that his statement supporting the Führer, published in the Austrian newspaper, helped Schrödinger obtain permission to travel to Berlin but there must have still been some risk with his attendance. It was his first visit back to Berlin since 1933 and the city was still reverberating with the success of the very recent Anschluss. The conference was attended by over 300 participants and Max von Laue, Planck's closest colleague still in Berlin, gave the main talk on superconductivity and magnetism.

The organisers worked hard to avoid any political connotations. Peter Debye had moved from the ETH in Zurich to become Director of the Physics Section of the Kaiser Wilhelm Institute and he chaired the proceedings. There was a highly original one-act humorous play written by the Potsdam astrophysicist Walter Grotrian in which the actors included Debye, Sommerfeld and Heisenberg and quantum theory was a theme.[33] There was a magnificent white-tie banquet in the Goethe-Saal of the Harnack-Haus for the participants and guests.

The symposium finished with the award of the gold Max Planck Medal for 1938 to Louis de Broglie.[8,34,35] Due to illness, de Broglie could not attend and the medal was presented to the French Ambassador on his behalf. Eventually the Planck Medal was awarded after the war in 1948 to Max Born who had been considered for the award, together with Schrödinger, back in 1934.[8] He thought long and hard as to whether he would return to Germany to receive it.[36] Four days after the symposium Schrödinger visited Planck and they talked together for three hours. Schrödinger, however, said nothing about his own situation.[37]

On the day after the Planck symposium Schrödinger and Anny visited Lise Meitner for afternoon tea.[37] Even though she was Jewish, Meitner was still able to work at the Kaiser Wilhelm Institute in Berlin and had been protected by Peter Debye and some other colleagues. She had been collaborating with Otto Hahn on the ground-breaking research that would lead to nuclear fission. Like Schrödinger, Meitner was of Austrian birth and had studied under Exner in Vienna. She had even served on the Italian front in a medical capacity in the First World War. She was the first female Professor of Physics in Germany and was well known in Berlin. As an Austrian national she was now, after the Anschluss, a citizen of Greater Germany and this was making it very difficult for her to leave the country. She had even been denounced by the Nazi Kurt Hess, an organic chemist from her Kaiser Wilhelm Institute for Chemistry, who stated: "The Jewess endangers the institute."[37] The rigid dogmatism of Kurt Hess also extended to science and he did not accept the macromolecular structure theory of polymers.

Meitner consulted with colleagues of Hans Kramers about a move to the Netherlands but that country was being swamped with refugees. Then an offer of a position came from Manne Siegbahn in Sweden. He had won the Nobel Prize in 1924 for experimental research on X-ray spectroscopy and was starting up a nuclear physics institute in Stockholm. Through fear of being prevented to depart, she told very few colleagues that she would escape from Germany. Kurt Hess attempted to inform the authorities but she quietly crossed the border to the Netherlands on 13 July 1938 with help from Debye, Paul Rosbaud and others, and in due course went on safely to Sweden.[37,38] It would be ten years before she would again step on German soil when she attended a commemoration for Max Planck in Göttingen.

Following her escape from Germany, there was a move to bring Meitner to Oxford. However, Lindemann wrote on 13 September 1938 to Sir Charles Ellis, Wheatstone Professor of Physics at King's College London:

> Miss Meitner is certainly one of the most distinguished women physicists in Germany and I very much hope if she would have to leave that country that she will find a congenial post somewhere. But I should not be prepared to press very strongly that we should invite her to come to Oxford... I hope you will not think me unsympathetic to Miss Meitner, whom I know and like, but whose plight seems to be more contingent than actual.[38]

Schrödinger's co-winner of the Nobel Prize Paul Dirac had also made his own daring visit to another country in political upheaval. In the summer of 1937 Dirac went to see his close friend and former colleague Pyotr Kapitsa in his summer house in Bolshevo near Moscow. This was just before one of Stalin's great purges. Within a year, five scientists who Dirac visited on this trip had been executed. Kapitsa, however, had been able to set up his laboratory again in Moscow and applied magnetic methods to achieve very low temperatures. He had just achieved the first observation of liquid helium which would win him the Nobel Prize in 1978, and Dirac nominated him several times for the Prize. Kapitsa survived the purges but he was not to meet with Dirac again for 29 years.

At the University of Vienna, it has been estimated that more than 2,700 Jewish or dissident members (professors, lecturers, students and administrative employees) were expelled in 1938.[39] This included 322 out of a total of 763 professors. Of these, 73% were on racial grounds and the remainder for political reasons. The process of expulsion was based on the same system that had been applied in Germany after Hitler came to power but was made successful through the active participation of university officials and professors who were already sympathetic to National Socialism. The expulsions of the academics happened much more quickly in Austria in 1938 than was the case in Germany after 1933. The process produced a frantic atmosphere in which Schrödinger was caught up.

Just three days after the Anschluss, Fritz Knoll, who was a Professor of Botany at the University of Vienna and a member of the Austrian National Socialist Party (NSDAP), was tasked with "looking after the regional leadership's interests at the University of Vienna". He then took over the Rectorship from Ernst Späth, a distinguished organic chemist who had achieved the first synthesis of mescaline. Knoll was also made the President of the prestigious and select Academy of Sciences in Vienna of which Schrödinger had been a corresponding member since 1928.[40]

Within a further three days, Knoll had replaced all the departmental deans at the University of Vienna with his own appointments. Viktor Christian, who was an ancient orientalist and editor of *Race Studies Magazine*, was appointed Dean of the Faculty of Philosophy, through which Schrödinger had been appointed in 1936 as an Honorary Professor. Christian was also a member of the Austrian National Socialist Party and eventually became an officer in the SS. The University of Vienna was put under the overall control of the German Reich Ministry for Science, Education and National Culture led by Minister Bernhard Rust, who had been a follower of Hitler from the early 1920s.[41]

On his return to Graz from the Planck birthday celebrations in Berlin, Schrödinger received a letter from Dean Viktor Christian dismissing him from his post in Vienna:

> On the basis of the decree of the Austrian Ministry of Education of 22 April 1938, No. 12474/I/1b, your authorisation to participate in instruction in the Philosophical Faculty of the University of Vienna is withdrawn.

> You are, therefore, to refrain from any professorial or other duties falling on you within the scope of your former appointment or especially assigned to you.[42]

Some 68 years later in Alpbach, Schrödinger's daughter Ruth told me that this dismissal was particularly hurtful to Schrödinger as he had been so proud to be a student at the University of Vienna. She said that when Schrödinger returned to the University as a fully appointed Professor in 1956 he asked for a copy of the official University document ordering his dismissal in 1938. She showed me the document which had a typed list of the many professors to be expelled and right at the end someone had added in handwriting "and Dr. E. Schrödinger".

Schrödinger's friend Herman Mark was also one of the 65 members of the Philosophy Faculty at the University of Vienna to be dismissed on April 22. Concern on his status was being raised in several letters being circulated in England. On the very day of the Anschluss he had been arrested by the Gestapo who were familiar with his friendship to former Chancellor Dollfuss. They took away his passport but he was released. However, being a skilful chemist, he had over previous months used some of his savings to buy a quantity of platinum which he had formed into several coat hangers. He bribed a local Nazi lawyer to retrieve his passport. He then drove himself and his family over the border to freedom in Switzerland with their clothes on the platinum coat hangers.

Mark then emigrated to the USA and took up an appointment at the Polytechnic Institute of Brooklyn, and established the first institute in the USA dedicated to polymer research. He was a leader of the rapidly expanding polymer science community in the USA and started the *Journal of Polymer Science*. He was subsequently awarded the US National Medal of Science by President Carter and also the Wolf Prize, an award considered by many to be second only to the Nobel Prize.

Schrödinger's closest friend from his student days and regular correspondent, Hans Thirring, was also dismissed from his Professorship of Theoretical Physics at the University of Vienna. He was not Jewish but due to his links with Herman Mark, and others disapproved of by the Nazis, he was not held in favour by the new regime. He was, however, well-connected in Vienna and, during the Second World War, managed to obtain a serious of short-term jobs in the industry with companies including Elin AG, Siemens and Halske. After the war he was reinstated to his former Professorship and was instrumental in bringing Schrödinger back to Vienna for the second time. He became active in the Socialist Party of Austria and was elected to the Austrian Parliament.

Back in Oxford, the Governing Body of Magdalen College was considering the status of Schrödinger's Fellowship which was soon due to finish on 2 October 1938. The matter was clearly uncertain and the College Actum of 15 June 1938 stated:

> That the question of the re-election of Dr Schrödinger to a Supernumerary Non-Stipendiary Fellowship be referred to the Fellowship Committee for further report to the October Meeting.[43]

Following Schrödinger's dismissal from the University of Vienna it would not be long before the same occurred from the University of Graz. On 26 August 1938, the Ministry of Education in Vienna wrote:

> Measures on the basis of the ordinance for renovation of the Austrian professional civil service. To Herr. Professor Erwin Schrödinger.
>
> On the basis of Sect 4. Paragraph 1 of the ordinance for renovation of the Austrian civil service of 31 May 1938, RGBI.I S. 607, you are dismissed. The dismissal is effective as of the day of arrival of this notice. You have no right to any legal recourse against this dismissal.[42]

Escape

The formal description of the ordinance was: "The regulation is the central instrument for the national reorganization of the public service. Section 3 concerns the removal of Jews, Jewish mongrels and officials with Jewish ethnicities from the public service. Section 4 orders the retirement or dismissal of politically unreliable civil servants." For the first time the alarm bells were truly ringing in the ears of Schrödinger and Anny. In her usual direct style Anny described what happened next:

Well, when the Nazis came to Austria, my husband got several invitations to foreign countries. He was not allowed to get the telegrams himself. They were brought to the university and he was called to the university and he was told: "Of course, you have to refuse. You can't go to Brussels or so." So it was really absolutely like a prison. Only through de Valera we were secured because de Valera knew that we were in danger there and he let us know that there was the possibility for an institute for advanced studies in Dublin which he wanted to create if my husband said, in principle, "Yes, I will come," or "no."

It was absolutely sure he could not write to my husband because everything was censored. So he asked Whittaker to ask Born; Born wrote to our friend Professor Richard Bär in Zurich. He told a Dutchman, who came to Vienna, about the possibilities. We were not in Vienna. He didn't come to Graz, he came to my mother and told her this important thing. She was afraid to take such a very important message without having anything written down, so he wrote down in just a few lines that de Valera wanted to create an Institute for Advanced Study and whether he would come, in principle.

This little piece of paper my mother sent to Graz, you know. We saw it and we read it three times and then destroyed it, put it into the fire, and told nobody about it at all. I went with my car and with Thirring as far as Munich. Thirring went with me and I went to Switzerland to Constance; I met our friends there and I told them: "Yes, in principle, he will come. But nothing should be done that will let anybody know that we are going away." We never were allowed to go away. My friends wrote to Born and Born told Whittaker and Whittaker told de Valera and that was finished. He never spoke to de Valera; he never knew de Valera — nothing at all. When this (missive) came he was perfectly sure that he must leave Austria at once.

A very good man here at the government — he came to Vienna and he spoke to him and said: "Oh, but it's easy for you with your name to get another job in the industry, or somewhere." My husband said: "No, for a theoretical physicist that is not so easy, and after all, all the Jews, their papers were invalid, displaced. I have to go to a foreign country to find my living again." And he (the official) said: "They won't let you go to a foreign country. Have you still got your passport?" But that was really well-meant. It was not a threat. That gave my husband a shock because he never thought he could be prevented to go to a foreign country.[44]

Schrödinger's daughter Ruth also told the author of this book some 70 years after this event that Schrödinger had to leave all of his personal documents behind when he fled from Austria. This included his priceless notebooks from 1925 and 1926 that indicate how he came to invent the Schrödinger equation. Together with his hugely valuable Nobel and Max Planck gold medals these were locked up in an inauspicious filing cabinet and left at the University of Graz.

Éamon de Valera was a very prominent Irish politician who had been a commander in the 1916 Dublin Easter Rising. After being prominent in the Sinn Féin and Fianna Fáil political parties, he had been the Taoiseach (Head of Government) of Ireland since 1937. He had graduated from the Royal University of Ireland and taught mathematics at several schools. Unusually for a politician he was very enthusiastic about research in mathematics. He had heard about the Institute for Advanced Study in Princeton and was impressed by the recruitment there of Einstein and Weyl. With the distinguished refugees leaving Germany and Austria in large numbers he had the vision of creating a similar Institute for Advanced Studies in Ireland through recruitment at the top level.

Ireland had a distinguished history in mathematics. At Trinity College Dublin in the mid-1800s, William Rowan Hamilton had reformulated Newton's classical equations into a general and useable form which became known as Hamilton's equations. Indeed, part of Schrödinger's equation included the Hamiltonian describing the kinetic and potential energy of an atomic or molecular system. Schrödinger also preferred to follow in appointments from great men of science such as Planck and Boltzmann and he would have considered Hamilton in this category.

Schrödinger had been very disappointed when he was not appointed at the Institute for Advanced Study in Princeton in 1934 and the imaginative idea of such an institute in Dublin appealed to him. Thus de Valera in 1938 considered Schrödinger to be a prime candidate to come to start his Institute for Advanced Studies in Dublin. Schrödinger had to leave Austria at once and the possibility of being able to go to Dublin was a great morale booster for him.

In addition, Sir Edmund Whittaker, who Schrödinger had met one year before in Rome when they were made founding members of the Pontifical Academy of Science, and also when Schrödinger was being considered for the Chair in Edinburgh, was a close friend and advisor to de Valera. He had been Professor of Mathematics in Trinity College Dublin and also the Royal Astronomer of Ireland. He had discussed the idea of the Institute for Advanced Studies in detail with de Valera and was a good choice to act as a go-between from De Valera to Max Born and other quiet and careful contacts.

Anny then continued the story of how she and her husband left Graz on 14 September 1938, taking the day-long train ride to Rome. At that time, most fortunately, it was possible to cross the border from Austria to Italy without a visa:

> So in three days we packed everything — three suitcases. We had nothing more. We left everything in Graz behind, took a return ticket to Rome, because everybody knew us in Graz. We didn't dare to take a taxi so that they knew what we have got with us. With my car I brought the luggage to the station and then I brought it back to the garage and I said that they should wash the car. I never saw the car at all again of course. I never saw the other things either because they were confiscated. And with ten marks in our pocket we left Graz.[44]

Once the Schrödingers arrived in Rome they headed for the Vatican where Schrödinger had been received in such a welcoming way at the Pontifical Academy of Science just one year before. Erwin and Anny had no money but Schrödinger contacted Enrico Fermi who he had met at the meeting of the Academy. Anny continued:

> We didn't have the money to pay the porter in Rome. We went to Rome because my husband was a member of the Papal Academy. There Fermi told us: "Don't write from Rome because it is already dangerous. It might be censored." From Vatican City, where the Papal Academy is in the Vatican gardens — wonderfully situated — he wrote three letters. One was to Lindemann to tell him that we left Graz; one to our friends in Zurich to get us some money because we had to borrow money from Fermi; and the third one to de Valera, who was the President of the League of Nations at that time. That was a Saturday when we posted the letters in Vatican City.

On Monday morning we went again to the Academy and after half an hour's time came a Diener (servant) and told us that his 'excellency' was wanted at the telephone. My husband didn't turn around. After having been thrown out of Graz — "Yes, yes," the man said, "You are meant." The Papal Academy uses the term "excellency". The Irish — not the ambassador, but the next man in line, not Botschafter but Gesandter, was at the telephone and he said de Valera rang up this morning from Geneva and told him to do everything for us to bring us as soon as possible into Geneva.

We should be at the legation in the afternoon and de Valera would phone; so we were there and this was the first time my husband heard de Valera speaking. He said that he was very glad that we were out of Austria and we should come to Geneva to discuss a few things but as soon as possible we should go on to England or Ireland because there was such a great danger of war in '38. The Munich Conference, you know.[44]

The Munich conference was going to be held at the end of September 1938. It involved the leaders of Britain and France meeting with Hitler and Mussolini. The question was the transfer of the German-speaking Sudeten part of Czechoslovakia to Germany. After much discussion the transfer was agreed with an international commission to consider the future of other disputed areas. On his return, the British Prime Minister Chamberlain famously announced, "Peace for our time." Lord Halifax did not go with Chamberlain to Germany and informed the House of Lords that the agreement made was a "lesser of two evils".

In the meantime, Richard Bär in Zurich had contacted Francis Simon in Oxford who wrote to Schrödinger on 20 September 1938:

Early today I had a letter from Bär which brought the news about you, and then your letter to Lindemann. He thinks it best to ask President Gordon of Magdalen to inform the Embassy in Rome so that you can obtain a visa. The French will then let you pass without anything further. I am writing this only hastily, since Lindemann apparently will not answer right away. He is now very busy, as you may imagine and — besides — frightfully depressed. So hopefully we shall see you soon, all else by word of mouth.[45]

It is unclear if Schrödinger received this letter but Simon was unaware at that time on the approach from de Valera. Anny continued the description of their movements:

> So de Valera tried to get everything ready for us and again we couldn't take any money out of Italy. He gave us a pound each and gave us first class tickets and off we went. I was quite happy already. I felt already safe, but my husband didn't feel safe at all. In Domodossola they looked at our passports and the luggage they hardly looked at all. But the passports were all right.
>
> And before we came to Iselle — Iselle is the point at the one end of the Simplon Tunnel — a carabinieri came into our compartment and we had to leave the compartment with all our luggage. He had a piece of paper with our name written on it. I nearly thought, "I'm dying." It was really the fright of my life. We had to take off everything. He was separated from me with the luggage and I was in another place. I couldn't speak Italian; there was a woman who looked through my things — saying: "Put everything toward the X-rays." They X-rayed every single bit of my things. I had my handbag, my tooth, everything. After about half an hour, or three-quarters of an hour, the train had to wait; it was an express train — we were allowed to enter the train again.
>
> Then he was very nervous, but he knew what the case was. They had looked at our passports and we had visas for all of Europe because de Valera had said that if we can't go through France, we must go through Spain or Portugal or something. We had everything. We were asked if we had some money. We said that we had one pound. Then they thought that we had to smuggle something, because one can't go through Europe on one pound.[44]

So, finally, after another 24-hour train trip, they arrived in Geneva and Schrödinger met with de Valera. It seemed the Italian authorities were not communicating at that time with the German or Austrian ones:

> De Valera was very, very pleased when we arrived. He was already in full dress because they had a banquet in the evening, but he received us and was very kind and we stayed for three days in his hotel and then rode on.[44]

Anny subsequently wrote on 18 October 1938 from Oxford in vivid terms to Lise Meitner about their escape:

> There was a blackout in Switzerland. On the other side of the Rhine one could see the bright lights of Germany. Trains were overflowing, and greatly delayed. In France we passed airports with countless airplanes standing ready. All bridges and tunnels were under military guard. Then in England in Hyde Park bomb shelters were being feverishly prepared, anti-aircraft guns pointed to the sky, everyone had gas masks. In Paddington we heard reports about the Four Power Conference in Munich. That day 80,000 children were evacuated! On the 28th we reached Oxford.[37]

On 28 September 1938 a telegram was received at the home of Francis Simon at 10 Belbroughton Road, Oxford stating: "Leaving 6.05 for Oxford — Schrödingers".[45] While she was in Oxford from 1933–36 Anny Schrödinger had got to know the other exiled wife Charlotte Simon quite well. Her daughter Kathrin recalls Anny turning up at their home to cry on her mother's shoulder during this period.[46] Simon wrote to Born on 6 October 1938 with more details of the surprise return of the Schrödingers:

> Schrödinger and wife (Anny) are here and are behaving so strangely... I want to give the background because I don't know how much you know about the previous details. After relocating to Austria, we feared the worst for them, as we knew that he had publicly committed himself to be quite anti-Nazi. We urged Lindemann to do something and he turned to Halifax, who on his part consulted Berlin. The desired success, namely allowing a visit to Oxford, did not materialize, but Schrödinger himself attributes this initiative to the fact that he was not imprisoned.
>
> In the meantime, his deliberate letter appeared in the newspapers, which caused a great stir here — with derisive remarks such as "he sold his soul" and quotes of the saying of the King of Hanover: "Professors are as good as whores." We always told everyone that he did this only under duress, even if he claimed the opposite. We continued to do so when we knew that he, unfortunately, did it voluntarily for personal reasons and wanted to make his peace with the Nazis.

Now they suddenly left (Austria) a few weeks ago, quite unexpectedly. They drove to Rome and Schrödinger from there wrote to Lindemann and other acquaintances… He told Lindemann that he wanted to come back — he was supposed to be here once a year anyway. He had originally applied to the German government to give lectures here; he did not want to spoil things with them. Lindemann then tried to get a visa (to enable him to pass through France), and Schrödinger probably would have got back that way. Meanwhile the crisis intensified and when Lindemann received a telegram that they would come very soon he was very concerned that they would have trouble here in the event of a war, especially after the letter.

They appeared on Wednesday evening 8 days ago and they came to visit us. After we had talked for a long time, I mentioned that it would be desirable if he would make it clear to his acquaintances that the letter had been written under pressure. First, he asked, which letter? Then he became very abusive and excited. He said he was here in a land of freedom and what he was doing was nobody else's business. He flatly refused to talk about this point at all… He was grateful to me for pointing this out to him, but I said he should accept anyone who mentioned this matter… We told him this because if war broke out tomorrow he could have a lot of trouble…

But he emphasised again that what he was doing was nobody else's business. I told him, as an example, that his affair with ICI was a major concern of ours. I don't know whether you know that he wanted to sue ICI from Graz, which allegedly gave him a few pounds too little to buy a couple of faucets in his house. The matter was very important for the other emigrants, especially those who were dependent on the ICI funding. It has been put in order by Lord Melchett personally paying the difference to avoid a major fuss.

The next day he was at Lindemann's, who must have said something similar. In any case, he has not shown himself to Lindemann or to us since then. When my wife called Mrs. Schrödinger after a few days, she didn't answer at first. She eventually replied that other people were so nice to them, they didn't talk about the letter at all and understood their situation so well — only we and Lindemann thought it necessary to mention this point.

He has a guilty conscience but does not want to spoil it with the Germans and apparently also does not have the heart to admit that he

> has done wrong... All in all, one has to say that he behaves like a spoiled boy, also with his belated sexual complex... He is negotiating with Ireland and will probably get appointed there. Incidentally, they forgot to take his Nobel Prize money from him — he still has it in Sweden.[47(t)]

It was becoming clear that, due to his letter seemingly supporting the Führer, Schrödinger was not nearly as welcome in Oxford in 1938 as he had been in 1933. Lindemann, in particular, was not able or willing to do more. The Schrödingers were fortunate to be able to stay in the spacious house of J.H.C. (Henry) Whitehead at 22 Charlbury Road, Oxford. Whitehead and Schrödinger had run their joint seminar in Oxford some three years before and had become good friends.

Schrödinger had been elected to a Fellowship at Magdalen College for five years from 3 October 1933. However, his status after 2 October 1938 was unclear. Schrödinger dined in the Great Hall of Magdalen on 1 October as was his right as a Fellow for one more day. The evening must have gone well as he also allowed himself to be weighed in the Senior Common Room after dinner. He was 10 stone 8 lbs and had lost one pound from a previous weighing on the evening of 9 November 1933 when he heard he had won the Nobel Prize.

Schrödinger raised with President Gordon the matter of his grant that had been agreed by Magdalen College two years previously providing that he resides in Oxford and takes part in the teaching of the University. Following his flight from Austria and Italy, Schrödinger had no personal resources except his Nobel money in Sweden and that was not readily available. The grant was then considered by the Governing Body on 12 October 1938 and the College Actum stated:

> That the report of the Fellowship Committee be approved; that a grant of £100 be made to Dr Schrödinger subject to his residing and taking part in the teaching of the University during the current term, and that this expenditure be provided for by transferring £100 from the Fellowship fund to the Grants Fund.[43]

There was no reference, however, to a renewal of Schrödinger's Fellowship. A non-tutorial Fellowship at Magdalen had to be associated with a funded post in the University, and Schrödinger did not now have such a position. At the same meeting, there was also a precautionary statement on air raid shelters, no doubt instigated by the recent Munich crisis:

> That the question of constructing permanent shelters or trenches in the Grove or elsewhere be referred to the Sub-Committee on Air Raid Shelters.[43]

George Gordon, by now Vice-Chancellor of the University of Oxford as well as President of Magdalen, then wrote on 14 October 1938 to Bursar Forster to emphasise the urgency of the grant to Schrödinger:

> When will Schrödinger be paid his money? I fancy he is pretty hard up. He will certainly give some instruction in the University as soon as it can be arranged. Can you proceed as if this can be assured? I shouldn't like him to be left with none of it until the end of term.[43]

The reluctant Forster then wrote to Schrödinger on 17 October 1938 to say:

> I enclose a copy of the College Order 8 of October 12, 1938. Would you kindly let me know as soon as you have arranged with the University authorities as to the instruction you will be giving in the University to enable me to pay the first instalment of this grant.[43]

From Henry Whitehead's home in Oxford, Schrödinger had written in some detail on the welfare of his colleagues from Graz in a letter to the Society for the Protection of Science and Learning on 10 October 1938:

> Having left Graz (Austria) a few weeks ago, may I be allowed to give you information about the displaced scientists which might perhaps not be fully known to you, particularly on account of the method which the authorities now usually follow and which, if not intended, is at any rate very fit for causing confusion of the evidence available. The method consists in taking comparatively mild decisions at first (*e.g.*, granting a substantial pension), then waiting a few months until the news has spread

and has been requested by you or by similar institutions, and only then pronouncing a severe sentence (*e.g.*, dismissal without pension).

Precisely this method was followed in the case of Professor Victor Hess, Physics Nobel Prize 1937, who was in Graz. After 9 days imprisonment in the early days after the Anschluss he was first suspended, then granted a comparatively good pension. I have reliable news that he has lately (about Sept 15) been dismissed without any pension. His case is aggravated by that he was compelled to deliver his Nobel Prize money to the Reichsbank about a fortnight before his dismissal. When I last saw him, he and his wife were without their passports (he is Aryan, his wife Jewish).

Otto Loewi, Graz, Pharmacist and Physiologist, Nobel Prize 1937, Jewish, had been in prison for 9 weeks. He was then pensioned. Then his two sons (who had been imprisoned longer) were made to sign that they would have to leave the country "voluntarily" before September 1st (I believe). Then he had to sign the same, the date being, as I am told, December 31st. Of course, he also had to deliver his Nobel Prize money. His pension ceases by his "voluntary" exile. What will be left of his fortune in Germany will be of no use to him.

Walther Schwarzacher, who was Professor in Graz, was dismissed August 31. His field is Forensic Medicine, of which he is one of the foremost students. He has a wife and three sons between 9 and 16, no fortune whatever. He lives now in Salzburg in a small cottage of his relatives and by their aid. He is Aryan and (as usually in that case) the greatest difficulty for him will very probably be to get out of the country. Once in, say, the U.S.A. he would, to all probability, very soon be placed for his subject is of greatest practical importance and I am told there are few first class representatives thereof.

For completeness sake I mention my own case, Erwin Schrödinger, physicist, Nobel Prize 1933. Without any previous hardship I was dismissed the same time as Schwarzacher (August 31st). Being an Aryan, I think, I was lucky to escape a few days later. Though I had to leave all my things behind, I had not yet been forced to deliver my Nobel Prize money (by a mistake I believe). By lucky circumstances my case is, as you see, not very severe, it ought, for the time being, not to worry you.[18]

Subsequent written reports confirm the accuracy of these details on the different colleagues of Schrödinger. Victor Hess was in particular danger as he had given advice to the government of the previous Chancellor Schuschnigg and his wife was from a Jewish family. His first detection of cosmic rays via balloon flights which he personally conducted in 1912 won him the Nobel Prize. He was then in the University of Vienna and the young Schrödinger did some theoretical analysis of his results.

After being dismissed as Director of the Physics Institute in Graz, Hess and his wife managed to escape eventually to Switzerland and then on to the USA where he was appointed to a Professorship at Fordham University. One year after the nuclear bomb was dropped on Hiroshima he made the first test for radioactivity in the USA from the 87th floor of the Empire State Building in New York. He did not return to work permanently in Austria.

Otto Loewi also managed to escape in September 1938 with his family after paying over his Nobel Prize award. He initially stayed with Henry Dale, with whom he had shared the Nobel Prize for Physiology or Medicine. In London, he met up again with his old friend, Sigmund Freud, who had also managed to escape from the Nazis. He moved to Brussels and then, on the outbreak of the war, worked in Oxford. Like many of the scientific refugees from Germany and Austria, he finally settled in the USA and had a research Professorship in Pharmacology at the New York Medical University. Following his death in 1961, Austria put the face of Loewi on the 4-schilling stamp.

Walther Schwarzacher stayed with his family in Salzburg and did not leave Austria during the Second World War. He was able to return to his Professorship afterwards. The letter also shows Schrödinger's own particular concern about pensions. Ever since his family was impoverished in the financial crisis of the early 1920s this had been a particular priority to him and he constantly brought the matter up in his negotiations for positions.

The new general Secretary of the Society for the Protection of Science and Learning was David Thomson and he replied to Schrödinger on 11 October 1938:

> Very many thanks for your letter of October 10th and the details which you have given therein regarding colleagues. We have been in touch of

> course with Professor Otto Loewi who is now in this country and we are also in touch with Professor Hess. We had not heard of the plight of Professor Walther Schwarzacher. We are glad to learn that you yourself were more fortunate than your colleagues in being able to take out your Nobel money with you.[18]

With just the small grant from Magdalen to live on in Oxford, the Schrödingers were living almost hand to mouth. However, de Valera remained very enthusiastic and provided funds for Schrödinger to travel to Dublin to meet him together with Arthur Conway, Professor of Mathematics at University College Dublin and the Minister for Education T. O'Derrig. De Valera explained that it would take about a year to get the legislation passed through the Irish Parliament to set up the Institute for Advanced Studies. Schrödinger also had an important private meeting with the Minister for Justice. This was described in the official history of the Dublin Institute for Advanced Studies:

> It was arranged for Schrödinger to have an hour-long interview with Mr. Duff, Department of Justice, regarding the admission to Ireland of his wife and a third party, Frau March. Schrödinger left Ireland, having met again with the Taoiseach and having been assured that if he (Schrödinger) had any difficulty he was to call him (De Valera). As Frau March had a family, it was unusual (to say the least) that Schrödinger did not anticipate that there would be difficulties associated with her visa. Schrödinger did not trouble De Valera with his own or his wife's visa — "But in the case of our friend, Frau Heldegrund March from Innsbruck, you advised me, Sir, to apply to yourself."[48]

The request concerning "Frau March" was highly unusual but de Valera did not want it to stop Schrödinger coming to Ireland. After the meeting with Schrödinger, however, Mr Duff reported to the Secretary to the Irish Government Maurice Moynihan that "it was extremely peculiar for a request to be made for facilities for this lady to abandon her family for the somewhat inadequate reason that she would be of assistance to Schrödinger in establishing a home here".[21]

De Valera also had an interest in Celtic Studies and constructed an Act to "make provision for the establishment and maintenance in Dublin of an Institute for Advanced Studies consisting of a School of Celtic Studies and a School of Theoretical Physics, to authorize the addition to such institute of schools in other subjects and to provide for matter incidental or ancillary to the matter". The strange mix of Celtic Studies and Theoretical Physics was to produce some criticism from the opposition members in the Irish Parliament that would delay the passing of the Act.

Schrödinger returned to Oxford from Dublin but was still not sure what he would be doing for the next year. Francis Simon wrote about the situation to Born on 20 October 1938:

> I met Schrödinger once in college. He is especially mad at Lindemann who is not willing to see him again. Yet Lindemann is certainly the Englishman who has done most for him.[47]

As was agreed in the terms of his grant from Magdalen College, Schrödinger needed to take part in the teaching of the University. Accordingly, he gave four lectures in November 1938 with the introduction:

> The subject for these lectures was chosen mainly because it is the thing on which I have been working in the last year, together with two pupils. It is the investigation of waves which are propagated not in ordinary space but in close spherical space... The waves which we are going to investigate are intended to explain and describe the particle structure of nature.[2]

He stated that this model is a basic representation of the Universe and then went on to derive the mathematics of extending the spherical polar representation of three-dimensional Cartesian coordinates to four dimensions. He also said: "Astronomical evidence points to the possibility or even probability that the world is a huge but finite box."[2] He was to develop these ideas further in a paper which he published a few months later when in Belgium.

Paul Dirac had heard that the Schrödingers were having housing difficulties and wrote to invite them to stay in his home at 7 Cavendish Avenue in Cambridge. On 27 November 1938 Schrödinger wrote back on Magdalen College notepaper to Dirac to take up the kind offer.[49] Anny had been made the godmother to Dirac's daughters and she and Dirac's wife Margit ("Wigner's sister") often corresponded, both being refugees with common interests in England and married to brilliant but complicated Nobel Laureates.

Like many other scientists Dirac was keen to discuss Schrödinger's surprising statement supporting the Führer. By now, Schrödinger was finally getting the message on the serious concern from the scientific community and explained to Dirac that he had been forced to make the statement and had done so as ambiguously as he could. This was accepted by Dirac who made his view on the matter clear in the obituary he wrote for Schrödinger in 1961.[27]

Word was getting round amongst the other refugee physicists about Schrödinger's plight. Rudolf Peierls wrote to Hans Bethe on Christmas Day, 25 December 1938, with a note of sarcasm:

> Schrödinger is wandering around England (he will presumably get a job in Dublin) and he is surprised when people ask him to recant his declaration in the Graz newspaper. Surely they must have known that, if he wrote something like this, it wasn't meant to be taken seriously.
>
> Besides that we are getting a number of letters from people who want to come to England and have had to tell most of them that it is nearly impossible.[50(t)]

Belgium

The careful plans that Schrödinger had made in 1935 and 1936 with the Austrian authorities and with President Gordon to have positions simultaneously in Graz, Vienna and Magdalen College had failed. Things were now very uncertain for him and Anny. However, as has frequently been the case in Schrödinger's story when he was unsure about his future, he then received a very welcome letter. This was from the Fondation

Francqui in Belgium offering him a Visiting Professorship at the University of Ghent. He at once accepted.

As was often his method to announce the next move in his career, Schrödinger arranged for *Nature* on 31 December 1938 to report:

> Professor Erwin Schrödinger has been appointed by the Fondation Francqui as a Visiting Professor for the next six months to a "Chaire Francqui" in the University of Ghent, Belgium. His address is Laboratory of Physics, Plateaustraat 22, Gand, Belgium.[51]

As soon as he arrived in Belgium, Schrödinger wrote on 18 December 1938 to Bursar Forster at Magdalen College. He often wrote to Magdalen on what might seem to be very minor matters, but these can be important with the conventions at the College:

> I should like to give a pound to the Xmas box of the servants (which is not too much in my case because I give vast trouble by forwarding letters, *etc.*). Would you be so kind to give it for me and could you deduct it from the second instalment of my grant?
>
> If that is impossible or no more possible I shall see and let you have it back in some other way. The best address to reach me now (in case it were necessary) is 11 Rue d'Egmont, Bruxelles. Or alternatively, Laboratory of Physics, Plateaustraat 22, Gand (Gent, Ghent). The latter is my office but I am not there on Sunday and sometimes also for a few weekdays in succession (especially during the Xmas vacation).
>
> If my second instalment has not yet been given it would simplify matters if you could kindly give direct orders to Barclays Bank (Old Bank), High Street, to put it down on my account. I thank you for the trouble I cannot avoid giving you.[43]

Bursar Forster then wrote back with a touch of irony on 21 December 1938 saying:

> It is extremely generous of you, but I cannot help feeling that £1 is far more than your proper share, since even those of us who live in the College are limited to this figure. Personally, I cannot see that there is

> any reason whatever for you to contribute since practically speaking you have never been here throughout the year, and forwarding is nothing. In any case I have already paid your cheque for £50 into your account at Barclays, Old Bank, High Street. With kindest regards and the very best wishes for Christmas and the New Year to you both from my wife and myself.[43]

However, on 24 January 1939 Forster received a cheque for £1 from Barclays Bank. He wrote back to Schrödinger:

> I have to thank you for your cheque for £1 received from Barclays Bank today. I am presuming that this is for the Servants' Fund as you suggested and am paying it over to the Vice-President accordingly. I think it is extremely good of you.[43]

When he was not worrying about the £1 cheque, Schrödinger was giving lectures in Belgium on the fundamentals of wave mechanics. Ilya Prigogine came as a refugee with his family from Russia to Belgium after the Russian Revolution and he recalled attending a lecture and meeting Schrödinger in 1939.[52] He also met with Schrödinger again at the 8th Solvay Meeting in 1948. Prigogine went on to win the Nobel Prize for Chemistry in 1977 for his contributions to the understanding of non-equilibrium thermodynamics.

While in Belgium Schrödinger published a paper in the Dutch journal *Physica* on "The Proper Vibrations of the Expanding Universe".[53] This paper has been highly cited and is thought by several cosmologists to be his most important contribution to their subject. The paper extended some of the ideas he had introduced in the short lecture course he had given in Oxford a few months before. In Belgium he had met with Georges Lemaître at the Catholic University of Louvain who had proposed the theory of the expanding universe. He had also met Lemaître two years previously when they were both inducted as founding members of the Pontifical Academy of Sciences.

Schrödinger also published a shorter paper which he submitted to *Nature* on the "Nature of the Nebular Red Shift".[54] The paper contained some simple estimations on an expanding universe related to recent observations by Edwin Hubble from the Mount Wilson Observatory in California that Schrödinger had visited some 12 years before. In these works, Schrödinger demonstrated his continuing interest in cosmology. Ever since his papers on entanglement and on Schrödinger's Cat, published in Oxford four years before, he had been unable to publish much substantial research. These new papers, however, demonstrated that whatever was happening and wherever he was, Schrödinger could still publish interesting works.

In the summer of 1939 Schrödinger and Anny moved to De Panne, a seaside resort in Belgium. There they were joined by Arthur March with his wife Hilde and daughter Ruth. March then returned to Innsbruck

Schrödinger in De Panne, Belgium in 1939.

leaving Hilde and Ruth behind. As always, Schrödinger continued to work but he was short of resources. He wrote to his friend Hans Thirring to ask if some German books could be sent to him, especially the trusted *Methods of Mathematical Physics* by Courant and Hilbert.[2]

On 19 July 1939, Schrödinger wrote in German from De Panne to his close friend Einstein in his usual personal style to attempt to explain what had happened to him over the previous year:

> It was ugly of me that I didn't thank you for your lovely letter from November (concerning Kottler). So now I'll catch up on it. Incidentally, I unfortunately did not find out what happened to the poor man. Hopefully he has managed to leave the hated ground after all — even the first step outside is like a whole nerve cure. A poor friend of mine who was dismissed at the same time as me, Walther Schwarzacher, is still there, with his wife and three boys. He is a forensic doctor...
>
> Thank you for your very appropriate congratulations on ending the romantic intermezzo. It was superfluous. I was of course aware of a certain danger when I returned to Austria. But until the end I didn't expect the fortress to be handed over without an empire of the sword. A few days before that I was with the head of the section in the Ministry and said to him: if you put a shotgun in my hand, I will be happy to defend myself, but what feelings do I leave with the signature of the same gentleman under the ordinances a few weeks after being thrown out after the transfer! I hope you did not seriously denounce my subsequent, certainly quite cowardly, behaviour. I wanted to stay free — and couldn't do it without gross hypocrisy.
>
> I spent the first ten months of emigration quite well, even as a little vagabond, first in Oxford, then in Ghent. At the moment I am a bit in a pinch. Not so much that I ask you to do something for me, but I would like you to orientate yourself. Namely, since an application that was smuggled to Graz in March or April of last year as a secret slip of paper, I am waiting for a position that de Valera has promised me at a new Institute for Advanced Studies in Dublin. He is terribly nice to me personally and in letters, but the implementation is postponed from six months to six months. First he wrote to me in Oxford: it will probably not be before January (1939). Then (when I had rejected a Brussels application for six months because of this appointment) he wrote: Oh,

see if you can still get it — because they are only free in summer. And so I came to Ghent for the period. Well I thought it would be autumn at the latest. But now he writes: yes, it will probably not be before January (1940)! Furthermore, there was opposition on the first submission in the Irish Parliament (which is unusual at first reading, as de Valera himself adds).

Should one still have confidence there? For his sake, I took it, for a politician, even taken him absolutely, as a highly decent, splendid person. But can I continue to answer for the inertia with which I have relied on it so far and have not seriously looked for other options? The stupid thing is that the matter with Ireland has got around quite a bit in a long time, so that anyone who thinks of me and inquires will probably be told: no, Schrödinger is well supplied, he will get a new institute in Dublin. I now have to do something about it — and so I talk to you in detail about something that is of itself so uninteresting (which you, please, will forgive me) is one of the first steps towards it.

It's a shame that I had to fill so much of this letter with my uninteresting personal matters. But writing (I mean about things like I mention above) is so terribly difficult, after all. If this letter reaches you on your sailing boat, I wish you much rest and comfort there. I am here on the lovely Belgian coast and with lovely, childishly happy people. If only one could be a little more light-hearted and think less about what will become of oneself. Holidays are okay, but holidays that don't have a definite ending are a strange thing. Greetings from your sincere devotee. E. Schrödinger.[25(t)]

Schrödinger's correspondence with Einstein always had a personal touch which brought out his deepest anxieties. Schrödinger mentions Friedrich Kottler who had worked on relativity and corresponded with Einstein. He was from a Jewish family and held a Professorship at the University of Vienna. He was dismissed at the same time as Schrödinger. With the help of Einstein and Pauli, Kottler eventually emigrated to the USA and worked for Eastman Kodak in Rochester.

It was clear from the letter that Schrödinger was still uncertain about his appointment in Ireland at this time. There was opposition to de Valera's proposals to the Irish Parliament to establish the Institute for Advanced Studies which was causing a delay. Schrödinger mentioned this uncertainty to Einstein just in case the possibility of another suitable post came up and Schrödinger still had hopes for an appointment at the Institute for Advanced Study in Princeton.

Einstein then replied, from Point Peconic, Long Island on 9 August 1939, referring to the discussions of five years previously when Schrödinger was hoping for an appointment to the Institute for Advanced Study at Princeton:

> I was very impressed by your letter, especially since I also thought that you had already found a secure appointment. Some time ago Weyl and I worked on Flexner so that he should send you a call to our institute. He excused his negative attitude with the lack of money, which back then was not honest, but since then has unfortunately grown into a serious argument. The change goes back to a couple of quite foolish appointments in the field of economics, which have put the institute in a somewhat precarious financial position.
>
> We had the impression at the time that Flexner was opposed to you. Maybe it was consideration for Princeton University whose appointment you have turned down. But we don't know. Incidentally, there is reason to expect that a change in the management of the institute will soon be imminent, which would have a significant and favourable influence on the situation, apart from the financial weakness that the successor will of course have to take on. In any case, I want to point out your true situation, just as the opportunity will present itself for me.[25(t)]

It was considerate of Einstein to reply but there were other things very much on his mind at the time. Just one week before his letter was sent to Schrödinger, Einstein had written one of the most momentous letters of all time on 2 August 1939, also from Point Peconic. This was to President Roosevelt at the White House urging him to initiate research that could lead to a nuclear bomb:

> Some recent work by E. Fermi and L. Szilard, which has been communicated to me in manuscript, leads me to expect that the element uranium may be turned into a new and important source of energy in the immediate future. Certain aspects of the situation which has arisen

seem to call for watchfulness and, if necessary, quick action on the part of the administration. I believe therefore that it is my duty to bring to your attention the following facts and recommendations:

In the course of the last four months it has been made probable — through the work of Joliot in France as well as Fermi and Szilard in America — that it may become possible to set up a nuclear chain reaction in a large mass of uranium, by which vast amounts of power and large quantities of new radium-like elements would be generated. Now it appears almost certain that this could be achieved in the immediate future.

This new phenomenon would also lead to the construction of bombs, and it is conceivable — though much less certain — that extremely powerful bombs of a new type may thus be constructed. A single bomb of this type, carried by boat and exploded in a port, might very well destroy the whole port together with some of the surrounding territory. However, such bombs might very well prove to be too heavy for transportation by air.

The United States has only very poor ores of uranium in moderate quantities. There is some good ore in Canada and the former Czechoslovakia, while the most important source of uranium is Belgian Congo.

In view of this situation you may think it desirable to have some permanent contact maintained between the Administration and the group of physicists working on chain reactions in America. One possible way of achieving this might be for you to entrust with this task a person who has your confidence and who could perhaps serve in an unofficial capacity. His task might comprise the following:

a) To approach Government Departments, keep them informed of the further development, and put forward recommendations for Government action, giving particular attention to the problem of uranium ore for the United States;
b) To speed up the experimental work, which is at present being carried on within the limits of the budgets of University laboratories, by providing funds, if such funds be required, through his contacts with private persons who are willing to make a contribution for this cause, and perhaps also by obtaining the co-operation of industrial laboratories which have the necessary equipment.

> I understand that Germany has actually stopped the sale of uranium from the Czechoslovakian mines, which she has taken over. That she should have taken such early action might perhaps be understood on the ground that the son of the German Under-Secretary of State, Von Weizsäcker, is attached to the Kaiser Wilhelm Institute in Berlin where some of the American work on uranium is now being repeated.
>
> Yours very truly, Albert Einstein.[55]

President Roosevelt replied to Einstein from the White House on 19 October 1939:

> I want to thank you for your recent letter and the most interesting and important enclosure.
>
> I found this data of such import that I have convened a Board consisting of the head of the Bureau of Standards and a chosen representative of the Army and Navy to thoroughly investigate the possibilities of your suggestion regarding the element of uranium.[55]

Einstein's letter set in motion the chain of events that established the extraordinary Manhattan project on developing nuclear weapons. This brought together in the USA so many of the scientists, on both sides of the Atlantic, who are mentioned in this book.

The summer of 1939 was a very tense time for the whole of Europe and Schrödinger was waiting more and more anxiously for the call to come to Dublin. He wrote to Born on 22 August 1939 asking about a possible position in India, where Born had spent some time, and also mentioned that he had been offered a permanent position at the University of Ghent by the Rector there.[25] His temporary position at Ghent was only for six months so it had terminated.

However, as was the case with Schrödinger a year before in Austria, a dangerous problem occurred. On 1 September, German troops invaded Poland and on 3 September Britain and France declared war on Germany. At that time, Belgium was remaining neutral but it was very vulnerable and had a border with Germany. Schrödinger still had German nationality and that at once caused a difficulty with the authorities in Belgium. With the announcement of war, de Valera realised he had to act very quickly and informed Schrödinger he should come to Dublin as soon as possible even though his Institute for Advanced Studies had not yet been approved by the Irish Parliament. A temporary professorship was being arranged for him. However, there was the challenge of travelling to Ireland in wartime. As Arthur March had returned to Austria, Schrödinger was also responsible for Hilde March and her daughter Ruth. Ruth was of course also the daughter of Schrödinger, although she was not aware of this at the time.

Schrödinger wrote to ask assistance from Professor Frederick Donnan. He had been involved in the early stages of the Academic Assistance Council and setting up the ICI funding that had supported Schrödinger and the other scientific refugees in Oxford. Donnan was an Irish chemist and head of the Chemistry Department at University College London. In this period Schrödinger was having a lively correspondence with Donnan on whether physical principles could be applied to biology which was to lead to his "What is Life?" lectures in Dublin four years later.[56,57] Donnan then wrote to Lindemann on 19 September 1939:

> Has Schrödinger written to you? He wrote to me about a week ago saying that he was staying at 7 Sentier des Lapins, La Panne, Belgium and was anxious to come to England en route for Dublin, together with two ladies, also Austrian refugees, and one child (British born). He seemed to anticipate difficulty with regards to permits and visas, and asked if I could help in any way.
>
> I wrote to the President of Magdalen concerning this matter, but have had no reply. I suppose that it is only necessary that some person or persons in this country should guarantee him the necessary immigration or other authorities. I am writing to you as, naturally, you are the chief person who helped Schrödinger when he came to England in 1933.[10]

Donnan also wrote on the same day to Tess Simpson, the Assistant Secretary for the Society for the Protection of Science and Learning:

> Some time ago I received a letter from Professor Schrödinger from Belgium, where he is at present staying. He had been as you probably know holding a Visiting Professorship for a few months at the University

of Ghent, but he said in his letter that he had been idle for the past four months and wished now to come to England en route for Dublin, together with two ladies, also Austrian refugees like himself, and one child (British born).

He seemed to anticipate difficulty with regard to permits and visas, and asked me if I could give him any help. I wrote some time ago concerning this matter to the President of Magdalen College, Oxford, since Schrödinger is an Hon. Fellow of that College, and told him about the state of affairs and asked if he could help, but I have received no reply.

I presume that it is necessary for some person or persons to get in touch with the immigration or other authorities to guarantee Schrödinger, so I am writing to know whether your Society could take such action, as no doubt you have full particulars concerning his movements and perhaps helped him when he came to Oxford in 1933. I was not personally concerned with his coming to England, although I have met him a number of times since 1933, and am very anxious to help in any way I can.[18]

The tireless Tess Simpson then wrote to Schrödinger in Belgium on 21 September 1939:

Today we have a letter from Professor Donnan saying that he heard from you some time ago that you wished to come to England on your way to Dublin. I do not know what your intentions now are; I should think it would be difficult, or at any rate, a lengthy procedure to obtain a permit now. However, if you are anxious to come to England first, please let me know and we shall see what we can do.

Professor Donnan's letter mentions two ladies and a child. I am afraid that we could only help if they were your immediate relatives. Would you let us know about this?[18]

These were the first letters of an official nature in Britain that referred to "two ladies and a child". This did of course refer to his wife Anny as well as Hilde March and her daughter Ruth, who was also Schrödinger's daughter although she was not stated as such at the time.

Schrödinger had been offered an Honorary Degree at the University of Ghent which was due to be presented to him at a ceremony on

9 October 1939. However, before that could take place, de Valera's previous promise to assist with travel permits and visas came through. Schrödinger had written several letters to him explaining his concern about a position in Ireland and appropriate visas for his party.[21] Some nine years later, in April 1948, Schrödinger was able to travel from Dublin to Ghent to receive finally his Honorary Degree.[6]

The party of Schrödinger and "two ladies and a child" were now able to travel through England with the necessary permits. Schrödinger himself said: "Perhaps Lindemann pulled a few strings on that occasion too, despite the rather unpleasant encounter we had had a year before. He was after all a very decent man, and I am convinced that as his friend Winston's advisor in matters of physics he proved invaluable in the defence of Britain during the war."[57]

They left Belgium on 4 October and arrived in Ireland two days later. Anny subsequently wrote from Dublin on 18 October 1939 to Lise Meitner describing the long trip of the party of four, first taking a ferry from Ostend to Dover.[58] The distinguished German writer John Hennig, who had come over the English Channel on the same ferry with his family, recalled the scene at the immigration office:

> The husband's name was called out first: "Schrödinger". "Forgive me if I ask a stupid question," I said when he came back, "but would you be the great Schrödinger?" "Whether or not I am great, I don't know," he answered. "I am a physicist, and like you, I am travelling on to Dublin." It must be a good omen, I thought, to emigrate in such good company.[21]

Hennig also wrote on a tragic incident when a Jewish person on the cross-channel ferry was denied entry to the UK and at once took his own life.[21]

After arriving in England, the Schrödinger party then travelled to London, leaving from there at 9 pm to Liverpool, going on to Holyhead in Wales and then taking another boat at 4 am to Dublin.[58] It was tense and exhausting travel in wartime, but Ireland was probably the safest country in Europe at the end of 1939.

CHAPTER FIVE

To Dublin and Final Days in Vienna

Ireland

Ruth March had been born in Oxford in 1934, moved to Austria at the age of two and was now arriving in another strange country as a five-year-old. She was in the unusual position of having two women looking after her, her mother Hilde March and her kind and doting godmother Anny Schrödinger. She would also be living in a house in Dublin with Anny's husband Erwin. While in Austria from 1936–38 Ruth had seen Erwin and Anny many times. Hilde's husband Arthur March was still a Professor back in Innsbruck and would not leave Austria during the Second World War. All the evidence suggests he was happy with this unusual arrangement and Ruth and Hilde would be living in what was a safe country.

It has to be noted that many children in Britain were being evacuated from their homes at this time of the war to live with families in safer parts of the country or to places further away such as the USA and Canada. So in the chaotic time of 1939, the arrangement was not so unusual. Ireland was very much a Catholic country but was tolerant with Schrödinger. He was also coming recently from a largely Catholic country of Austria and had been made a founding member of the Pontifical Academy of Science. This all helped with his assimilation into Ireland even though he did not follow a religion.

Schrödinger found a house at 26 Kincora Road at Clontarf on the northern shore of the bay at Dublin. Initially he rented the house but he bought it outright four years later. After a succession of nine moves in 20 years, with appointments in Jena, Stuttgart, Breslau, Zurich, Berlin, Oxford, Graz, Ghent and Dublin, he did not anticipate that he would be living in Ireland for as long as 17 years. The Institute for Advanced Studies was still being debated in the Irish Parliament and his position, at least initially, was a temporary one.

With the uncertainty of his appointment, there were parallels to Schrödinger's academic insecurity when he first came to Oxford some six years previously. In that case, he had an unclear promise from Lindemann that a Professorship would be found for him in due course once one of the senior professors in an appropriate field at the University of Oxford retired or died. However, in the period 1933–36 such an opportunity had not arisen. Schrödinger was relying on the considerable influence of President de Valera to come good this time in Dublin.

Upon his arrival in Ireland, Schrödinger was being funded on a temporary basis by University College Dublin, where he agreed to give several lectures, and the Royal Irish Academy to which Schrödinger had been elected back in 1931. Irish politics, however, was volatile and de Valera could easily have lost his position of Taoiseach and his proposed Institute for Advanced Studies.

Nevertheless, Schrödinger's reception in Dublin was much more enthusiastic than had been the case in Oxford. In Dublin he was an academic superstar who was head and shoulders above other academics in the country in terms of his international reputation. In contrast to his introduction to Oxford in 1933, he was invited to give many public lectures which were enthusiastically received, advertised and reported in the national Irish press. His first talk was at the University of Dublin with the general title "Some thoughts on causality". He had developed a popular style of lecturing in almost perfect English and in general terms with no mathematics. It was hard to find lecture halls large enough for his audiences. This was a significant contrast with Oxford where only a handful of students had attended his lectures. The *Irish Independent* on 4 November 1939 reported:

Prof. Erwin Schrödinger, Nobel Prize Winner who formerly held an academic chair in Vienna, and who stated he was dismissed from his post by the Nazis, began a course of lectures on the latest form of the Quantum Theory, intended for advanced students in physics, at University College Dublin, yesterday. Prof J.J. Nolan presided and the attendance included Mr de Valera. The course will last for several months.

Prof. Schrödinger, who is an honorary member of the Royal Irish Academy, said that the great Danish physicist, Niels Bohr, and others, guessed, on the one hand, the general law governing sequences of energy levels and, on the other hand, the general law by which the latter determine the radiation of the atom. Both laws are obtained from Max Planck's original idea by straightforward generalisations, provided the known empirical facts about line spectra were carefully attended to.

The lecturer paid a tribute to the work of the late Sir William Rowan Hamilton, Astronomer Royal and former President of the Royal Irish Academy. He had been deeply impressed by his work, he said, long before he had come to Ireland.[1]

However, the uncertainty of a temporary position once again weighed heavily with Schrödinger. Therefore, when he received a letter from the Vice-Chancellor Amarnath Jha of the University of Allahabad in India offering him a permanent Professorship he took the approach very seriously and clearly wanted to keep all options open. On 30 December 1939 he replied to the Vice-Chancellor in a positive tone:

I thank you very much for kindly considering, even under the present circumstance, the possibility of an appointment as Professor to your University.

Since I left my fatherland Austria in September 1938, I have not entered a permanent position and I am without one still. Though I never encountered any hardship, I should be glad to reach definitive settlement again. I should therefore consider the offer of a permanent chair at your University very seriously indeed.

I am not quite sure about what steps would have to be taken to enable me and Mrs Schrödinger to travel to India whilst the war is going on. Formally we are considered enemy aliens in Great Britain and France. Actually we could not venture the journey without being assured of the full protection by the authorities of these countries. Only a route that practically excludes the possibility of an encounter with enemy (I mean: with German) forces would be possible for us.[2]

Schrödinger was well aware that Max Born had enjoyed his extended visit to Bangalore in India four years before and had very nearly taken up a permanent position in C.V. Raman's institute. The Vice-Chancellor then wrote from Allahabad on 19 January 1940 to David Thomson at the Society for the Protection of Science and Learning, which had been so much involved previously in assisting Schrödinger:

> It has been suggested to me that Prof. Erwin Schrödinger who received the Nobel Prize for Physics in 1933 and who is Austrian by birth and succeeded Max Planck at the Berlin University in 1927 and is now in Dublin at 45 Victoria Road would be a valuable addition to our Physics Staff. I had written to him on the subject of his coming out to India and I enclose herewith a copy of the reply that he has sent.
>
> I have asked the Government of India to help me in removing the political difficulty to which he refers; but I should like to know if your Society could get into touch with the Government at Whitehall and arrange that he might come out to this country. I shall be most grateful for any assistance that your Society might be able to render.[2]

Gertrud Bing at the Warburg Institute was delegated with the complicated matter of arranging travel for Schrödinger to go to India in wartime. She had been concerned nearly two years before after Schrödinger's statement seemingly supporting the Führer had been published in the Austrian press. The Transcontinental Travel Service wrote to her on 17 February 1940:

> With reference to the proposed journey of Dr Schrödinger to Allahabad, we have to inform you that it will be first necessary to obtain the permit from the authorities in India, and after this has been received he will also require an English transit visa in order to come to London to embark. There are regular sailings from the U.K. to Bombay, and the fares are approximately as follows: 1st class £80.00, 2nd class £58.00, by the P & O Line and similar lines serving Bombay.[2]

Gertrud Bing then wrote on 27 February 1940 to Tess Simpson at the Society for the Protection of Science and Learning:

> Enclosed is the information you required from Dr Schrödinger. Mr Döry asks me to add that there are no direct sailings from Eire to Bombay, so that it will be necessary for Dr. Schrödinger to sail from the United Kingdom. On the other hand, there will be no difficulty whatsoever about the obtaining of a visa once the permit from the authorities in India is available.[2]

Tess Simpson then responded on 28 February 1940 to the Vice-Chancellor in Allahabad:

> I apologise for not having replied sooner to your letter of January 19th, but I have been making enquiries on the question of getting Professor Schrödinger to India, and have now got this information:
>
> It is first of all necessary to obtain permission for Professor Schrödinger to go to India from the authorities in India themselves. After this permission has been received, he will require a British transit visa in order to embark in London, as there are no direct sailings from Eire to Bombay.
>
> This is where we could be helpful, and there will certainly be no difficulty — but we must first have the consent of the authorities in India. There are regular sailings from the United Kingdom to Bombay and the fares are approximately £80 first-class and £58 second-class.[2]

Sir Girija Shankar Bajpai, Education Secretary of the Government of India, New Delhi then wrote on 25 March 1940 to the Vice-Chancellor of the University of Allahabad:

> I write to say that the Government of India would have no objection if Professor Schrödinger were invited by you to take up the post of Professor of Physics in the Allahabad University. I may add that, if the Professor should come, there would be no objection to his wife accompanying him.[3]

There is no mention in any of this correspondence about an additional lady and a child. In the meantime, however, Schrödinger had informed de Valera about the approach from the University of Allahabad and that, rather surprisingly given the ongoing war, the arrangements were being made for him to travel to India by boat with his wife. This helped to get things moving in Dublin and on 19 March 1940 the *Irish Independent* newspaper reported:

> Mr de Valera was admitted to membership of the Royal Irish Academy, being formally welcomed by the President Dr A.W. Conway. The President announced that Dr Erwin Schrödinger had been appointed Professor of Theoretical Physics at the Academy from April 1. Dr Schrödinger, a native of Vienna, is a member of numerous scientific academies including the Pontifical Academy of Science. He was elected an Honorary Member of the Royal Irish Academy in 1931, and won the Nobel Prize in Physics in 1933.[1]

It is relevant that several press reports around this time frequently referred to Schrödinger being a member of the Pontifical Academy of Science, the message that did him no harm in this Catholic country. De Valera had been pleased with the positive reaction to Schrödinger in Dublin. He agreed that an announcement describing progress in the establishment of a new institute could be placed in *Nature* (on 6 April 1940), a publication that had reported several times on Schrödinger's movements:

> At the election meeting of the Royal Irish Academy in Dublin on March 16, it was announced that Dr. Erwin Schrödinger had been appointed Professor of Theoretical Physics in the Academy as from April 1. The funds for the Professorship are being supplied to the Academy by the Irish Government. Prof. Schrödinger has been giving a course of lectures on wave mechanics at University College, Dublin, since November last. This course has had a large attendance from members of the two Dublin Colleges. It will be continued now in the Royal Irish Academy. It is believed that the institution of this Professorship is intended as a temporary measure, pending the setting up of an institute for theoretical physics in Dublin, in which Prof. Schrödinger will have a permanent appointment.
>
> The recent publication of the text of a Bill making provision for the new institute shows that Mr. de Valera does not intend the financial

> He had by an ill wind, which had certainly blown us good, come to our shores. Schrödinger's name would ever stand out as one of the great names of Mathematical Physics, taking rank with Newton, Laplace, and Hamilton. His late Holiness Pope Pius XI, in reconstituting the Pontifical Academy of Science, appointed him one of the original 70 members of the Academy.[1]

Wolfgang Pauli, another brilliant Austrian theoretical physicist, had been working at the ETH in Switzerland but had automatically become a German citizen after the Anschluss. On the outbreak of the war he hoped to be made a Swiss citizen but this was denied to him. He was from a Jewish family and this put him in danger if he was forced to leave Switzerland.[5] However, the Institute for Advanced Study in Princeton had had their eye on Pauli for some time and offered him a Visiting Professorship. To get to the USA he had to travel from Geneva to Barcelona in Spain and then on to Lisbon for a boat across the Atlantic. His visa to the USA only came through after France had been invaded by Germany but on 31 July 1940 he and his wife managed to take the train across southern France to Spain which was a dangerous route at that time. Schrödinger wrote an affectionate and somewhat envious letter in English to Pauli on 1 November 1940:

> I had news from you today via H. Weyl though I knew that you had reached Princeton, for some time. And I thought we might at least try to resume correspondence, however disgusting it may be to know that the crossing of the ocean takes 12 hours and the remaining 3 weeks delay are caused by censorial precautions.
>
> First of all, I congratulate you on your move. I suppose you feel very happy in Princeton, owing mainly to the fact that you have your wonderful wife with you — well it may startle you for the moment that I attribute all importance, crushing importance to this fact. But in the second moment you will agree that it is a terrible milieu unless you have a very good friend quite near. Otherwise, how could you stand, *e.g.*, Mrs Eisenhart! I do hope that you will find occasions of unobtrusively annoying her.
>
> On the other hand there are also very charming persons in Princeton society, *e.g.* — and more than *e.g.* — Mrs Ladenburg. But instead of

anticipating reports from Princeton let me report from here. In the whole nothing has changed. The repercussions of the war in this country are hitherto so slight, so absolutely negligible, that I occasionally catch hold of my nose to make sure that I am not just dreaming.

The Dublin Institute for Advanced Studies has come into existence about a fortnight ago. And two days ago four Professors have been appointed, three in the School of Celtic Studies and one (my humble self) in the School of Theoretical Physics. And there you are. L'état c'est moi. I am the one and only member of the latter School, my activity is devolved to discussing with the Registrar the qualities of my writing table to come (which must not have a middle drawer, lest I should be prevented from crossing my legs) and to writing post-cards to booksellers and publishers, in order to beget a library.

People are as nice to me as nice can be. Newspaper articles extoll my fame, graciously discuss my private life in detail. No wish could I utter with respect to the School that would not immediately be cared for. A remark that I must have made a long time ago recently came back to me in this form: the Prime Minister stressed the necessity to provide for an afternoon tea, to invite the members of the Institute in an informal way, *etc. etc.* So nothing is missing except the others, except this that I am still quite alone. Oh, could I get you here! That is nonsense now, I know. But perhaps a little later.[6]

Then on 21 November Schrödinger was appointed Director of the School of Theoretical Physics at the Institute for Advanced Studies. This was essentially the first senior post that Schrödinger had taken involving administration. Despite the lonely caveats in his letter to Pauli, this gave him the opportunity to make new appointments and to invite, in due course, his scientific friends from other countries to give lectures in Dublin. It is clear, at least initially, that he enjoyed holding this position.

During his 17 years in Ireland, Schrödinger was hugely popular with the public and with the press. As time moved on he chose more and more general titles for his public lectures such as "What is Life?", "Science at Play" and "Fun in Science" and they were sell-outs. Like the Director of the Royal Institution or the President of the Royal Society in London, he had become the public face of science in Ireland. Every new prize, election to an academy, new appointment or Honorary Degree — and there were many of them for Schrödinger — was reported in complementary terms in the *Irish Press*.

De Valera himself had founded the *Irish Press* newspaper which always reported very positively on the Taoiseach and his colleagues. Accordingly, Schrödinger was regularly asked by the *Irish Press* to comment on almost any subject from animal rights to the role of the wives of businessmen. Even when he went on cycling holidays to Kerry or Connemara it was reported in the press. It seems that any movement he made or any comment he expressed was reported. There were non-sensational reports of his life at home with Anny, Hilde and Ruth with no difficult questions asked. After the Second World War he was also in regular demand to speak on highly serious topics such as the nuclear bomb or how to deal with German or Austrian Nazis.

However, the continual praise from the press and his most senior colleagues connected with the Institute for Advanced Studies, including Arthur Conway, Professor of Mathematics and President of University College Dublin, and Albert McConnell, Professor of Natural Philosophy at Trinity College Dublin, at times served to his disadvantage, especially in heated scientific discussions in correspondence with colleagues such as Albert Einstein or Max Born. It is helpful for academics to have close colleagues who can make positive but constructive comments and Schrödinger certainly had this when in Zurich and Berlin. He would have benefited from a colleague such as Pauli who, as a new colleague of Einstein at the Institute for Advanced Study in Princeton, was a sounding board for Einstein's ideas.

Walter Thirring, a distinguished theoretical physicist who was the son of Schrödinger's great friend Hans Thirring, first met Schrödinger in 1937 in Vienna when he was ten years old. He went on to undertake research with Schrödinger in Dublin. He went back to Austria and was in due course appointed to the same Chair of Theoretical Physics that his father once occupied at the University of Vienna. On Schrödinger's work after his great papers of 1926, Walter Thirring said: "It is strange that he had in fact many missed opportunities partly because he did not have enough appropriate collaborators to follow up his ideas, to recognize their importance."[7]

Schrödinger had become an impressive self-publicist. One of his press releases said:

> A course of lectures on the latest form of quantum theory intended for advanced students was started yesterday in U.C.D. by the Viennese E. Schrödinger, Honorary member of the Royal Irish Academy, honoured by the Nazi government with pensionless dismissal, without notice, from his academic chair in Austria.[8]

Time and time again the pension aspect for himself, and especially for his dependable wife Anny, would be raised by him and this would eventually be a key factor when he left Dublin and returned to Austria. Schrödinger submitted several papers to the *Proceedings of the Royal Irish Academy*, a journal which would have been hardly read outside of Ireland. His first paper on "A method of determining quantum mechanical eigenfunctions and eigenvalues" proposed a technique for factorising his Schrödinger equation and making it easier to solve in some cases.[9] This was a rare example in the 1940s of him linking directly back to his great papers of 1926.

On 20 April 1941, Schrödinger wrote to Paul Dirac in Cambridge with a request for photographs of the leading Cambridge physicists to display on the walls of his new Institute. He was apparently oblivious to the desperate situation back in England:

> We try to put a little picture gallery of physicists to adorn the walls of our very nicely furnished and equipped, but still a little soberly looking, new little place. From Cambridge physicists we would, in the first place, wish to have: yourself, Rutherford, J.J., Aston, Eddington and anybody else you might suggest. Would you help me at least in some of these cases? The Institute will be all too glad to meet any expenses which occur in providing and forwarding the copies. There is no ceremony because we are not a Department of the Government.
>
> The starting of the Institute in times like this has been a daring plunge, at least as far as the School of Theoretical Physics is concerned. Of course, I have every reason to be grateful for it. Being appointed Director of the School, I am assiduously directing myself, when as yet

> there is nobody else to be directed. The other wing, the Celtic Studies, consisting of three senior professors are more fortunate, not only on account of their plurality but for a prospective assistant professor who they are going to direct to deliver courses on phonetics — and who, by the way, is a handsome and attractive young lady!!
>
> Well, if you can do anything to add to my picture gallery I shall be infinitely grateful.[10]

Schrödinger was soon able to make an appointment to a junior professorship at his Institute but, sadly for him, it was not a "handsome and attractive young lady" as mentioned in the letter to Dirac. It was his former colleague Walter Heitler. After his brilliant work with London on the chemical bond, done in Zurich in 1927 not long after the publication of the Schrödinger equation, Heitler had gone to Göttingen to work with Max Born. He was from a Jewish family and had to leave Germany in 1933. The Academic Assistance Council helped him find a research post in Bristol University where the newly appointed Professor of Theoretical Physics was Nevill Mott.

Mott had been at the Cavendish Laboratory in Cambridge where he had developed the quantum theory for collisions between atoms, ions and electrons. He had been a keen member of the Kapitsa Club. He had written a classic book on atomic collisions with Harrie Massey who was at University College London.[11] Mott then changed direction in his research to apply wave mechanics to magnetic and disordered systems in the condensed phase. He was eventually awarded the Nobel Prize in 1977 for this work together with Philip Anderson and John Van Vleck. At Bristol, Mott had recruited a group of talented theoretical physicists who had been forced out of Germany. In addition to Heitler, this included Hans Bethe, Herbert Fröhlich and Klaus Fuchs.

In Bristol, Heitler had written a well-received book on the *Quantum Theory of Radiation*.[12] He was also moving into the field of quantum electrodynamics. As a German citizen he was interned on the Isle of Man in the summer of 1940 along with many other refugees. He never published jointly with Schrödinger but was highly valued in Dublin and eventually became the Director of Theoretical Physics at the Institute for

Advanced Studies when Schrödinger wanted to direct all his energies to unifying relativity and electromagnetic theory. Following Heitler's move to Dublin, Schrödinger wrote in English on 5 October 1941 to Max Born:

> Having had Heitler here for some time, I feel I must thank you particularly for recommending him so strongly. He is a most valuable asset in every respect. When scientifically at least equal to his "milk-brothers" London (I mean that brothership with respect to their first great achievement), as a man and as a teacher he is incomparably better. Indeed, he has a marvellous gift of understanding the difficulties and objections of another person.[13]

Fritz London had been the assistant to Schrödinger in Zurich and Berlin, and both had then been supported by ICI in Oxford, but it is clear from this remark that their interaction was not perfect. Heitler took British citizenship and this qualified for him to be nominated for Fellowship of the Royal Society to which he was elected in 1948.[14] He also managed to extract from Germany his mother, sister Annerose and brother Hans, who had been interned in a concentration camp. They all lived with Heitler in his house at 21 Seapark Road which was near to the Schrödingers in Clontarf. Brave Annerose had set up a Jewish orphanage in Freiburg Germany from 1935–38 and stayed in Ireland for the rest of her life.

Another appointment to the Institute was Peng Huanwu. He was a rare scientist in Europe in the 1940s who was born in China. He had come to work with Max Born in Edinburgh for his PhD. He then collaborated with Heitler in Dublin on the theory of new particles including mesons. After the war he returned to China and eventually led their nuclear and hydrogen bomb programmes. It is remarkable that J. Robert Oppenheimer, who famously directed the US nuclear bomb research; Peng Huanwu, who did the same for China; Heisenberg, whose research in Germany in the war involved examining the possibility of a nuclear bomb; and Klaus Fuchs, who gave key secrets away to the Soviet Union on nuclear weapons, had all worked with the pacifist Max Born. In April 1964 Born wrote a letter to Oppenheimer saying:

> I have followed your public career with deep interest and sympathy, not only because you proved to be a leader of men and a most efficient administrator, but because I felt that you were burdened with a responsibility almost too heavy for a human being.[15]

Born was particularly critical of Lindemann's role as the science advisor to Winston Churchill who advocated bombing of German cities:

> I knew that once Hitler was in power there was no way of getting rid of him without defeat in war. The idea of breaking Hitler by killing ordinary people, women and children, and by destroying their homes seemed to me to be absurd and detestable. But it was just this idea which Lindemann suggested and forced through against the objections of better men.[16]

By 1940, and in the spirit of Einstein's letter to President Roosevelt, many of the refugee scientists based in Britain had suddenly turned their own research direction to assisting the war effort against Germany. Many of those in the USA would soon do the same. Their hatred for the Hitler regime was so great that they did not hesitate to act in this way even if it was to lead to the near-destruction of their home country. Many of the prominent British scientists and refugee physicists were involved in the key UK research that led to the nuclear bomb. Even Paul Dirac made a contribution in formulating the details of isotope separation using centrifuges. Schrödinger, however, who was working in neutral Ireland with a close connection to President de Valera, was not involved at all in any war-related research. Arthur March also declined to undertake research linked to the German war effort and between 1939 and 1945 published seven papers in Innsbruck on his fundamental ideas of the smallest length and on space, time and natural laws.

Magdalen College back in Oxford was aware of Schrödinger's recent appointments but many Fellows who had briefly got to know him had taken leave to join the forces or undertake urgent work for the Government. The published Summary of Events of Magdalen College for 1939–40 included the following items:

> The President was re-appointed Vice-Chancellor. His portrait was painted for the College by Mr William Coldstream.
>
> Sir Charles Sherrington was elected an Honorary Fellow of the College.
>
> Sir Robert Robinson received the Hon. Degree of D.Sc. from the University of London.
>
> Sir Kenneth Clark, formerly Fellow, was appointed Director of Films and then of Home Publicity in the Ministry of Information.
>
> Dr E. Schrödinger, formerly Fellow, received the Hon. Degree of D.Sc. of the University of Ghent, and was appointed Professor of Theoretical Physics at Dublin.
>
> The following Fellows of the College are absent on National Service, some with the Forces of the Crown, others on Government work at home or abroad: Messrs Driver, Weldon, Johnson, Denholm-Young, Opie, Moullin, Mackenzie, Austin, Bazell, Morris, Hunt, Cheney, Waterhouse, Urmson and Rees.[17]

Several of these former colleagues of Schrödinger played important roles in the war. Sir Godfrey Driver found his language skills useful when he was put in charge of Arabic and Hebrew publications for the Ministry of Information. Harry Weldon became the Personal Staff Officer to Sir Arthur Harris who was head of the British Bomber Command. Weldon had to apply all of his philosophical expertise in justifying to the public that area bombing was a means of shortening the war. The physicist Patrick Johnson, who had attended Schrödinger's lectures in Oxford, became the Deputy Scientific Advisor to Field Marshal Montgomery. Eric Moullin joined the Admiralty Signals Establishment and his papers on magnetrons were useful in the development of key radar devices. Redvers Opie was made the UK Treasury Representative in Washington DC. Following on from this, he helped set up the International Monetary Fund and the World Bank.

William Mackenzie, who had reported to Lindemann on his meeting with Schrödinger in Austria just after the Anschluss, became secretary to several key technology committees in Whitehall covering topics such as anti-submarine devices, aircraft production, operational research and radar. He worked closely with Lindemann and also initially with Henry

Tizard. At the end of the war Mackenzie was commissioned to write the history of the Special Operations Executive which had been established to conduct espionage, reconnaissance and sabotage in occupied Europe and Southeast Asia. John Austin, who was an expert on the Philosophy of Language, made a vital contribution to the D-Day invasion of Normandy by providing to the invading forces a notebook containing a comprehensive and meticulous picture of northern France, including coastal defences, road networks, industry, topography, supply lines, infrastructure, human terrain and enemy troop concentrations. For this he was awarded the O.B.E., the French Croix de Guerre and the US Legion of Merit. In addition, John Morris, who was a formidable Law Tutor, was responsible for building up the force of unique landing craft used in the invasion of Normandy.

The historian Christopher Cheney joined the MI5 Security Services but he was said to find the MI5 files less gripping than medieval manuscripts. Sir Ellis Waterhouse, who was an art historian, was in Athens when the war broke out and worked in the British Legation in Greece and Crete. He escaped on the last British boat from Crete after the island was invaded. He then went to Cairo to work in the Intelligence Corps. At the end of the war he was commissioned to retrieve works of art stolen by the Nazis. He discovered that a painting which had been acquired for the voluminous collection of Hermann Göring, and which was attributed to Vermeer, was a fake. James Urmson, who was a Fellow by Examination in Philosophy, was involved in the frantic withdrawal of British forces from Dunkirk. He then fought in North Africa and Italy where he was taken as a prisoner of war. He was awarded the Military Cross.

Sadly, Schrödinger would not be hearing again from President George Gordon at Magdalen College. Gordon had written dozens of letters and worked so hard for Schrödinger while he was a Fellow at Magdalen from 1933–38. He had become Vice-Chancellor of Oxford University while still being President of his College. This was a very difficult job in war time with the mobilisation of most students and Fellows, and with confidential research programmes being conducted in the University. The strain was too much and Gordon died on 12 March 1942. However, some of the most prominent Magdalen Fellows, such as the Nobel Prize winners Charles Sherrington, Robert Robinson, John Eccles and Peter Medawar, continued to correspond with Schrödinger.

George Gordon was replaced as President of Magdalen by Henry Tizard who had once been a student in the College and a physical chemistry researcher in Oxford. On the advice of his research supervisor Nevil Sidgwick, who had befriended Schrödinger when he was in Oxford, Tizard had gone to work with Nernst in Berlin before the First World War and had overlapped there with Lindemann. In 1908 he had hoped to attend Planck's lectures in Berlin and said:

> I intended to put down my name for Planck's lectures but postponed doing so because of the cost. Later on I was told that the lecture room, which held 400, was already crowded. How different from Oxford, I thought, where a Professor of Mathematical Physics could count himself fortunate if he kept an audience of half a dozen.[18]

Tizard then researched with Townsend in the Clarendon Laboratory at Oxford on the motion of electrons in gases. In the First World War, he did important research on fuels for aircraft and subsequently proposed the octane rating to classify petrol. He eventually became Rector of Imperial College, London and in the 1930s and 40s chaired key government science committees on topics that would soon become crucial for the war effort such as radar and nuclear fission.[18]

Tizard had a power struggle with Lindemann in 1940 on who should give science advice to the new government of Winston Churchill and lost out. Therefore, the appointment to the Presidency of Magdalen in 1942 seemed timely for him. Despite the fact that he was the first scientist to head an Oxford College, as was proudly proclaimed in *Nature*,[19] there is no evidence that Tizard corresponded with Schrödinger at any stage. He did not find being President of Magdalen easy and the Fellows in social science and humanities were rather uncooperative. He also had difficulties in remembering the names of Fellows. His Vice-President Bruce McFarlane noted, in a letter to a friend Dame Helena Wright, that Tizard called him MacNab, Vice-Chancellor and Prime Minister in one 24-hour period.[17]

After just four years, an untypically short period in post for a President of Magdalen, Tizard resigned and returned to giving advice to the Government, becoming the Chief Scientific Advisor to the Ministry of Defence. Tizard was given an Honorary Degree by Edinburgh University in 1946 and there he met Max Born. Almost his last action as President of Magdalen was to invite Born to give the prestigious Waynflete Lectures in the College. Tizard was replaced as President of Magdalen in 1947 by Tom Boase who was an Art Historian. Magdalen had to wait until 1968 for its next scientific President, the physicist James Griffiths who had been appointed a Fellow by Examination in 1934 only after the crucial intervention by Schrödinger at a College Meeting.

The refugee scientists back in Oxford did very important work in the early 1940s that started off key technology on nuclear weapons. In Berlin in 1938 at the Kaiser Wilhelm Institute for Chemistry, Hahn and Strassmann had detected the production of barium from the bombardment of uranium by neutrons.[20] Just before this, their close collaborator Lise Meitner had departed Germany for Sweden. From reading the Hahn and Strassmann paper, Meitner and her nephew Otto Frisch realised that the uranium nucleus was splitting in two and in doing so would give off a large amount of energy. This was the basis for the nuclear bomb.

Frisch also had to leave Germany and eventually went to work with Rudolf Peierls in Birmingham. It had been realised that the key isotope of uranium for nuclear fission was uranium-235 and Frisch and Peierls calculated the critical mass needed for producing the chain reaction needed for a nuclear explosion. They concluded that just a few kilograms of the isotope would explode with the energy of thousands of tons of dynamite. This was communicated in a memorandum to Henry Tizard who was then Chairman of the Committee for the Scientific Survey of Air Warfare.[18] This initiated a major secret research effort in the UK dubbed the Tube Alloys project and administered through a new "MAUD" committee chaired by George Thomson, whom Schrödinger had met on his visit to Cambridge back in 1928. A key aspect was to develop an efficient method for separating the uranium-235 from the dominant uranium-238 isotope.

Peierls had become friendly with Francis Simon when they were both in Germany and knew he was a highly innovative experimental physicist. Peierls suggested to Simon he should direct his team in Oxford to consider experimental methods to separate uranium isotopes.[21] Despite the fact that many of the key players in Simon's group in Oxford had come from Germany they were still to be trusted in this research. With the other German refugees Kurti and Kuhn, an experimental procedure was developed and tested for separating isotopes by diffusion of gaseous uranium hexafluoride. By passing this material through a fine membrane screen, the slightly lighter uranium-235 would diffuse more quickly and, by repeating this process many times, the lighter isotope could be separated out.[21]

Simon and his group, together with Schrödinger, had been funded by ICI and this link with their Chemicals Division enabled Simon to obtain sufficient quantities of the uranium hexafluoride and, with their Metals Division, appropriate perforated membranes. These early experiments of Simon and co-workers were promising enough for the technique to be taken up subsequently on a massive scale in the USA in a plant at Oak Ridge to produce the uranium-235 used in the nuclear bomb dropped on Hiroshima in Japan in 1945.

Given the progress on nuclear research, there was a concern that the German scientists might be making similar progress. Simon surveyed the German physical chemistry scientific literature to see if there were any clues and Peierls, together with his assistant Klaus Fuchs, did the same for physics. They came to the conclusion that it was unlikely that significant progress was being made in Germany.[22]

However, a number of physicists left behind in Germany, including Heisenberg, were undertaking research in this direction. Following a meeting with Hitler on 6 May 1942 the German Armaments Minister Albert Speer organised a lecture on 4 June 1942 by Heisenberg in the Harnack House in Berlin to examine progress on German atomic research. Leading military representatives were present including the armaments chiefs of the three branches of the armed forces, General Fromm of the Army High Command, Field Marshal Milch and Admiral

Witzell. Several scientists, including Otto Hahn, attended in addition to Heisenberg. Speer wrote:

> After the lecture I asked Heisenberg how nuclear physics could be applied to the manufacture of atom bombs. His answer was by no means encouraging. He declared, to be sure, that the scientific solution had already been found and theoretically nothing stood in the way of building such a bomb. But the technical prerequisites for production would take years to develop, two years at the earliest, even provided the programme was given maximum support... I had been given the impression that the atom bomb could no longer have any bearing on the course of the war.[23]

However, the suspicion of the work in Germany did much to accelerate the research on nuclear weapons in the UK and USA, which did not abate during the period of the war. In his Nuremberg Trial after the war, upon being asked if Germany was close to producing a nuclear bomb, Speer said:

> We had not got as far as that, unfortunately, because the finest experts we had in atomic research had emigrated to America, and this had thrown us back a great deal in our research, so that we still needed another year or two in order to achieve any results in the splitting of the atom.[24]

Speer was fortunate to avoid a sentence of death at the judgement of his Nuremberg Trial. He was sentenced to 20 years in prison.

Lise Meitner's position in neutral Stockholm proved useful to the allies opposing the Nazis in the Second World War.[25,26] Through a Norwegian Physicist Njål Hole, who was working in the same laboratory as Meitner, and also her contact Paul Rosbaud, the Springer publisher who had helped her leave Germany and somehow subsequently remained in Berlin, any information on the progress of German nuclear research was scrutinised in great detail. Both Otto Hahn, who continued to work on nuclear research in Germany, and Max von Laue visited Meitner in neutral Stockholm in 1943 and 1944 respectively. Any news on the place of work and research activities of Heisenberg and Hahn was of particular interest to the allies. Following Niels Bohr's dramatic escape to England in 1943, and his subsequent contributions in the research team working on the nuclear bomb, a confidential message was sent to Meitner suggesting that the same arrangement could be made for her. However, Meitner had volunteered in the First World War as an X-ray nurse technician and following this traumatic experience she refused to be involved in research linked to military activities.

Many of the threatened scientists and promising PhD students in Germany had managed eventually to find safety in other countries after the Nazis came to power and the same was true, to a certain extent, with those who left Austria. However, some scientists from other countries which were suddenly invaded in the war were not so fortunate. Henri Abraham and Eugène Bloch were both distinguished directors of the Laboratoire de Physique at the École Normale Supérieure in Paris and were from Jewish families. They had made important contributions to understanding radio waves. Following the invasion of France both tried to escape but were arrested and sent to Auschwitz where they were murdered. Georges Bruhat had also been Director of the same Laboratory. He was arrested by the Gestapo as a suspect for providing assistance to the French Resistance. He was sent to the Sachsenhausen concentration camp where he perished. After the war, the Three Physicists Prize was set up by the École Normale Supérieure in memory of these three directors.[27] The Prize has been won by several notable physicists mentioned in this book including Nevill Mott and J.R. Oppenheimer.

Back at Magdalen College in Oxford, C.S. Lewis was contemplating about his former colleague. In 1942 he wrote:

> To explain even an atom Schrödinger wants seven dimensions: and give us new senses and we should find a new Nature. There may be Natures piled upon Natures, each supernatural to the one beneath it, before we come to the abyss of pure spirit.[28]

It is interesting to speculate on the view Schrödinger might have had on this comment. Schrödinger's ideas had a huge influence on fields beyond physics such as chemistry and biology. It seems this influence was also extending to theology.

A particularly attractive aspect for Schrödinger in his position as Director of Theoretical Physics at the Dublin Institute for Advanced Studies was that he could invite, even in wartime, his close academic friends who were working in England and Scotland to come to neutral Dublin for an extended visit and give lectures. In July 1942 Paul Dirac and Arthur Eddington made the trip from Cambridge to Dublin for two weeks of talks and receptions. Dirac was amazed to find that "there is plenty of food here — ham, butter, eggs, cakes — as much as one wants". He also said to the bemused *Irish Press* on 14 July 1942: "This is my first visit to Ireland and I won't say a word about it until I look around."[1] On the front page of the *Irish Independent*, close to photographs of the visiting scientists Dirac and Eddington, there was a report on "Soviet Reserves now in Action in Stalingrad".[1]

The distinguished visitors were surprised that de Valera attended their lectures. They had never seen anything similar to that from a leading politician in England. De Valera even took them on a personal joyride around the local countryside. It was a very different atmosphere to the blacked-out and bombed-out Britain of 1942. Dirac must have liked Dublin and the Institute for Advanced Studies as he was to visit to lecture again in 1945, 1953 and 1963.

Ernest Walton also attended these lectures. Originally from Waterford in Ireland, he and John Cockcroft had discovered artificial nuclear disintegration at the Cavendish Laboratory at Cambridge University in 1934, while Schrödinger was in Oxford. A member of St John's College Cambridge, Walton had come across Dirac and Born who were Fellows there at that time. Walton had been made a Fellow of Trinity College Dublin and was awarded the Nobel Prize for Physics in 1951 with Cockcroft.

In Dublin, Eddington lectured on a topic that had become close to Schrödinger's heart: "The Combination of Relativity Theory and Quantum Theory". Schrödinger still was not happy with how the number of particles in the universe came into Eddington's cosmological theory. Dirac also dealt with combining the two theories but in the context of quantum electrodynamics. Just six weeks after Dirac returned to Cambridge, his second daughter was born who was named after his mother Florence. She had accompanied him to Stockholm to receive his Nobel Prize together with Schrödinger and, as is described in Chapter 2, had written in such glowing detail of all the celebrations.

Max Born also took the trip over from Edinburgh to lecture at the Institute in July 1943. He had some reservations and wrote to Frederick Donnan on 14 May 1943 to say:

> I think it distasteful to be "neutral" in a conflict like this war with the Nazis. I do not wish to give the impression of consenting to this attitude by being the guest of de Valera. On the other hand, I think it is important to keep science out of and above all political struggle. And personally I wish to discuss things with Erwin Schrödinger. So my soul is divided.[29]

Paul Ewald came at the same time and his subject was the theory for determining the structure of crystals from X-ray patterns. Ewald had taken Schrödinger's position when he moved on from Stuttgart in 1921. His wife was Jewish and they left Germany in 1933, moving first to Cambridge and then to Belfast where Ewald was appointed Professor in Mathematical Physics at the Queen's University. One of his academic colleagues there was Frank Boys who moved subsequently to the University of Cambridge where he became a leading pioneer on extending Schrödinger's wave mechanics to realistic calculations on molecules.

As time went on, and unlike Schrödinger, his old friend Max Born was moving more towards applicable theoretical research that could be linked directly to laboratory observations. In Dublin, he lectured on crystal dynamics. The lectures were reported in the *Irish Press* on 16 July 1943 with a picture of Max Born and de Valera. The headline of the newspaper was "Russians on Offensive in Orel Salient" and there was a detailed report on the battle for Catania in Sicily.[1] The tide was turning in the war. Max Born found Ireland to be "a strange world, quite separate from the big waves which shake the ships of the other nations. But they are a loveable people altogether".[30]

Schrödinger's main research interest during his time in Ireland was to attempt to unify Einstein's relativity theory with Maxwell's equations for electromagnetic radiation. He read his first paper on the topic to the Royal Irish Academy in January 1943. He would have a correspondence with Einstein on this subject which would get more and more heated and would finally develop into a bitter public row that took place in the press. A few months after Schrödinger's first paper on this research the *Irish Press* reported on 28 June 1943 that a new theory by Schrödinger had been confirmed:

> The Royal Irish Academy will hear today of an important development which provides strong confirmation of Professor Erwin Schrödinger's new theory of physical fields. It is the reaction of the compass needle recording variations in earth magnetism intensity which has unexpectedly afforded proof of Prof. Schrödinger's great theory. It has done so in much the same way as the movement of the stars gave Einstein confirmation of the validity of the Theory of Relativity, which Prof. Schrödinger's new theory supplements and, in a measure, replaces. Einstein's discovery of the deflection of starlight beams as they pass the rays of the sun on their way to the earth, and his correct calculations in regard to the position of the stars showed that the stars and celestial bodies generally were obeying his laws, not those of the older astronomy.
>
> Prof. Schrödinger's Unitary Theory of Physical Fields was worked out by him in the School of Theoretical Physics of the Dublin Institute for Advanced Studies of which he is Director. It was first announced in the Irish Press early this year. Since then it has attracted world-wide attention and he has received inquiries and comments from many eminent physicists in various countries.[1]

Life

At the same time that he was working on his new unified theory Schrödinger gave four public lectures at Trinity College Dublin, under the auspices of the Dublin Institute for Advanced Studies entitled "What is Life? The Physical Aspect of the Living Cell".[31] The publication of these lectures was to make him famous in biology. The lectures addressed the problem "how can the events in space and time which take place within the spatial boundary of a living organism be accounted for by physics and chemistry". The lectures introduced an "aperiodic crystal" based on chemical bonds that contained genetic information.

At that time the mechanism for biological inheritance in terms of molecular structure was not known. Schrödinger was largely thinking on his own from first principles and had limited interaction with active chemists or biologists on this subject. This was to lead to subsequent criticism from some of the leaders in the field of molecular biology.

He also introduced confusing new phrases that were not used by other scientists. For example, "aperiodic crystal" would have been much better described as a biopolymer, and Schrödinger's close friend when they were in Vienna, Herman Mark, had been an early pioneer in the polymer field. It should be noted, however, that at this time genetic material was thought to be proteins made up of amino acids, rather than nucleic acids. Schrödinger emphasised "negative entropy" as the driving force for stability of his "aperiodic crystals" when many biochemists already realised that entropy was only half the story and free energy was the important quantity that also included total energy. After the first publication of the lectures Francis Simon wrote to Schrödinger to correct him on the matter of free energy and he did add an addendum in future reprints of the lectures to refer to this key point.

Schrödinger's "little book", however, influenced several young scientists at that time who were to play a central role in the revolution of molecular biology.[32] These budding scientists were surprised to see that the great Schrödinger was applying the principles of physics and chemistry to biology. A good example is Francis Crick, who with James Watson in 1953 famously deduced the double helix structure of DNA and its hereditary implications. Crick was given the *What is Life?* book by his physics professor Harrie Massey at University College London while doing rather uninspiring physics research on the viscosity of water.

Reading Schrödinger's book helped inspire Crick to change his research field and start to think about the fundamentals of biology and whether heredity could be explained by molecular physics. He said: "The book conveyed in an exciting way the idea that, in biology, molecular explanations would not only be extremely important but also that they were just around the corner."[33] James Watson had similar views and said: "This book very elegantly propounded the belief that genes were the key components of living cells and that, to understand what life is, we must know how genes act."[34] After Crick and Watson had published their famous paper on the double helix structure of DNA in *Nature*, the grateful Crick wrote to Schrödinger in Dublin on 12 August 1953:

> Watson and I were discussing how we came to enter the field of molecular biology, and we discovered that we had both been influenced by your little book *What is Life?*.
>
> We thought you might be interested in the enclosed reprints — you will see that it looks as though your term "aperiodic crystal" is going to be a very apt one.[35]

Maurice Wilkins, who published the first X-ray diffraction images of DNA with Rosalind Franklin, was awarded the Nobel Prize for Physiology or Medicine in 1962 with Watson and Crick. He wrote:

> Schrödinger used the language of physicists and that stimulated me, as a physicist, to persevere with his book and its introduction to genetics, and to decide that this was the general area that I wanted to explore as a 'biophysicist'.[36]

It was not just physicists who appreciated Schrödinger's book. Irene Manton, who was a leading authority on cell biology, said: "When a great physicist takes the trouble to explain in simple language some of his matured thoughts on topics of general interest outside his own subject, it is an event for which one cannot be too grateful."[37]

As always, the *Irish Press* was keen to report, on 6 February 1943, on Schrödinger's latest ideas:

> Crowds were turned away from the Physics Lecture Theatre yesterday for a lecture on "What is Life (the physical aspect of the living cell)". This was his first public appearance since our announcement of the completion of his new theory which supplements Einstein's theory of

> relativity. The lecture was the first of three statutory lectures. He was given an ovation by the packed theatre, and so many people desiring to gain admission were disappointed that the lecture is to be repeated on Monday.
>
> Prof. Schrödinger said that the whole problem turned on the question: How could the events in space and time which occurred within the boundary of a living organism be accounted for by physics and chemistry and on the obvious impotence of present day chemistry and physics to account for these events?[1]

An obscure paper that had inspired Schrödinger's series of lectures had been published in Berlin by Timofeev-Ressovsky, Zimmer and Delbrück back in 1935.[38] Delbrück had worked with Lise Meitner in Berlin and knew Schrödinger when he was working there. Their paper described a model for genetic mutation based on the principles of atomic physics and stable molecules. Delbrück himself hated the Nazis and had emigrated to the USA to work on biological problems. It seems his citation by Schrödinger in the little book impressed a selection committee at the California Institute of Technology in Pasadena who offered him a Professorship. There he became a world leader in phage genetics which won him the Nobel Prize in Physiology or Medicine in 1969.[39]

There have been mixed reactions over time to the science described in the "little book" but its inspirational effects are without doubt. The eminent geneticist J.B.S. Haldane wrote a long and detailed review in *Nature* which started:

> As a result of the war, many scientific workers have been too busy with the applications of science to keep up even with the development of their own branch. Schrödinger, as an exile in neutral Eire, has found the leisure to study another, namely, genetics, which he describes as "a new branch of science, easily the most interesting of our days". I wonder if posterity will find crossing-over as interesting as exchange energy, or mutation as atomic transition. However this may be, every geneticist will be interested in Schrödinger's approach to his or her science.[40]

While his conclusion was also positive:

> The book is one to which one comes back again and again. I have lent it to several genetical colleagues, and the verdict has been uniformly favourable. We may disagree with details, or even with fundamental principles, but we cannot stop reading it before the end. Unfortunately it contains only 93 pages. There are 93 elements but 738 isotopes have so far been described. Perhaps we may hope for a book of 738 pages of biology in general. Schrödinger may not doubt that it will find readers.[40]

In a letter of 11 January 1945 to Schrödinger, Haldane also stated, "I am in favour of physicists trespassing into biology and conversely."[41] Some of the leading experts in molecular biology, however, were not so positive. Linus Pauling, who had travelled to Zurich in 1926 hoping to work with Schrödinger and had subsequently been a Nobel-winning pioneer in molecular wave mechanics, had changed direction into studying the structure of biological molecules. He wrote:

> We might now ask the following question: To what extent, aside from his discovery of the Schrödinger equation, did Schrödinger contribute to modern biology, to our understanding of the nature of life? It is my opinion that he did not make any contribution whatever, or that perhaps, by his discussion of "negative entropy" in relation to life, he made a negative contribution...
>
> When I first read this book I was disappointed... Schrödinger's discussion of thermodynamics is vague and superficial to an extent that it should not be tolerated even in a public lecture.[42]

For many years, Max Perutz was one of the leaders of the field of molecular biology. He won the Nobel Prize for Chemistry in 1962 for his work on the structure of haemoglobin. Like Schrödinger he came from Austria and was a student at the University of Vienna. He studied with Schrödinger's friend Herman Mark who encouraged Perutz to go and work in Cambridge in 1936 with the crystallographer J.D. Bernal. In this way, Perutz missed the Anschluss and stayed for his PhD in Cambridge where he also worked with Lawrence Bragg.

In the Second World War Perutz worked on a project to create an ice platform for aeroplanes to land in the mid-Atlantic. This original idea was never realised but after the war with the support of Bragg he obtained funding to establish the Molecular Biology Unit at the Cavendish Laboratory in Cambridge. Many brilliant scientists joined this group, including Crick and Watson. The Laboratory of Molecular Biology which Perutz founded in Cambridge has had a remarkable 26 researchers who have gone on to win the Nobel Prize. Perutz criticized strongly Schrödinger's bold statement:

> Living matter, while not eluding the laws of physics as established to date, is likely to involve other laws of physics hitherto unknown which, however, once they have been revealed, will form just as integral a part of this science as the former.[31]

The author of the current book recalls Perutz lecturing in Cambridge in 1987 with a very negative view on Schrödinger's book:

> The apparent contradictions between life and the statistical laws of physics can be resolved by invoking a science largely ignored by Schrödinger. That science is chemistry.[43]

Perutz was also very critical of Schrödinger for not realising the ability of catalysts such as enzymes to drive biological processes by reducing the potential energy barriers for chemical reactions. His conclusion was made in no uncertain terms:

> Sadly, however, a close study of his book and of the related literature has shown me that what was true in his book was not original and most of what was original was known not to be true even when the book was written. Moreover, the book ignores some crucial discoveries that were published before it went into print.[43]

Perutz did mention one positive aspect that he considered came out of Schrödinger's book and that was the citation of the work of Delbrück and co-workers which was hardly known before. Despite misgivings such as these, Schrödinger's "little book" has been reprinted many times since and has been examined and discussed by generations of biologists and geneticists. The book was published by *Cambridge University Press* in the UK. There are reports that the materialistic aspects of the book did not go down well with some members of the Catholic Church in Ireland and so Schrödinger looked for a publisher elsewhere.[44]

More than anybody else, de Valera was responsible for creating the state of Ireland and played the leading role in that country for a remarkable period from the Easter rising in 1916 to his retirement as President in 1973. He had done well to keep Ireland neutral in the Second World War. However, this policy lead to a major aberration on his part. On 3 May 1945, following the announcement of the death of Adolf Hitler, his own *Irish Press* reported:

> Taoiseach and Minister for External Affairs, Éamon de Valera, accompanied by the Secretary of External Affairs, Joseph Walshe, called on Dr Hempel, the German minister, last evening, to express his condolences.[1]

De Valera's action led to fierce criticism from countries round the world. He is still remembered by many today who considered this to be a serious lack of judgement.

The end of the Second World War was a huge relief and Schrödinger celebrated with a major colloquium at the Dublin Institute for Advanced Studies in July 1945. As shown in the picture, he brought back his close scientific friends Paul Dirac and Max Born to lecture. He also wrote by hand on 14 May 1945 to his personal friend Patrick Blackett inviting him to come and emphasising the famous Irish hospitality:

> I hasten to answer your letter of 9 May, which I have just received with intense joy. Oh do come, that would be a delight!
>
> I am enclosing a short prospect. As you can see from it, the time is the fortnight from 5–18 July (which contains two weekends "in the interior"). Should you only be able to come for part of the time, I would try and put the lectures you are most interested in in that part. Apart from the lectures and discussions we always have a jolly good time during these meetings, with official and unofficial invitations and dinners, almost a bit too many for my taste, but one need not go to all of them.

> If I can do anything to facilitate your travel or in ordering hotel accommodation, please tell me. The British Representative's office is always very kind and helpful. The hotels are not very good and pretty expensive, but that cannot be helped. If you take your bike with you, some private quarters near me (about 3 miles from the centre) might be preferable. Unfortunately my little house is over-crowded.[45]

In the war, Blackett had been on Tizard's committee which initiated key nuclear-related research and he played a major role in insisting this information was shared with the Americans. He was Director of Operational Research with the British Admiralty. He wrote back to Schrödinger to say that he was still doing work with the Admiralty and sadly could not get leave to attend the colloquium in Dublin.

Colloquium at the Institute for Advanced Studies, Dublin, July 1945. From the left: Born, de Brun, Dirac, de Valera, Conway, Schrödinger, McConnell and Heitler.

Schrödinger had also invited the Hungarian astrophysicist Lajos Jánossy who was a colleague of Blackett at Manchester University. He was an expert on cosmic rays and had studied under Schrödinger in Berlin before leaving Germany to work with Blackett. He must have impressed the people in Dublin with his lectures as he was shortly after appointed to a permanent post at the Institute for Advanced Studies.

A fee of one guinea was charged to attend the lectures on crystal optics, statistics of radiation, quantum electrodynamics and cosmic radiation. Schrödinger was quite strict on who should attend, saying: "Would newcomers not known to us, please indicate by reference to their curriculum or to their previous studies that they are likely to profit from lectures and discussions concerned with questions on the frontiers of knowledge of physics."[45]

Bomb

Then, on 6 August 1945, the nuclear bomb was dropped on Hiroshima. Many researchers not involved in the highly secret scientific efforts in the UK, Canada and the USA were astonished with the news. All of the scientists who had been involved with developing the understanding of the atom and the nucleus had contributed in some way directly or indirectly to the nuclear weapon. Many of the refugee scientists had played crucial roles in different ways including Einstein, Szilard, Peierls, Frisch, Wigner, Bethe, Fermi, Franck, Teller, von Neumann, Fuchs, Simon, Kuhn, and Kurti. Niels Bohr, who escaped from occupied Denmark in 1943, should also be included in this list. Without his contributions it would have taken many more years for the bomb to be developed.

The President of the USA who made the decision to drop the first nuclear bomb was Harry Truman. President Roosevelt, who had received Einstein's crucial letter in 1939, had died just four months before and Truman was automatically promoted from Vice-President. The witty and insightful Nobel Prize winner and theoretical physicist from the California Institute of Technology, Richard Feynman, made an interesting comparison: "Schrödinger was the Harry Truman of Physics, not obviously destined for greatness, but one who certainly lived up to the demands when his time came."[46]

Always in the back of the minds of the refugee scientists was the unclear role Heisenberg had played in the war including his role in attempting to develop a German nuclear bomb. As was vividly portrayed in the play *Copenhagen* by Michael Frayn,[47] he had visited Niels Bohr in 1941 in occupied Denmark and had made some ambiguous statements on the recent research progress on nuclear fission.[48] In 1943, he did the same with Hans Kramers, Paul Erhenfest's successor, during a visit to Leiden and other cities in occupied Holland.

Heisenberg interacted with the most infamous Nazis. In December 1943, he lectured in Cracow at the invitation of his former school friend Hans Frank, the Governor of occupied Poland.[48] Heisenberg stayed at the Wartenburg Castle and Frank presented him with the Copernicus Prize. This visit was just weeks after the liquidation of the Cracow and Warsaw

Jewish ghettos under the direction of Frank. In 1946, Frank was executed by hanging after his trial at Nuremberg.

After the war, the British authorities interned, in Farm Hall near Cambridge, the leading physicists who had remained in Germany during the war and listened secretly to their discussions. This group not only included Heisenberg but also Otto Hahn who had just been awarded the Nobel Prize for Chemistry in 1944 for his breakthrough in nuclear fission. Schrödinger's Austrian colleague Lise Meitner had undeservedly been omitted from this award. This was possibly because she had been working in the laboratory in Sweden of Manne Siegbahn who was on the Nobel awards committee and there had been some friction between them. Von Laue was also interned in Farm Hall but he was not involved with research related to the Nazi war effort. The Farm Hall tapes seemed to suggest that Heisenberg and colleagues had failed to realise that it was possible to produce a nuclear weapon from a relatively small mass of uranium-235.[48]

Being the most celebrated scientist in Ireland who had won the Nobel Prize explicitly for his contribution to "the discovery of new productive forms of atomic theory" Schrödinger was asked for his opinion on the nuclear bomb. As usual, he responded straight away to the *Irish Press* on 8 August 1945, just two days after the bomb was dropped, in a report titled "Prof Schrödinger's Views on the Bomb". He wished to make clear it should be called a nuclear bomb, not an atom bomb:

> It appears that the discovery has caused even mere astonishment among scientists, physicists in particular, who are in a position to realise its full implications, than it has among the general public. "If the claims made for the atomic bomb are really true," said Professor Schrödinger, "its invention is a tremendous event in the history of mankind. It has been difficult to explain its importance in non-technical language. The release and use of atomic energy, as it has been called, has been a sort of scientific sea-serpent, always cropping up everywhere, but on investigation proving a myth.
>
> They are now talking about the release of atomic energy but in view of the claims made that is obviously quite wrong. It is not a question of the energy that binds the atom but of the energy which binds the

nucleus. These are two very different things: ordinary atomic energy may be involved in ordinary explosions. The nucleus is 100,000 times smaller than the atom but the energy which unites it is immeasurably greater.

Uranium would be probably the easiest substance in which to cause breakdown of the atom nucleus because of all materials its nucleus is the largest and least stable. One of the difficulties of breaking up the nucleus by artificial means is that it is inconceivably small. In uranium, it has reached the limit of stability… Probably what the British-American scientists have done is to contrive a means of giving the uranium nucleus an extra push which made it get through the breaking up process rapidly."[1]

A few months later on 1 February 1946, the intrigued *Irish Press* wrote a report called "An Atom Man at Home". The article featured a photograph of Schrödinger playing chess with his daughter Ruth. He spoke in a matter-of-fact way about Arthur March, domestic matters and the threat of a Russian bomb:

Dr Erwin Schrödinger, the man who has the atomic theory and nuclear fission at his fingertips, was playing chess with Ruth March when we arrived at his home at Clontarf. Ruth is the daughter of Dr. Arthur March, Professor of Theoretical Physics at Innsbruck, Austria, Dr. Schrödinger's friend and colleague. Ruth, now aged 12, has had the 'flu and, convalescing, pitted her wits against the Great Brain in the Ancient Game. That was her first illness since she arrived in Ireland six years ago.

Dr. Schrödinger greeted us, went back to the game, and when he had allowed Ruth to beat him, told us: "She and her mother, Mrs. March, were with us in Belgium when the war broke out. We brought them with us. We hear from Dr. March regularly now. He had an anxious time during the war."

Ruth showed us her dolls, told us she was going to the Holy Faith Convent, Clontarf and then we went to find what the Schrödinger family looks like when at home. Schrödinger works, alone. Mrs. Schrödinger does not worry about the atomic theory. She knits cardigans for her husband and cooks grand Austrian food out of Irish ingredients. Mrs. March helps her and they converse in German or English at will. Only Ruth

Schrödinger with his daughter Ruth in Dublin, 1946.

and Dr. Schrödinger know anything about Gaelic yet. Burschi completes the family. They got him as a collie pup in the Wicklow Mountains and his German (Austrian) name means "Laddie" to you and means chops and caresses to Burschi.

Dr. Schrödinger, who has retired from the directorship of the School of Physics of the Dublin Institute for Advanced Studies, now does much of his work at home. He still is a Senior Professor at the Institute. He recently told the Royal Irish Academy how "one of the nuclei to undergo the reaction (fission) is split by a chance hit of an erratic particle (neutron) and, on fission, gives birth to a number of particles of the same kind (neutrons). Each of these may score an efficient hit on some other nucleus, and will do so, unless it is removed earlier, either in escaping from the reacting body or in being caught in a non-splitting hit. ..."

"Doctor," I asked, "have they atomic bombs now that would eliminate all Ireland? "Oh, I would not say so," said he, quite unabashed, "but watch out if they find some substance like lead, for instance, that disintegrates in the same way as uranium." "That self-disintegration of uranium," I said, "how does it work?" "Quite simply explained," said Dr. Schrödinger. "If there is too great a quantity of uranium together the explosion occurs. So you must have two portions of uranium, neither of them being large enough to explode by itself, but both together being over the limit. Bang them together with sufficient force and —"

He threw up his hands in despair and turned to show us one of the lamp-shades he makes in his spare time.

Schrödinger is a man of peace. The atomic theory on which he works, in collaboration with the great scientists of the world, must be directed towards peace, he says. He wants a world of people who play games with children, chop sticks in the garden, and have dogs like Burschi.[1]

With the war over, Schrödinger was beginning to think about returning once more to his home country of Austria. Although he had been treated very badly when he was last in Austria eight years before, a pension for himself and Anny was always very much on his mind. De Valera had managed to remain as the Taoiseach for the whole period of the war and the Dublin Institute for Advanced Studies was under his protection. However, after receiving a letter from the Austrian Government inviting him to return to his home country Schrödinger wrote to de Valerā in formal terms on 28 April 1946:

Your Excellency.

Excuse me bothering you by this letter. But since it is you to whose personal protection I owe my salvation from the wreck of Central Europe, you must be the first to whom to tell that the Austrian Government has written to me, asking to resume academic activity in Austria, more particularly at the University of Vienna. They add that the interruption of eight years, caused by my dismissal in 1938, will be included as active service for the purposes both of my advances sent in salary and for the imputation of old-age-pensions to myself and to my widow (which amounts to reaching the top scale in all three respects).

The main reason why I have to consider this call very seriously is the question of pensions. I understand that at the Institute it is not yet definitively settled for myself. But much more serious is the prospective destitution of my wife if I die before her (as is probable for general grounds). To face it calmly, in order to spend and end his own life in more favourable conditions and more according to his predilection, would be seen a nasty choice for a man to take.

I had always been protected in this respect by a States pension to the widow, up to my removal in 1938. To take out an insurance then was virtually impossible. Owing to the age I have reached the terms would have been impossible, prohibitive. I am afraid that if I definitively turned down my Government's call, this would be legal ground for my forfeiting all claims for pensions in Austria.

Believe me to be, Sir, your respectfully and gratefully devoted servant. E. Schrödinger.[6]

With the Russian army gathered around Vienna, Schrödinger did not want to return to Austria at this time but he wished to keep all options open and draw the attention of de Valera to the pension issue. In his own brief biography Schrödinger commented:

As early as 1946 I had been offered an Austrian chair again. When I told de Valera about it he urgently advised me against it, pointing to the unsettled political situation in Central Europe. He was quite right in that respect. But while he was so kindly disposed towards me in many ways, he showed no concern for my wife's future should anything happen to me. All he could say was that he wasn't sure what would happen to his wife in such a situation either.

So I told them in Vienna that I was keen on going back, but that I wanted to wait for matters to return to normal. I told them that because of the Nazis I had been forced to interrupt my work twice already and start all over again elsewhere; a third time would certainly put an end to it altogether.[49]

His wife Anny also commented on the challenges of his possible return to Austria:

After the second war, Austria wanted to have him back, you know. Already in '46 there was a state president who was very much interested in physics (Dr. Renner). He tried to get him back but my husband was absolutely right to tell him, "No, I can't come back as long as the Russians are here, because I can't go into a country occupied by Russians. This is impossible." This was quite right. Many people were taken away. They got a beautiful position in Russia, but after all, they were taken away.

Well, it took ten years before we could go back. Then we were old enough and we liked to go back to our own home country, of course. Everybody was so kind here. The government was so kind and everybody was enthusiastic that he came back. He still lectured a year and a half or two years and gave public lectures and he got all the honours. I remember how often he was a member of an academy.[50]

A few months later Schrödinger arranged for his English-speaking Aunt Rhoda to visit Dublin for an eye operation. She commented publicly on "Her Memories of Ruined Vienna" to the *Irish Press* on 27 May 1946:

Mrs. Rhoda Arzberger, aunt of Dr. Erwin Schrödinger, Senior Professor in the School of Theoretical Physics, Dublin Institute for Advanced Studies, arrived in Dublin by air on Saturday, having travelled to London from Vienna in a bomber. She has come here on the recommendation of Prof. Linder, Viennese eye specialist, to undergo an eye operation by a Dublin specialist.

When I saw her at the home of Dr. Schrödinger in Kincora Road, Clontarf, on Saturday, writes an Irish Press reporter, she sighed as she looked out on the sunlit garden and remarked that it was wonderful to be in such a quiet and peaceful place after her harrowing experience in Vienna. She said that Vienna was in terrible ruin. The hospitals were appalling. The Russians, she said, had stripped the city of everything they could lay their hand on.

Until the Russians came, she said, "there were large stocks of food but they were quickly seized, and since then the position was appalling. Peas and potatoes were the principal items of food available but the potatoes were now becoming scarce. It seemed as if the whole world had gone crazy."

Her husband died about a year ago, and for ten days his remains lay in a room in their house because they could get no facilities to bury him. Mrs. Arzberger's mother was English and her father's ancestors were English. She herself was brought up in England, and speaks English perfectly. She never visited Ireland before but says that, from what she has already seen of it, it lives up to Dr. Schrödinger's descriptions of its beauty and charm and she looks forward to recuperating here.[1]

It is clear from his aunt's comments that a return to Austria for Schrödinger was not sensible at this time. The letter to de Valera about his position in Ireland shows that Schrödinger was once again nervous about his future. Quite often when this happened his feelings became known elsewhere and new and unexpected offers came in. This was true, for example, when he first arrived in Dublin and was approached by Allahabad University in India. So on 7 August 1946 he received yet another letter from Tess Simpson at the Society for the Protection of Science and Learning:

> I have had a long conversation with Professor Albert Eckstein who is Professor of Paediatrics at the University of Ankara. He used to be at the Düsseldorf Medizinische Akademie. Professor Eckstein has come to this country for a visit of a few weeks and was commissioned by the Dean of the University of Ankara to contact a small number of scientists over here. I understand that the University of Ankara have already been in touch with you but Professor Eckstein was asked to follow up this matter personally, and to repeat to you some of the aspects of the offer made to you.
>
> The University of Ankara is new, and is only now starting up certain scientific departments. They invite you to come over for one year, or longer, or even permanently. The salary they offer from 1st January 1947 is 1,400 Turkish lira a month. (The lira is 5.33 to the £ sterling) plus free passage for yourself and your family if desired. You have permission to send out to any other country you wish one-fifth of your salary in the currency of the country concerned. The language need offer no difficulty as where required an interpreter will be provided.
>
> Professor Ekstein, who has been in Ankara since 1935, would like to provide you with any information you would like to have from him. He will communicate to the Dean of the University of Ankara any message you care to give him.
>
> I was so glad to see you and Mrs Schrödinger in London last week; I do hope you found your stay in England useful and amusing.[2]

Moving to Turkey in 1946 was not such an unusual possibility as it might seem. Following Hitler's rise to power, Turkey was the most popular destination for refugee scientists outside of the UK, USA and Switzerland. Albert Eckstein himself was an expert on children's illnesses and played a major role in reducing infant mortality in Anatolia in

Turkey. In addition, Erwin Finlay-Freundlich was a distinguished astronomer from Potsdam who in 1933 moved to found the Astronomy Institute at Istanbul University. He had done astronomical observations which were used to test Einstein's theory of relativity.

Schrödinger considered the offer seriously and had a correspondence with the officials at the University of Ankara. In the end on 27 March 1947 he declined the offer in a letter to Professor Dikmen at the Office of the Dean:

> I thank you very much for your letter, which you kindly addressed to me on behalf of the Faculty of Science of the University of Ankara, offering me a chair of physics at this Faculty. I wish to explain that I sincerely appreciate your offer and regard it as a great honour.
>
> Speaking quite frankly and without concealing the egoistic point of view, I would say that, quite apart from the attractive task of being charged with a sort of pioneer work on what is really of the oldest ground of human culture, I am also quite sure that I should spend the rest of my life much more happily in Asia Minor than in the detestable fog and rain of the North of Europe.
>
> However, I am sure you will understand that the political situation which has become ever more pointed since your letter was written does make my decision at this moment a rather delicate one — please do not misunderstand me. I truly wish your country to receive the help that is in prospect, and I consider it in the interest of Europe.
>
> Apart from this momentary difficulty, which may soon be resolved, I am afraid I have to add this. I really do not think I could ever accept a post which is, albeit formally, granted only from year to year and of which the superannuation condition cannot be settled before I have held the post for five years.
>
> If you reflect calmly you will, I am sure, understand this without further explanations. If you do wish to get any person worth speaking of from abroad you will, I am afraid, have to get these conditions appropriately amended.[6]

It is interesting that Schrödinger refers to his preference for fine weather with some similarities to his response when he was approached about a Professorship at the University of Edinburgh in 1936. However, the climate did not seem to concern him so much when he took up new appointments in both Oxford and Ireland. Furthermore, the pension aspect was again very important for him. Schrödinger also refers to the political situation in Turkey at that time. There was then a stand-off between the Soviet Union and the USA on whose navy would have rights to move through the Dardanelles and the Bosphorus. Schrödinger had already previously made more than one foolish move to a country with an imminent and potentially dangerous crisis and he did not want to do the same again.

In the summer of 1947 Schrödinger was also considered for the prestigious Chair that Arnold Sommerfeld had held at Munich and had been transferred to the Nazi Wilhelm Müller in 1939 despite the strong interest of Heisenberg and objections by many physicists. After the war, Müller had been dismissed from the post. However, Schrödinger was not willing to return to Germany at this time due to the uncertain political situation which still prevailed there.[51]

After the war there were several trials of people closely associated with the Nazi regime. This included some prominent scientists. The Nobel Prize winner Johannes Stark had given his name to the Stark effect, the splitting of spectral lines in electric fields, which Schrödinger had so rigorously explained with his wave mechanics in 1926. However, Stark was one of the leaders of the Deutsche Physik and frequently spoke and wrote against Jewish scientists holding academic positions. In 1947 he was put on trial and sentenced to four years of hard labour. Von Laue and other physicists, including Schrödinger, had the view that this was too harsh a sentence.[52] This may have influenced Stark's appeal and in 1949 his prison sentence was transformed into a fine. Wilhelm Müller was also sentenced to prison originally but this was similarly turned into a fine on appeal.

Ruth March, who had been born in Oxford in 1934 and baptised in Magdalen College Chapel, had moved to Austria with her mother at the age of two and, accordingly, had no memories of her time in England. This book is very much about her as well as her father. She wrote in very affectionate terms about her life as a young girl in Dublin:

I spent my formative years in Dublin between 1939 and 1946 together with my mother, my godmother Anny Schrödinger, Lina the maid, Burschie the dog, and Erwin Schrödinger, who is actually my father, although I did not know that at the time.

The school I attended was the Holy Faith Convent at Clontarf. I loved this school. I really don't know why my father sent me there. There were other schools just as near, a National School and small private schools, so the distance cannot have been the reason, for it would not have taken me any longer on my bicycle to reach them. Maybe my father thought this school was as good as any, maybe it was recommended to him; and what he couldn't foresee was that his child would contradict him and tell him she preferred the nuns' view to his.

When my father sent me back to Austria after the war — he stayed on till 1956 — one of the reasons was surely the nuns' great influence. I thought at the time and for many years to come I could never forgive him. Only when I met and married the man who is my dear husband and realized I could never have met him otherwise, did I come to terms with being so far away from Ireland.

For seven years Clontarf was my home, and a farm near Aughrim in the Wicklow Mountains was where I spent my summer holidays. And when asked what I would like to be when I grew up, I said I wanted to marry an Irish farmer. For me those seven years were, apart from the terrible death of my beloved dog, quite uneventful.

The war, the cause of my being in Ireland, did not leave a great impression on me. I feared maybe having to wear those gas masks; the dog always howled with the test-sirens and I remember the night of the bombing of the North Strand. I heard about, but did not see the effects. I was sometimes invited to my school-friends' houses and they came to my parties. I am still in touch with many of these friends. It was a very happy time for me.

During all these years my father, my mother and my godmother lived their lives, which did not affect mine. All I learned from them was that the Nazis were evil, so I refused to speak German with the excuse it was the Nazis' language. My godmother spoiled me extensively and sought my affection. She called herself my second mother and I did not object and thought my mother far too strict. My mother received letters from her husband, who had stayed behind in Innsbruck. There was

> often very little left of them, because the censors did not want any news spread in Ireland or anywhere else.
>
> My mother had always been fascinated by the theatre, and needless to say Dublin offered her great opportunities… My father, too, would ride his bicycle to town. He never went by bus to the Institute in Merrion Square. He nearly always worked after I had gone to bed, frequently up to the early hours of the morning and slept well into the following day. He got very angry if we disturbed him. So we had to be very, very quiet. I had to practise the piano in our neighbour's house or our friend's house opposite.[53]

With the end of the war, Schrödinger was able to make his first visit to England for seven years and attended the Isaac Newton Tercentenary Celebrations at the Royal Society starting on 16 July 1946. This event had been delayed by three years due to the war. Niels Bohr and Lise Meitner also attended and the only scientist still working in Germany who was invited was Max Planck. Planck's last years in Berlin in the war had been tragic when his son was executed after being implicated in the attempt to assassinate Hitler. His house was destroyed in an air raid but he escaped from Berlin and he was essentially homeless for a period. He was to die just one year after his visit to London.

Following the trip to London, in which he also visited Tess Simpson at the Society for the Protection of Science and Learning, Schrödinger made his first visit for seven years to the European mainland and attended an "East meets West" Eranos conferences in Ascona and also the bicentenary meeting of the Swiss Natural Research Society in his old haunt of Zurich.

Ruth and Hilde were now able at last to return to Innsbruck and saw Arthur March again for the first time in seven years. Naturally for a child who had been away for so long, Ruth found the return home to Austria difficult.[53] Ruth's doting godmother Anny, who had no children of her own, also spoke very favourably indeed about her life in Ireland. She said:

> We were very, very happy when we at last arrived in Dublin. It was seventeen years we stayed there. It was really wonderful, peaceful. Really, he liked it very, very much. He was very interested in Gaelic as well,

you know. He tried to learn Irish, but it is so complicated. You have no grammar, no nothing. At the last he gave it up. But he liked it very much.

It was especially nice in Dublin because there he had no duties whatsoever. The whole scientific coming together. The main thing was to teach in the morning and to hold a colloquium or seminar in the afternoon. Never real lectures but always a conversation. He had a very nice room there and they could come and speak to him whenever they liked. They appreciated it very much.[50]

Most of Schrödinger's letters which still exist were either accurately typed or neatly handwritten. Anny commented on his working methods saying that he worked until 11 or midnight and:

He never had a secretary, even if he could have had one, he wouldn't have liked to dictate. He preferred to write the things on his little typewriter. He wrote it himself. Only in Dublin he had a secretary for the institute, but not for his private work. He never used it. For his manuscripts he always had it handwritten as you see.[50]

With Ruth and Hilde living back in Innsbruck Schrödinger was interested in the possibility of returning there himself and wrote in German to Arthur March on 16 October 1946:

Thank you very much for your suggestion. I seem to have suggested to them that they could just as easily put me in Innsbruck as a Professor with extra status as anywhere else — and I would probably prefer that... The main problem today is always that under normal circumstances one would say: well, one simply sells the house here and buys a corresponding one at the new place. But first of all, the latter is unlikely to be possible, secondly, there are a number of currency exports and imports in between permissions and an often fantastic range of fluctuation between official and unofficial exchange rates...

The complete lack of titles of our Austrian colleagues whom I met in Zurich made an indelible impression on me. They had a little "pocket money" for the three days of the congress. I also gather from other, partly printed, publications that incomes at the moment, measured in any old value, are impossibly small, that they only get an incomparably higher internal rate through extensive import control, but that "luxury articles" are outside the realm of reasonable accessibility. What I am now wondering is, on a four-week summer vacation in Selva or the like, what is such a luxury item and is it likely to stay for a long time?

You will perhaps laugh when you see this one thought occupying such a large space. I think a lot of other things too, of course. I love the Tyrol in general, especially dear old Innsbruck, which I have known since childhood. It is definitely closer to the mountains there than in a Viennese city apartment, for example.

The Minister has suggested alternate appointments in Vienna and in Innsbruck (semester by semester). Reading is now technically a bit difficult for me but I have to first think through things properly. The present letter is my first reaction to the ideas.

Many greetings, and to Hilde and Ruth and also from Anny. Your old friend.[6(t)]

Then on 24 October 1946 Arthur March responded. It can be seen that his correspondence with Schrödinger was, as usual, helpful and very friendly. There was no animosity at all about Hilde or Ruth:

Dear Erwin!

Thank you very much for your letter. I should of course have asked you first before I proposed the plan to the Minister. But when I happened to meet (Minister) Hurdes in Alpbach, where a school was taking place, and since Hilde had told me that you would rather be in Innsbruck than in Vienna, I explained the advantages to him…

I do believe that you will feel very comfortable here… Last year I made the most of the favourable opportunity to gain first-class staff for the vacant positions. We have set up various colloquia in which physicists, chemists, natural scientists and philosophers take part and which are very stimulating.

Unfortunately, however, the living conditions never seem to be such that I can advise you to relocate in the near future. We live very well personally, but that is only possible because last year I was lucky enough to get a nice apartment. But it also costs us constant effort to create what is necessary for life; and certain things that are not absolutely necessary for life, but which a mentally working person cannot forego, such as tobacco and beverages, are barely available in sufficient quantities...

It will be better one day, but one can only recall the problems after the First World War. The ministry will not want to wait that long and the question is what could be done about it. Isn't it possible to have brief guest roles in the meantime? I would imagine that you would take a leave of absence for 2 or 3 months and hold lectures here during this time. I think the Ministry would be happy with that for the time being. Hess wants to do it, after a letter I received, and will only spend a visiting professorship here, because he does not want to lose his American citizenship...

Ruth is once again quite sick with the doctors. A fifth has been consulted already and is unable to say what she really lacks. She does not come out of a mild fever and could not be treated for lack of a diagnosis. But she is quite amused and receives the many visitors with the charm of a film diva.

With the warmest of greetings also to Anny. Arthur.[6(t)]

This correspondence with Arthur March and the Austrian authorities continued and Schrödinger was able eventually to take a sabbatical in Innsbruck during 1950. The letter from March also refers to Alpbach which is a beautiful small town not far from Innsbruck. This soon became Schrödinger's favourite place and home in Austria.

In the late 1940s there were several new arrivals at the Dublin Institute for Advanced Studies. Neville Symonds worked with Schrödinger on theoretical biology. He eventually became a Professor of Biology at the University of Sussex. There he lectured in 1972 to undergraduates, including the writer of this book, on the fundamental principles of genetics and emphasised the importance of Schrödinger's *What is Life?* book in bringing people like him into biological science.

Another arrival was Cornelius Lanczos who had once worked with Einstein and had emigrated to the USA where he was eventually forced to leave due to his communist links. He became well-known in developing efficient numerical analysis techniques and was eventually to take over the Directorship of the Dublin Institute for Advanced Studies. Heitler left Ireland to return to Zurich in 1949 and Lajos Jánossy also departed to go back to Hungary. He seemed to do this in a hurry and there was some discussion in the *Daily Mail* as to whether he may have been too close to rogue physicists such as Klaus Fuchs.

In the meantime, Schrödinger was continuing his major project on attempting to unify Einstein's theory of relativity with Maxwell's theory of electromagnetic radiation.[54] On 28 January 1947 he spoke to the Royal Irish Academy on what he thought was a breakthrough in his research. As usual, the *Irish Press* reported in enthusiastic terms on his lecture:

> Twenty persons heard and saw history being made in the world of physics yesterday as they sat in the lecture hall of the Royal Irish Academy, Dublin, and heard Dr. Erwin Schrödinger declare, simply and slowly: "I have the honour of laying before you today the keystone of the Affine Field Theory and, therefore, the solution of a 30-year old problem the competent generalisation of Einstein's great theory of 1916 . . ." The Taoiseach was in the group of professors and students. And, in that twenty, four races were represented — White, Black, Brown, Yellow.
>
> Dr. Schrödinger, the Austrian-born Professor of the School of Physics in the Dublin Institute for Advanced Studies (he resigned the directorship of the School some time ago to allow himself more time for research) confined himself to the language of the physics classroom in his explanation of the theory, which summed up (he told me) relates the Einstein Theory to Electro-Magnetics while it had hitherto expressed gravitation only and did not comprise anything else.
>
> The Einstein Theory, he said, was "purely geometrical", now electromagnetism must be included. The extension of the general theory of relativity had been "the aim of Einstein and others, including myself, for thirty years," said Dr. Schrödinger. During his lecture, he said, having placed his solution on the blackboard: "I am prepared to see some of my mathematical friends among you shaking their heads and saying "If it is as simple as that, why did not the fool try it before? Is not that square root the simplest, the most suggestive Lagrangian, that anybody would of course try first?"...
>
> When I asked him if he himself was quite confident that the theory was really the solution of the thirty-year-old problem of the generalisation of the Einstein Theory, he said: "This is the generalisation. Now the Einstein Theory becomes simply a special case. Just as the parabola

described by a stone thrown directly upwards is a special case in the general idea of a parabola. And," he added, "I believe I am right. I shall look an awful fool if I am wrong," referring to the fact that he had discovered that the solution had an "affine basis". Notwithstanding that Einstein had held otherwise, he said: "I beg my younger Fellows of the Institute to take this as a lesson — never rely on an authority in science. Even the greatest genius can be wrong — whether he has one, two Nobel Prizes, or none."

But it was the opening statement of Dr. Schrödinger's lecture that still had most point for one of the "men in the street". It was: "The nearer one approaches truth, the simpler things become." Because, in contrast to that, the closer one is to history, the harder it is to write it down.[1]

The news of Schrödinger's "breakthrough" was then reported in many other newspapers and magazines around the world. Einstein, however, was not happy and felt this was an attack on his own theories. He sent a statement to the *Irish Press* which was published on 24 February 1947 with the headline "Einstein's Comment on Schrödinger claim":

Professor Einstein, in an important declaration in New York yesterday, in what he called general remarks about the present state of physics, denied that the foundations of physics are determined at present and said that as a rule discoveries of that importance could only be reached through the untiring labour of several generations of scientists...

"This is purely a mathematical problem. Prof. Schrödinger's latest attempt must be considered as such an effort. It can, therefore, only be judged on the basis of its mathematical qualities but not from the point of view of "truth" and agreement with facts of experience. Even from this point of view it can see no special advantages over the theoretical possibilities known before.

I want to stress the following point: It seems undesirable to me to present such preliminary attempts to the public. It is even worse when the impression is created that one is dealing with definite discoveries, concerning physical reality. Such communiques issued in sensational terms give the lay public misleading information about the character of research. The reader gets the impression that every five minutes there is a revolution in science somewhat like a coup d'état in some of the smaller unstable Republics. In reality one has in theoretical science a process of development to which the best brains of successive generations add by untiring labour and so slowly lead towards a deeper conception of the laws of nature. Honest reporting should do justice to this character of scientific work."[1]

Sadly, Schrödinger took the bait and responded at once to Einstein through the *Irish Press* on 1 March 1947 in a statement entitled "Schrödinger replies to Einstein":

> Prof. Schrödinger stated in his lecture that he had solved the 30-year-old problem of the generalisation of Einstein's Theory of Relativity and the statement was commented upon by Einstein in an interview in the United States.
>
> Dr. Schrödinger states: "In his comment on my latest work, Professor Einstein has denounced 'such communiques issued in sensational terms' which 'give to the lay public misleading information', *etc.* I wish to state here that neither before nor after his comment was published, Professor Einstein can have seen one printed line published under my name on the subject in question. It is therefore not quite clear whom he decries.
>
> A meeting of the Royal Irish Academy can hardly be referred to as the lay public, nor can the Press, habitually admitted to these meetings, be blamed for reporting on them in such detail as they choose. And surely Professor Einstein is the last to dispute an academician's right of reporting to his Academy and giving his opinion freely.
>
> Or ought the academician to ask for exclusion of the Press whenever their faithful report might possibly displease anyone? To vent such displeasure in the daily papers is unusual in scientific circles. But if somebody finds it after all unavoidable, he ought to state frankly which published article or articles he accuses of not conforming with his own ideas of "honest reporting"."[1]

It is unfortunate that the deep friendship between two of the greatest scientists of the 20th century should have been marred in a public dispute in the open press as opposed to the more dignified medium of papers in peer-reviewed scientific journals. However, neither Schrödinger nor

Einstein held long-standing grudges and they did subsequently correspond again, as they had so many times before, in friendly terms.

In June 1947 Schrödinger received a letter from the Dean of the Faculty of Mathematics and Natural Sciences at the University of Berlin asking if he would be interested in returning as Emeritus Professor. On 24 June 1947, he replied:

> Dear Dean! I have to apologize for the fact that I am so late in replying to your kind letter for which I thank you very much. The years at Berlin University are among the happiest of my life. I keep an eye on the possibility of returning there, even if only as an emeritus. It was a great pleasure for me to be thought of, in addition, to reactivate me, when I first heard about it from a very gracious address from the Rector on June 10, 1946. Nevertheless, for such a decision I have to wait for the situation to be clarified and improved.
>
> In the middle of the years in which a scientist's work tends to develop in the most intense and normal way I was thrown off track twice by the unfortunate times when I left Germany in 1933 and Austria again in 1938, to withdraw from despotism. Both times I escaped, the second only with a suitcase, all of my belongings were left as they are today, as long as they have not been looted in the meantime. I state this to justify the fact that I could only undertake a third transplant in completely clear and secure circumstances, a renewed struggle with still adverse circumstances would consume too much of the rest of my strength...
>
> For the time being, I have to be satisfied with my ideal affiliation to your university. In the above-mentioned letter from the rectorate (dated 10 June 1946) it said that all members of the renewed university and confirmed members of all teaching staff will be issued a printed certificate, and that one will be sent to me. I have not yet received it and I am indeed very happy about it.[55(t)]

Then on 6 November 1947 Schrödinger was granted Emeritus Professor status at the University of Berlin.[55] Encouraged by the reaction of some leading biologists to his *What is Life?* book, Schrödinger was contemplating even more complex issues such as science related to the mind. He was very impressed by the book published in 1940 by Sir Charles Sherrington, the Nobel Laureate from Magdalen College,

Oxford on *Man on his Nature.*[56] On 25 September 1947 he wrote to Sherrington from the Dublin Institute for Advanced Studies in enthusiastic terms:

> Dear Sir Charles,
>
> I believe that the mind is attached somehow to new events in a living system. New inventions, new decisions are in the spotlight of mind. Once they are practised and memorised, they slowly disappear in the dark sink below the threshold. We notice this every day. The same course of activity that was originally intensely conscious gradually turns into reflex in proportion of its being well practised and "learnt by heart".
>
> I mean this ontogenetic fact to apply, very probably, to phylogeny. In man mind is attached to brain events, because only they are confronted with new situations. With us brain has become the organ of paramount biological importance; practically the only one that undergoes further development.
>
> These are old ideas of mine, which come to the surface under the spell of your magnificent chapter on the brain and its work in "Man on his Nature". This chapter is not just an excellent popular exposition of a difficult scientific subject. It is a sublime work of art (Kunstwerk) — that art which shapes an organic Whole not of colours on marble or human fate and feeling, but of the delicate material: thought, conceptions and ideas.
>
> Yours very sincerely.
>
> Erwin Schrödinger (Fellow of Magdalen 1933–38).[6]

Schrödinger eventually went on to publish his ideas on the working of the brain in *Mind and Matter* which he also gave as the Tarner Lectures at Cambridge University in 1956. In those lectures he stated: "I cannot convey the grandeur of Sherrington's immortal book by quoting sentences; one has to read it for oneself."[49] Sherrington had also been a founding member with Schrödinger of the Pontifical Academy of Sciences in 1936.

John Eccles was also a Fellow at Magdalen College at the same time as Schrödinger and Sherrington, and he went on to win the Nobel Prize for Physiology or Medicine in 1963 for his work on the synapse. Like these two great scientists, Eccles was elected to the Pontifical Academy of Sciences. In 1966 he organised a meeting of the Pontifical Academy on

"Brain and the Conscious Experience" which was dedicated to Sherrington and Schrödinger.[57]

In 1949 the BBC had remembered Schrödinger's previous broadcast in 1935 on "Equality and Relativity of Freedom" and invited him to do another. The BBC recording of this lecture is still available and gives a fine example of Schrödinger's ability to speak in clear and almost perfect English. This time the title was "Do Electrons Think?" and he again took the opportunity to refer to Sherrington:

> What is the present attitude of physiology to the question? We may take for it the word of its great master, Sir Charles Sherrington. According to Sherrington, there is no boundary between the animate and the inanimate. The same laws of physics and physical chemistry hold within the living body as outside. The most careful investigation of the physiological processes, in the nerves and in the brain, reveals no leverage whereby mind could take direct influence on matter: Mind, per se (that is, by its very nature), cannot play the piano — mind per se cannot move a finger of a hand.[6]

Born

The administration required of being Director of Theoretical Physics at the Dublin Institute for Advanced Studies was getting too much for Schrödinger and he had handed the job over to Heitler in 1946. Heitler then had a grand plan to appoint Max Born to the Institute. With Schrödinger and Born in post, the Dublin Institute would almost have rivalled the Institute for Advanced Study at Princeton where Einstein and Weyl were based. Born took the approach from Heitler very seriously and had a somewhat clandestine meeting with him in Cambridge in May 1947 to discuss details. Born described these negotiations in a long letter to his wife Hedi on 10 May 1947 and he told her that in Dublin "the heating with peat seems not too bad".[15] He also suggested she could write to Anny Schrödinger to see if she could stay with her for two or three days to see Dublin for herself. However, Born was approaching the age of 65 and would have needed to stay in post for ten years to obtain any kind of pension. For this reason he declined the approach.

Born had also been invited by President Tizard to give the prestigious Waynflete Lectures at Magdalen, Schrödinger's former College in Oxford. He agreed to stay for two months early in 1948 and he chose the title "Natural Philosophy of Cause and Chance" for his lectures.[58] The overall theme of the Waynflete Lecture Series was the boundary between science and philosophy, the subject at the heart of the Magdalen Philosophy Club. This visit to Oxford also gave him an opportunity to see more of his children and grandchildren who were based nearby.

Born's lectures were given in the Great Hall of Magdalen and were all sell-outs with standing-room only. Together with the new President of the College, Tom Boase, he had to push his way through the crowd to get to the podium. The main theme of the lectures was how chance can follow rules and why cause-effect relationships cannot always predict the future. He took examples from astronomy, electromagnetic fields, thermodynamics, the kinetic theory of gases and quantum mechanics to emphasise how physical theories are developed. It is unfortunate that Schrödinger had not been asked to give the Waynflete Lectures when he first arrived at Magdalen College in 1933. It would have been an ideal way to introduce him to the Oxford community of scholars.

On nearly every day of his stay at Magdalen, Born wrote long, detailed and affectionate letters about his colourful life in the College to his wife Hedi, who remained in Edinburgh.[15] Some of the letters were in English and some in German. He interacted with many of the Fellows who had known Schrödinger some 15 years before. He lived the life of a bachelor Fellow which enabled him to take part in more College activities than was the case for Schrödinger. He was well looked after. The President of the College Tom Boase invited him to dinner in his President's Lodgings and Born wrote about the event to Hedi on 20 February:

> On Wednesday evening there was a dinner party at our President, who is a bachelor. I was the only professor from our College, the others were all heads of Colleges: the Master of John's with his wife and two lady Principals, of Somerville and St Hugh's, very elegant women. I have never been at such a luxurious party in such a lovely house for the last 20 years! Two footmen serving incredibly good food, wonderful tapestries and old pictures on the walls, absolutely fantastic.[15]

The Fellows invited Born to the Magdalen Philosophy Club, which had also been attended by Schrödinger during his time in the College. Born said to Hedi:

> The local philosophers, Prof. Ryle and Weldon, and the physicist Griffiths have a discussion evening every Monday on philosophy, mainly about the foundations of science. They are convinced to be very modern and smart. They took me to the room of Griffiths where six students appeared, talking to us on a vague kind of philosophy about the meaning of scientific notions, *etc.* But it was quite clear that none of them, apart from Griffiths, know what scientists are doing. They all spoke with a low half veiled voice, a little mystical and dark. It was long and tedious. I took only little part. If philosophy is treated here everywhere in this dull, ritual, monotonal way I understand why they flock to my lecture.[15]

Some 14 years before, Schrödinger had intervened at a Magdalen College Meeting to propose James Griffiths for a Fellowship by Examination. Griffiths was clearly now moving up the hierarchy at Magdalen to be elected eventually President of the College. The chemist Leslie Sutton, who was the College Vice-President and who had previously much lamented Fritz London's departure from Oxford, invited Born to his rooms for a glass of beer with distinguished guests. This included Linus Pauling (who had overlapped with Schrödinger in Zurich in 1926 and was in Oxford for the year on an Eastman Professorship), Henry Whitehead (who Schrödinger had stayed with after his escape from Austria in 1938) and Sir Robert Robinson (who had just won the Nobel Prize for Chemistry and was President of the Royal Society). Born said that "the discussion was most lively and quite amusing. People here have a peculiar alertness and wit".[15] Sutton also introduced Born to the undergraduates studying Chemistry who impressed him more than the philosophy students. Following a discussion on chemical forces, he described them as "bright, young boys". Born also attended a service in the College Chapel and said: "It is very romantic, all dark except candles, choir boys and men in white robes, lovely gothic architecture, beautiful singing, but it does not agree with me, particularly the lessons".

On a busy 9 March, Born visited his former research assistant Klaus Fuchs at the nearby Harwell nuclear facility. Fuchs had given the impression to Born that he was bored by the work at Harwell. Born had replied that he could find a professorship somewhere to which Fuchs had given the ambiguous response: "he had a greater task at Harwell".[59] Afterwards, a military car then took Born back to dinner at Christ Church with Lindemann (now Lord Cherwell) and Heisenberg, another previous research assistant of Born, who was being permitted a rare visit to Oxford by the British military authorities.

Just four days after Born's visit, Fuchs had a meeting with his Russian contact Major Feklisov to whom he gave key information on the new American research on the hydrogen bomb.[60] Some two years later, Fuchs had another visitor at Harwell. This was the MI5 "gentleman interrogator" Jim Skardon and this resulted in a sentence for Fuchs of 14 years in prison for espionage of nuclear secrets.

Born attended Magdalen High Table frequently where he met C.S. Lewis who "looks and speaks neither religious or poetical but is a youngish fat little man with humanistic eyes and some wit".[15] Every morning Born walked around the extensive College garden paths known as Addison's Walk. There he often met Lewis and they once "had an interesting talk about the sun and the earth, and what is going on upon this planet, and our work and his writings which I do not know more than he does know mine, and about the budding green and the spring flowers. But suddenly the clock on Magdalen Tower chimed 10, and he spurted away to his pupils".[15]

Born, who like everyone in Britain at the time had a ration book, enjoyed breakfast in the Common Room in the "lovely cloisters" where "the rule is not to talk to each other and we have real egg and bacon. There are newspapers, I take the *Manchester Guardian*".[15] However, unlike Schrödinger, he did not take to the dessert after dinner in the Senior Common Room which he described as "lit only by real candles, eating some fruit and nuts, and port or madeira. I shall reduce this to a minimum as it does not agree with my stomach".

Like many visitors, he was also critical of the heating in the College "in my room with the electric fire on, only in my coat and the bedroom is so cold I need a hot water bottle". Born, however, often enjoyed playing Brahms, who had been a personal friend of his parents back in the good times in Germany, on the "fine College Bechstein pianos". On 4 March 1948, Born also wrote to Einstein about his visit to Magdalen, speaking favourably of the "very beautiful Oxford College with (relatively) good food".[61]

Born must have been impressed by Magdalen as he sent his son Gustav to the College to undertake graduate study for a D.Phil. degree. Gustav worked in the research group of Howard Florey at Oxford who, as a Rhodes scholar at Magdalen, had been a research student of Sherrington. During the war Florey had, with the German refugee Ernst Chain, famously discovered the clinical action of penicillin which was to win them the Nobel Prize for Physiology or Medicine in 1945. Gustav graduated with a degree in Medicine and as a doctor saw service just after the war in the very difficult situation in Hiroshima after the explosion of the nuclear bomb.

Gustav Born went on to have a distinguished career in Pharmacology. He was appointed to a Chair at Cambridge and was elected a Fellow of the Royal Society.[62] He was aged 12 when his father had to leave Germany and he had memories of meeting Schrödinger and Anny in the Tyrol in the summer of 1933. Right up until his death in 2018, Gustav Born spoke regularly and eloquently about the important work of the Academic Assistance Council in helping his father and many other scholars, including Schrödinger, during the crisis in the 1930s and why such organisations continue to be important in the present day.

One of the most distinguished elections for a scientist who is a citizen of Britain or the Commonwealth is to Fellowship of the Royal Society. This allows for FRS to be added after a surname. Max Born, Francis Simon and Rudolf Peierls had taken British citizenship and were elected as Fellows in 1939, 1941 and 1945, respectively. Klaus Fuchs had also become a British citizen. Shortly after Born's visit to see him at Harwell, Peierls and John Cockcroft, the Director of Harwell, nominated Fuchs to the Fellowship with the citation: "There is hardly a theoretical problem in the atomic energy field in which our knowledge has not been widened considerably by his work, or by work under his guidance and inspiration."[59]

In retrospect, the Royal Society must have been highly relieved that this nomination was not successful.

After the Second World War the Fellows also had discussions about electing distinguished scientists from overseas to Foreign Membership of the Royal Society, which was a particularly select group. This had already occurred with Einstein in 1921 and Planck in 1926, and the Fellows were keen to elect more highly distinguished theoretical physicists as Foreign Members.

Names of Nobel Laureates such as von Laue, Heisenberg and Schrödinger were considered towards the end of the 1940s. However, Max Born opposed Heisenberg, even though they had collaborated in their pioneering research on quantum mechanics.[30] Simon and Peierls also did not support Heisenberg as they still had concerns on his work with the Nazi nuclear programme.[30] Born had written to Simon during the last days of the war on 24 April 1945:

> Concerning Heisenberg I had further news about his bad behaviour. Two Belgian scientists were here, de Hemptinne and Junger, from Louvain. They told me amongst other things that H. has visited Holland during the first half of the war just after Stalingrad at the invitation of pro-German professors, and there he met Kramers (who did not belong to these but could not avoid meeting him). They spoke about the war and Heisenberg admitted that it was probably lost (Stalingrad!) but added "well, the next war we will certainly win".
>
> Dirac has entered H's name in the book for proposing Foreign members but without my signature. I have told Dirac my standpoint in very clear words in a letter. I have also written to Niels Bohr and described to him the "appeasing" or at least un-interested attitude of the R.S. and other people... All the old tricks are beginning again, because the wealthy people are frightened by the socialists. Whittaker openly says we must join hands with the Germans against the Russians. What a world![15]

However, as time went on, Born's position on Heisenberg softened. He wrote to Peierls on 13 June 1946:

> About 2 years ago, when I was a member of the Sectional Committee for Mathematics for the Royal Society, Dirac asked me to join him in

> proposing Heisenberg for election as Foreign Fellow of the R.S. After some consideration I refused and kept to this decision in spite of Dirac's persuasion who rightly said that Heisenberg's discovery will be remembered when Hitler is completely forgotten. I said I would reconsider the matter after the war when I had the opportunity of knowing exactly how far H. has collaborated with the Nazis.
>
> Meanwhile I have heard contradictory reports. He was certainly not behaving like Laue and Hahn, but it is said he tried to obstruct the development of nuclear explosives. I do not know whether this is true. My feeling now is that these fellows have got their punishment. They are not only having a hard life and little to eat but their conceit is badly shaken. In my letter to Dirac I wanted to show that I am prepared to reconsider the question but if you object and the people of the occupied countries object I shall not move any further.[15]

There was much discussion also on the case of Schrödinger. His statement that had been published in the Graz newspaper seemingly supporting the Führer was still fresh in the minds of the scientific refugees. Rudolf Peierls had become highly respected by the scientific community in the UK and USA after his memorandum with Frisch on nuclear weapons. Born wrote to Peierls on 22 May 1948 asking for support for Schrödinger's case:

> I got today information from the Royal Society about the recommendations of names for election of Foreign Members. Apart from Cripps, and two names unknown to me, there are Brouwer and Pauling. But not Schrödinger. Dirac and I have tried for years to bring him into the Royal Society. The main obstacle is the letter he wrote in Graz, in which he expressed his agreement with the Nazis (or something like that; part of it was, I think, published in *Nature*). On account of this silly document a group of people have prohibited his election. The formal difficulty is that he is not British; hence he cannot be an ordinary Fellow. And as he lives in Eire which is regarded as part of the British Empire, he cannot be a Foreign Member. Now the latter obstacle has been removed, as far as I know, by a special decision of the Council.
>
> When I heard that Heitler was to be elected I wrote some strong letters to members of the Sectional Committee, saying that it would

be an affront to Schrödinger, if Heitler would be in the Society and he not. Actually I have a very high opinion of Heitler, but still I think Schrödinger is of a higher order of magnitude. There is hardly any paper in theoretical physics in the world where not a Schrödinger equation is used. He has actually revolutionised our science. That he was not very lucky in recent years seems to me of less importance. Planck has also not done any fundamental work after 1900. Therefore I feel rather strongly in this matter. I should like to know what you think. It is a great pity that Dirac is not here; I think I could agree with him on some drastic action. But now I do not know what to do. I am quite aware of Schrödinger's shortcomings and of the enmity he has accumulated through his own behaviour. But I think that all this ought not to matter in the question of election to a purely scientific society.[22]

Peierls quickly responded to Born on 29 May 1948 in negative terms:

As regards Schrödinger, there is, of course, no doubt whatever about his merits as a scientist, and surely no member of the Royal Society Council can have any doubt that he is much more eminent than practically any Fellow or Foreign Member recently elected. The issue is, however, whether scientific distinction is the only factor to be taken into account in the election. As regards Fellows that is perhaps the case. Schrödinger, however, does not appear to be eligible as a Fellow and election of a Foreign Member is a rather different story. It is in many ways analogous to conferring an honorary degree, where the personal record is most certainly taken into account. I have understood the decision of the Royal Society to mean that this is their interpretation.

On grounds of personal record I think there is a very strong case against Schrödinger. The famous Graz letter is only one example, but it is bad enough. I would not pass it over by just describing it as "silly". Our distress at events in Germany surely was so bitter just because there were so many people who failed to understand the seriousness of the issues, and the importance of personal integrity. If a man with a world-wide reputation says what he knows not to be true for reasons of expediency, it is not merely silly. Admittedly most of us would say or sign such things under strong pressure or in acute danger. But we would expect to pay for it, and in any event would presumably disassociate ourselves from what we said as soon as it was safe to do so.

To my knowledge, Schrödinger has never troubled to explain this letter was written under pressure. (As far as I know it was not very severe pressure at that). About other similar things I only know from hearsay, such as his retaining his name on the books of the German Legation in Dublin practically as long as such Legation existed. It all adds up to a consistent impression of irresponsibility. At a time when scientists are so much in the public eye and when their words count more than ever, standards of behaviour must be particularly severe.

The only case you could make would be to claim that the election of Foreign Members should be based exclusively on scientific standing. But this would have to be applied consistently. Would you take the same view if Heisenberg's name were proposed? Or in the hypothetical case that Weizsäcker had done work of comparable importance, would you take the same line in his case?[22]

Born was annoyed and replied at once to Peierls on 1 June 1948:

In the matter of Schrödinger, I cannot agree to your standpoint. You will know that some of the Scientific Advisors of the British Military Government in Germany have no objections to collaborating with former Nazis, and that they even treat them better than those people who have stood firm against Hitler's regime. If you want to know details, ask Simon at Oxford. Under these circumstances, it seems hypocritical to put some weight on Schrödinger's letters from Graz.

I entirely agree that he has certainly not been under severe pressure; but a sensitive and excitable fellow like him must have felt even a slight terror or stronger than a more robust man. Anyhow, it is easy to be severe against such things if one, oneself, has not suffered under any terror. I am quite sure that future generations will find it very strange that S., though living in Ireland, is not a Fellow or Foreign Member of the R.S.[15]

Born would not give up and wrote on 20 June 1948 to Francis Simon:

Schrödinger has again not been elected Foreign Member of the R.S., though they have changed their statutes to make it possible. There must be strong opposition, and I found out that Peierls belongs to it. I think it is sheer hypocrisy, as "Nazi generals" and such folk are great favourites with our representatives at the German universities, like Fraser.

Schrödinger is a difficult fellow and he gets on people's nerves. But in the whole world's physical literature there is hardly one paper without a "Schrödinger equation". Later generations will be baffled by the fact that his is not connected to the R.S. and whatever the real reasons, the blame will be given to the R.S. It has happened before. I think our time is thoroughly philistine. I hate the Nazis like anybody but I can make a distinction between a little mad genius and a scoundrel. But I do not know what to do in the matter.[63]

Simon, however, also had reservations and agreed with Peierls. He replied to Born on 23 June 1948:

A few words about Schrödinger. I am not of your opinion that his treatment by the Royal Society is an outrage. Of course he should be elected a Fellow of the Royal Society and actually we had him first on the Sub-Committee's list for the last two years. It turned out, however, that he is not eligible because he has not got British nationality. He preferred to sit on the fence and see what the outcome of the war would be. This is hardly a point for which one can blame the Royal Society. I am quite sure that if the Germans had won the war that Schrödinger would have written another letter saying that he had been mistaken and that after all the Nazis were the right kind of people.

I myself suggested at the Committee that foreign membership would be the way out, and as you know, the election rules have been changed in order to make it possible to elect him. But people are not in a desperate hurry about it and I must say I agree with them. To qualify for foreign membership it is not sufficient just to have the scientific standing — there are at least ten times as many people who would qualify if this was so as could be elected. The person must have some additional merits and you must admit that Schrödinger has not made it very easy for them to see his specific merits.[63]

Simon then goes on, in rather strong terms, to recall Schrödinger's time in Oxford:

I do not think you fully realised how he behaved when he was in Oxford. Everything in England was wrong from the bicycle brakes and door knobs to more important things and only things in Germany were right. He freely commented on these things to people who wanted to hear them and those who did not. He was a menace to neighbours, not only because of his complicated matrimonial affairs about which he wanted everyone to know — actually he seemed to be very proud of it, but also in many other matters where he behaved absolutely ruthlessly. Thus, when we were looking for a theoretical physicist it never came into anyone's mind to ask Schrödinger.

You mention Peierls is working against him. I do not think that Peierls has any influence in this matter. I think it is much more probable that Robinson, who was at the same college as Schrödinger in Oxford, has all the information himself. I am quite sure Schrödinger will be elected a foreign member in a year or two and I think the delay is entirely his own fault.[63]

The letter from Peierls to Born of 29 May shows that Simon was incorrect in his comment that Peierls did not oppose Schrödinger's election. However, two pioneers of quantum mechanics, Douglas Hartree and Paul Dirac, eventually facilitated the election. Hartree had given his name to Hartree-Fock theory which is a powerful and widely used wave-mechanical method for calculating orbitals for many-electron systems. He had also been a pioneer on developing very early electronic computers in Manchester and Cambridge and had taken part in a major debate with Alan Turing on the prospects for artificial intelligence.

Hartree was Chairman of the Sectional Committee that made proposals to the Council of the Royal Society for election of candidates in Pure and Applied Mathematics. The Sectional Committee consisted of experts in the field but also included two members of the Royal Society Council who acted as observers. On 22 December 1947, Hartree had written from the Cavendish Laboratory at Cambridge to Dirac who was then on sabbatical leave at the Institute for Advanced Study in Princeton:

I am writing to you as Chair of the Royal Society Sectional Committee for Mathematics to ask your view on this year's candidates...

> Schrödinger was put top of the Sectional Committee's list last year, and I think the year before, but was not accepted by the Council; I had the impression last year that both Council members were antagonistic to his claims for consideration; the attitude being that one hunch, however good and however important the development from it, was not enough of a claim and needed more following up with sustained evidence of ability. My own view is the one the Sectional Committee took.
>
> The question of Schrödinger's standing will certainly come up again this year, and I would very much appreciate a statement of your views.[64]

Many scientists working in quantum mechanics would consider the "one hunch" sentiment implied here from the Royal Society Council members on Schrödinger's great contribution as completely outrageous. However, there is something of a brutal truth in it.

Hartree realised that support from Dirac would be essential to guarantee Schrödinger's election to the Royal Society. Nevertheless, the election did not happen in 1948 and Walter Heitler was elected FRS that year which, as Born states, is likely to have been disappointing to Schrödinger. However, in 1949 Schrödinger was finally elected as a Foreign Member of the Royal Society for "his contributions to the quantum theory". His name was put to the ballot of Fellows on 7 April. The election was formally confirmed and signed on 12 May by the Council headed by its President, Sir Robert Robinson of Magdalen College, Oxford.[65]

Foreign Membership of the Royal Society was the first time Schrödinger had been acknowledged in the 1940s in a significant way by a major British institution and he was pleased. On 18 May 1949 he wrote to Dirac from Dublin to thank him for his support:

> Many thousand thanks. It was awfully good of you to manage this difficult problem in this fashion. You really are very nearly a saint. Any non-saint would have got cross with me and done nothing more.
>
> I am very glad and grateful for this gracious form of clinching my sentimental attachment and true devotion to British Science and Learning.[64]

As was often the case with Schrödinger's communications, there was a touch of irony in the last sentence of his letter. His old friend Hans Thirring wrote from Vienna on 12 May 1949, the same day as Schrödinger's election, to say he had just heard on the radio that he had been elected "F.R.S." and conveyed his congratulations.[6] Schrödinger responded to correct him:

> For my part, thank you for your congratulations to the Royal Society — not F.R.S., by the way, but For.Mem.R.S. (I copy the strange abbreviation of the secretary's letter to me, as the foreign member should write it).[6(t)]

Also elected as a Foreign Member at the same time as Schrödinger was Max von Laue. Although he had stayed in Germany during the war, von Laue had shown his disapproval of the Nazis and had kept away from any war research, being, to a certain extent, shielded by his colleague Max Planck in Berlin. Laue had the firm support of the refugees who had previously been elected FRS such as Born, Simon and Peierls.[30] After the war, von Laue moved to Göttingen and was made Director of the Max-Planck Institut für Physik which had replaced the Kaiser Wilhelm Institut.

In due course, Schrödinger passed through London and on 1 February 1951 signed the Royal Society Charter Book, often called "the roll", which contains the signatures of Fellows and Foreign Members of the Royal Society from its Foundation to the present day. The first signature in 1663 is of King Charles II and then his brother who became King James II, and after that Prince Rupert. Then the very early Fellows have signed including Isaac Newton and Robert Hooke. Later on there is the signature of James Clerk Maxwell and, a few pages later, Erwin Schrödinger. So the inventors of the great equations, Newton, Maxwell and Schrödinger, have all signed the same book. On the same page as Schrödinger there is the very neat signature A.M. Turing, and also two Fellows of Magdalen College: the physical chemist Leslie Sutton and the Nobel-winning physiologist John Eccles.

Schrödinger was elected to many prestigious academic societies including those based in Berlin, Boston, Brussels, Dublin, Lima, Madrid, Moscow, Munich, Rome and Vienna but his election in these cases was not nearly so complicated as was the case for the Royal Society of London. Despite the initial opposition from Born, Simon and Peierls, Heisenberg, who, like Schrödinger, was also strongly supported by Dirac, was eventually elected a Foreign Member of the Royal Society in 1955, the same year as Lise Meitner. In 1968 de Valera himself was elected an Honorary Fellow for his establishment of the Dublin Institute for Advanced Studies.

The letter from Peierls to Born mentions Schrödinger's ambiguous interaction with the German Legation in Dublin during the war. Ruth Braunizer (née March) noted that her father needed to be careful not to upset de Valera.[53] An excerpt from Schrödinger's diary on 17 December 1940 sheds more light on this matter:

> Conversation with the German Ambassador Hempel... With friendly demeanour he is trying to get me to declare myself friend or foe of the Führer in no uncertain terms. "This is just between you and me. Feel free to speak your mind, Professor." This went on for almost an hour, painlessly monotonous and with only the slightest common deviations from the subject, the same question twenty times over. The events of 1933, 1936 and 1938 were discussed in accordance with the true facts, *i.e.*, that I officially left Berlin in 1933 due to poor health (and not because I was against the regime, as His Excellency was so kind to surmise), or that I was fired in 1938 for reasons I was not informed of. What was he actually trying to get at? That now I am a big shot here and one would rather see me as a German than as one thrown out and running for life? I spared no words to be blunt: I was not to blame if anyone got that impression. And clearly I had been thrown out of Germany.
>
> He explained the actual reason for his urgent questions. Should I be included in the list for the embassy receptions? Well, I said, I have been reprimanded for political reasons. Your Excellency does not know the real reason why and neither do I, for I was not told. So it would seem to me that we should let the matter rest until Your Excellency has received sufficient information. "The point is," he replied, "do you no longer want to have anything to do with us and do you want this breach

> to be a permanent one or would you rather it were straightened out?" My answer: "I'd really prefer a breach. Not a full one, though. I would be more than delighted to receive an official declaration to the effect that what happened was unfortunate, but that one realizes it was a mistake, for which one is sorry. Other than that (with respect to the question I was repeatedly asked) your Excellency may consider me as a good and decent German. Your Excellency may rest assured of that."[53]

Postwar

Throughout the war and afterwards, Anny Schrödinger in Dublin and Lise Meitner in Stockholm had been swapping many letters with news of their mutual friends, some of whom were tragically lost in the Holocaust.[66] The health and welfare of Max and Marga Planck were a mutual concern as was that of Stefan Meyer who had worked with Meitner and Schrödinger in the radiation laboratory in Vienna. Anny also often freely mentioned the challenges of her domestic situation in Ireland, friction with Hilde March and her close relationship with young Ruth March. They frequently shared their nostalgic memories of Berlin before 1933. They discovered that as they were in neutral countries during the war they could send letters by air mail through Lisbon, providing they were written in English. Anny also described to Meitner her husband's "signing of the roll" at the Royal Society in some detail and the historical links of the Society going back nearly 300 years to King Charles II.[66]

Of particular concern to Meitner and Anny in their letters was Peter Pringsheim.[66] He had been a pioneer in the science of fluorescence and phosphorescence and had interacted with Frank and Einstein. He had overlapped with Schrödinger in Berlin where he was a Professor of Physics. Pringsheim came from a famous Jewish family in Silesia who had made their fortune from coal mining and railways. This made him very vulnerable to the Nazis. His wife Emilia was born in Belgium and, as soon as Hitler came to power in 1933, they moved to Brussels where he was given a position at the Free University. Both Schrödinger and von Laue, as professors in the University of Berlin, gave written permission for Pringsheim to transfer his equipment from Berlin to Brussels.[67] In 1939 he and his wife were visited in Brussels by the Schrödingers during their sojourn in Belgium.[66]

When Belgium was invaded in May 1940 by the German army, Pringsheim was arrested. He was first sent to the internment camp at St. Cyprien and then to the notorious camp at Gurs in Southern France run by the Vichy regime. His brother-in-law, the famous German writer and Nobel Laureate for Literature Thomas Mann, who had moved to the then-neutral USA, helped Pringsheim obtain the offer of a position at Berkeley in the University of California. This enabled him to be given permission to leave France and he managed to take the transatlantic liner SS Excambion from Lisbon to New York in February 1941. After a short period at Berkeley, he joined James Franck in Chicago. His wife remained in Antwerp in Nazi-controlled Belgium and did not know where her husband was during the war. In due course, most of the Jewish prisoners and dissidents at the Gurs camp were sent to their deaths at Auschwitz. So this could have been the fate of Schrödinger if he had not escaped from Belgium after the declaration of war in 1939.

Two co-workers of Pringsheim in Belgium, the photochemist Fritz Duschinsky and Emanuel Oskar Wasser, who had previously worked with Ehrenhaft in Vienna, were also interned in France. They were subsequently transported to Auschwitz where they were murdered.[68] Duschinsky had written in 1933 to congratulate Schrödinger on his Nobel Prize.[6] His "Duschinsky Rotation" is a mechanism for electronic transitions arising from internal rotations of groups of atoms in molecules.

Schrödinger could not resist returning to Belgium to attend the 8th Solvay Conference in Brussels from 27 September to 2 October, 1948. The subject was elementary particles but it was a reunion of many of the great pioneers in quantum mechanics with Schrödinger, Bohr, Pauli, Dirac, Kramers, Peierls, Bloch and Oppenheimer all attending together with others who had made highly significant contributions to physics such as W.L. Bragg, Blackett, Richardson, Frisch, Meitner and Cockcroft. However, as in 1924, scientists from Germany could not attend since the English, American, French or Russian authorities occupying Germany would not provide the permits required. There still remained significant suspicion as regards the activities of Heisenberg, Hahn and others in the

Second World War. The emerging post-war leaders in theoretical physics such as Julian Schwinger and Richard Feynman, who were pioneering the new theories in quantum electrodynamics, did not attend the Solvay Conference although Oppenheimer did present some of their latest results. The heart of theoretical physics, and indeed much of science, had been transferred from Europe to the USA by the Second World War.

Shortly before the Solvay Conference, on 16 September 1948, Schrödinger returned to Oxford. He had dinner at Magdalen College with David Kendall who was a new Tutorial Fellow in Mathematics and an expert on probability theory. Kendall had become a close colleague of Schrödinger's friend Henry Whitehead who had been elected Waynflete Professor of Mathematics at Magdalen. Both Schrödinger and Kendall were recorded as being weighed after dinner in the Senior Common Room.[17] Schrödinger must have retained some affection for Magdalen to have returned some 15 years after he was elected to a Fellowship.

At the time he was corresponding with Simon about Schrödinger and the Royal Society, Max Born had just heard from von Laue that he had won the Max Planck Medal. This had been awarded to Schrödinger in 1937, and in the wartime of 1942 to Pascual Jordan. Jordan had published an important early paper with Born on the matrix representation of quantum mechanics but had been an early supporter of National Socialism. The medal had also been awarded to Walter Kossler in 1944 who had written the first paper on the converged electron diffraction beam technique. Kossler had been a student of the ardent Nazi physicist Philipp Lenard and had signed the notorious "Professors' Confession at the German universities and Colleges to Adolf Hitler" in 1933.

Max Planck had died in 1947 and Max Born wrote his Biographical Memoir for the Royal Society after his death. Born was a supporter of Max Planck and in the Memoir emphasised the quote from the diary of the infamous Reich Minister of Propaganda Goebbels: "It was a great mistake that we failed to win science over to support the new state. That men such as Planck are reserved, to put it mildly, in their attitude towards us, is the fault of Rust (the Minister of Education) and is irremediable."[69] However, given the political connotations of accepting the Max Planck Medal, Born was nervous and wrote on 20 June 1948 to Simon:

> The Deutsche Phys. Gesellschaft has given me the Planck Medal which I will receive at their autumn meeting in Clausthal-Zellerfeld. I have agreed to come although with some misgivings. I do not relish shaking hands with former Nazi generals and such like. But I think I could not decline without an affront to my friends like Laue (who informed me about the award).[63]

Simon responded on 23 June 1948 to Born saying:

> My congratulations on the Planck Medal. I believe you are right to go but I do not think you will have a very pleasant time. Laue also invited me to attend the Conference, but I do not think I will go.[63]

In March 1948 Heisenberg, after much discussion, was allowed by the British authorities in Germany to visit Oxford. He stayed in Christ Church where he met Lindemann and Simon. He also talked to Kurti and Mendelssohn. As is discussed earlier in this Chapter, Born was giving the Waynflete Lectures at this time at Magdalen College and he also had several discussions with Heisenberg.[15] It was reported that:

> Heisenberg thought that his visit had been a success. The time he spent there had been pleasing in every way. Born had been as friendly and nice as in the old days. Simon and Peierls had also been very hospitable but Heisenberg believed that they found it difficult to free themselves from the injustice they had suffered.[70]

Like several of the scientists who had remained in Germany, Heisenberg failed to understand the deep and bitter feelings of the Jewish scientific refugees. Many had lost family members in the Holocaust. Born calculated that as many as 35 of his own family, colleagues or teachers had perished.[15] In addition there was the clear evidence, well known to Simon and Peierls, that Heisenberg had worked closely with the Nazi regime on research examining the possibility of producing a nuclear weapon.[71]

Like Einstein, Simon was unforgiving with German scientists after the war and thought they should have done more to apologise. He wrote on 22 March 1951 to Karl Friedrich Bonhoeffer:

> In my opinion German scientists as a group lost their honour in 1933 and did nothing to get it back. I admit that you cannot say that everybody should have risked his position or life, but such risks are no longer necessary after the war. The least you can expect after all that happened was that German scientists, as a group, would state publicly and clearly that they regretted what happened. I did not notice anything of this kind.[72]

In the same year as Schrödinger's election to Foreign Membership of the Royal Society, Peter Medawar was elected a Fellow. He had been a student during Schrödinger's time at Magdalen College and had been promoted to the Fellowship there in 1936 where he taught Zoology. In the Second World War, Medawar had been an air-raid warden in Oxford with the famous writer and historian J.R.R. Tolkien. He had also done research on skin grafts with the aim of assisting airmen with severe burns. This started a major research programme in immunology that won Medawar the Nobel Prize for Physiology or Medicine in 1960.

Ten years after their elections to the Royal Society Medawar had some subsequent correspondence with Schrödinger following a review he wrote on Schrödinger's "Mind and Matter" lectures. Medawar wrote to Schrödinger on 25 May 1959:

> I was delighted to hear from you about my review of your book. I did indeed study it and with the most intense interest. I felt that I, in common with all biologists, owed you a debt of gratitude for *What is Life?*.
>
> Many years ago, by the way, we dined together in Magdalen in John Young's rooms, but I didn't become a Fellow of the College until after you had left.[73]

It is possible Medawar is referring here to the dinner organised by Arthur Tansley for Schrödinger to meet members of the Magdalen Philosophy Club. There would have been lively discussions at this dinner with the zoologist and conversationalist John Young and Peter Medawar, who himself became a major populariser of science. This may well have served to implant some biological ideas in Schrödinger's mind. The writer of this book recalls sitting next to John Young at a dinner at Magdalen College in 1989. Squid was on the menu and Young was the world expert on the nervous system of this cephalopod. During the dinner he gave to me a detailed and what seemed to be complete survey of the physiology of the squid.

Schrödinger always enjoyed mixing with top intellectuals and a visit to the hometown in West Wales of the great philosopher Bertrand Russell had been arranged for him with the help of Patrick Blackett. Schrödinger was visiting London and wrote in personal terms on 6 September 1949 to Blackett from the Society for the Protection of Science and Learning with a curious request concerning Hansi Bauer-Böhm. It is perhaps not surprising that this kind of request often led to rumours about his personal life:

> Thank you very much for your letter on August 23rd which I answer only now after just getting rooms reserved in Penrhyndeudraeth for 8–15th September. I am very grateful for your having announced me to the owner, because it is always nice not to arrive as a complete stranger; it also may improve the rooms you are given, particularly with single rooms. (I am in the company of an old friend of my wife, who is still on the continent and very much enjoying a marvellously prolonged stay made possible only by her being put up with friends half of the time.)
>
> I shall be grateful if you write me a line of introduction to your friends, provided that they do not take exception to my turning up with a lady other than Mrs Schrödinger (it is Mrs Hansi Böhm from Vienna, living now in London with her husband and children).[45]

Schrödinger corresponded quite frequently with Bertrand Russell who, like many philosophers, had a particular interest in the fundamental ideas of quantum mechanics. Unfortunately, Schrödinger was now beginning to have more serious medical problems. In addition to his tuberculosis, he was a heavy pipe smoker. He often had bouts of severe bronchitis and eye problems made worse by cataracts that needed two operations. He was also hospitalised by appendicitis. From the late 1940s onwards, this restricted his opportunities to take up many post-war offers to travel, lecture and attend conferences in places like Cambridge and Harvard.

In 1949 the Institute in Dublin had a visit from Leopold Infeld. He was a scientific refugee from Poland who had worked with Max Born in Cambridge on a non-linear theory for the electromagnetic field. This research had drawn the attention of Schrödinger who Infeld had visited in Oxford. He then went to work with Einstein in Princeton and this was followed by a Chair in Toronto. He returned to Warsaw in 1949 where he continued to work on relativity. Infeld was commissioned by *Scientific American* to write a popular article on the Institute for Advanced Studies in Dublin. His article had the heading "A noted theoretical physicist calls upon the remarkable little group of eminent scientists at Ireland's exotic Institute for Advanced Studies":

> Schrödinger has done much first-class work since 1928, but none of it of such a revolutionary character as that in the years 1925–28. The same thing can be said, though perhaps to a lesser degree, about Heisenberg and Dirac. Schrödinger is not only a great scientist; he is a most interesting and charming man — intelligent, witty, erudite. He admires Spinoza and good literature, and is himself an excellent writer. I saw him for the first time when he lectured, with spirit and artistry, in the Berlin of 1928. He was then at the peak of his fame; he occupied Max Planck's chair in theoretical physics, the greatest scientific honour in Germany.
>
> I saw him again in 1934 in Cambridge, England, when, on a Rockefeller Fellowship, I was working on the unitary field theory with Max Born, who had just left Nazi Germany. On Schrödinger's invitation, I went to visit him at Oxford and spent a delightful evening in his home. He was interested in the unitary field theory, and wrote an important, original paper on the subject. The distinctive mark of Schrödinger's genius has always been the originality of his thinking, his self-confidence and disregard for tradition. He showed a lack of political judgment, however, when he left Oxford to return to Austria in 1936.[74]

Infeld then went on to give his overall impression of the Institute and science in Dublin:

> Ireland is untouched by war or fears of war... Every scholar longs to be in a place like Oxford, Cambridge or Dublin which seems to be

> outside the world of trivial realities. In isolated Ireland, the Institute for Advanced Studies with its scholars, most of them fugitives from oppression, seems to be the most peaceful spot on earth.[74]

By 1950 the situation in some parts of Austria had improved and Schrödinger was now able to take a three-month sabbatical at the University of Innsbruck which he had started to arrange in 1946. Innsbruck was in the French zone and was more congenial than Vienna which was still surrounded by Russian troops. In Innsbruck he saw Hilde and the 16-year old Ruth for the first time in five years. He enjoyed his visit to his home country and wanted to arrange a permanent move back to Austria. His daughter Ruth told the writer of this book that around this time she and Schrödinger were sitting by a waterside when he commented that her feet in the water looked very much like his. He then told her for the first time that he was her father.

While on this visit, Schrödinger wrote to Einstein to help patch up their public quarrel on the unification theory. What better topic than to discuss the probabilistic interpretation of wave mechanics which they had been discussing on a regular basis for the last 25 years. On 18 November 1950 Schrödinger wrote from Innsbruck:

> It seems to me that the concept of probability is terribly handled these days. Probability surely has as its substance a statement as to whether something is or is not the case — an uncertain statement to be sure... No reasonable person would express a conjecture as to whether Caesar rolled a five with his dice at the Rubicon. But the quantum mechanics people sometimes act as if probability statements were to be applied just to events whose reality is vague.[75]

Einstein responded on 22 December 1950 with a reference to the paper Schrödinger had published in Oxford in 1935:

> You are the only contemporary physicist beside Laue who sees that one cannot get round the assumption of reality... They somehow believe that the quantum theory provides a description of reality, and even a complete description; this interpretation is, however, refuted most

> elegantly by your system of radioactive atom + Geiger counter + amplifier + charge of gun powder + cat in a box, in which the ψ-function of the system contains the cat both alive and blown to bits. Is the state of the cat to be created only when a physicist investigates the situation at some definite time? Nobody really doubts that the presence or absence of the cat is something independent of the act of observation.[75]

Max Born was keen to get an Honorary Degree for Schrödinger at the University of Edinburgh. He put the case to Professor R.N. Arnold, Dean of the Faculty of Sciences, on 7 January 1951 and stated: "He is one of the founders of modern quantum mechanics and his name is perhaps the most quoted of all living physicists since the Schrödinger equation appears in almost every paper containing any theory."[15] However, the proposal was not successful and perhaps some of the University authorities recalled Schrödinger's rejection of the approach from Edinburgh made 15 years before. Born was not to be put off and proposed Schrödinger again on 9 January 1953. He stated that this was the last opportunity to bring Schrödinger to Edinburgh before his retirement and also: "Although I am at present in a scientific dispute with him I still have the highest admiration for his skill and judgement."[15]

The new nomination was eventually successful. Schrödinger did not go to Edinburgh to get the degree in 1953 as the date coincided with Max Born and James Franck receiving the freedom of the City of Göttingen.[76] However, he was presented with the Honorary Degree of Doctor of Laws in 1955, although this was after Max Born had retired to Germany. Max Born himself was also awarded the Honorary Degree in 1957 by Edinburgh University and he would have been very proud to learn that his son Gustav was given the Honorary Degree in Medicine in 1982.

Schrödinger had been negotiating with Wilhelm Klastersky who was the trusted 70-year-old Cabinet Director reporting to the President of Austria. Klastersky had been imprisoned for a time by the Nazis after the Anschluss. Schrödinger had been hoping to organise a dual appointment at the Universities of Innsbruck and Vienna with an appropriate pension for himself and for Anny. Klastersky explained to Schrödinger that the "Federal President of Austria is urgently aware of the problem of your return to Austria".[6]

However, it did not look as if a position would become available in Innsbruck though one at the University of Vienna was possible. Then in 1955 the Russians left Austria creating the opportunity for Schrödinger to return at last. His great friend Hans Thirring was very much in favour of Schrödinger returning once again to a Vienna without major political uncertainties. Thirring was now himself involved in politics and consulted with the Ministry to make the arrangements.

Return

So on 23 March 1956 Schrödinger and Anny left Dublin by boat on the start of their journey to go back to Austria. De Valera and many others were present to see them off at the quayside. As usual, the *Irish Press* reported on the event and the headline was "One of our Age's Great Men of Science":

> When Erwin Schrödinger sailed from Dublin last night we said goodbye to one of the great scientists of our age. A notable figure of the revolution in scientific thought, which was led by Planck and Einstein, he will be remembered with these immortals. Not that his contributions to knowledge were merely developments of theirs — they were as original and in many ways as far-reaching.
>
> It is not easy to convey a complete picture of the man. A vivid personality with an acute mind and courteous manner. He likes controversy, the inquiring scientist and mathematician in him does not stifle an interest in humanity and literature nor those poetic qualities that go with great imagination. The power of intellect coupled with a capacity for dreams — these are the qualities of genius, and Schrödinger possesses them.
>
> When, in the 1920s, he gave the world his wave mechanics, he made a fundamental contribution to atomic theory. Schrödinger's equation and his theory of atomic processes have proved to be one of the most powerful means of gaining knowledge of atomic processes, and have conditioned the thinking of physicists in this field. With Heisenberg and Dirac, who developed different aspects of the subject, he is the cofounder of modern quantum mechanics, and as such will be remembered in history.

> While continuing to contribute to the science which he founded, Schrödinger in later years was interested in the problem of wedding relativity, electrodynamics and atomic mechanics into one theory. His main contribution in this sphere was made while he was in Ireland. His enquiring mind has also roamed in the fields of biology and psychology.
>
> He was the Senior Professor of the School of Theoretical Physics of the Dublin Institute for Advanced Studies. That School has now an international reputation, and Schrödinger's work contributed in great measure to this. We have had our own great scientists here: Hamilton of Trinity College was the greatest. Schrödinger confessed that Hamilton was his inspiration. While both will be remembered while physicists and mathematicians remain in the world Ireland will especially remember Schrödinger with Hamilton.[1]

In his happy 17 years in Ireland Schrödinger had not achieved great scientific breakthroughs comparable to his papers on wave mechanics in 1926 but he had still been influential. His little book *What is Life?* helped to accelerate the revolution in molecular biology and he did much to inspire the public understanding of science in Ireland.

Back in Austria Schrödinger was treated like a returning hero. He was the only Nobel Prize winner to return to a position in Austria after the war. There were many letters and telegrams welcoming him back again to his home country. He gave his inaugural address to a standing ovation at the University of Vienna from which he had been so curtly dismissed 18 years before. He was awarded the Prize of the City of Vienna which had first been given to Lise Meitner in 1947. It was presented by the Bürgermeister in the Festival Room of the Rathaus. The Erwin Schrödinger Prize was initiated by the Austrian Academy of Sciences. It was decided to make the first award to Schrödinger himself in the year of his return. This prestigious prize continues to be awarded to the present day to "scholars who work in Austria and who have achieved outstanding scientific achievements in the subjects represented by the mathematics and natural sciences".

Schrödinger's status as a Corresponding Member of the Academy of Sciences in Vienna had been cancelled in 1940. His membership had been re-designated in 1945 and on his return to Austria in 1956 he was

Erwin and Anny Schrödinger in Alpbach, Austria in the late 1950s.

made a Full Member of the Academy which had been renamed the Austrian Academy of Sciences. His former teacher in Vienna, Egon Schweidler, had been Vice-President of the Academy during the war period and it seems he had helped to prevent the Academy being joined formally with the German academies during that very difficult period.[77]

Schrödinger's return to Austria was instigating several other awards to him from the cities and countries which had let him down before. He was very pleased to hear from Otto Hahn that he had been awarded the most significant German honour from the President of Germany: the Pour le Mérite. This award, on the civilian side, had been given previously to some of the greatest scientists or mathematicians including Darwin, Faraday, Gauss, Planck and Einstein. It is also given for contributions to the humanities and, in the previous year, had been awarded to Hermann Hesse and Thomas Mann. There is also a military honour which sadly had been applied to the likes of Göring and Rommel, but this did not deter Schrödinger from accepting the award. Schrödinger was holding no major grudges, unlike Einstein who refused to have anything to do with Germany after he was forced to leave following the significant threats to his safety of the 1930s.

Lise Meitner and Erwin Schrödinger in Alpbach in 1952.

Anny said that Schrödinger valued this post-war recognition from Germany of the Pour le Mérite more than any other award except the Nobel Prize and the Max Planck Medal.[50] Schrödinger was not well enough to go to the West German capital of Bonn for the presentation so the award was presented to him by the German Ambassador in Vienna. He and Anny were also very pleased to hear that Lise Meitner, who many feel should have won the Nobel Prize with Otto Hahn for her work on nuclear fission, was given the Pour le Mérite in the next year. Meitner was also the first female member of both the Austrian and Germany Academy of Sciences.

Max Born had come full circle and had gone back to live in Germany at Bad Pyrmont with a pension from the German government. He had finally won the Nobel Prize in 1954, shortly after retiring from Edinburgh

University, for his "fundamental research in quantum mechanics, especially in the statistical interpretation of the wave function". Schrödinger always had reservations that Born was not awarded the Nobel Prize at the same time as Heisenberg and had written to Born on 13 January 1943 to say:

> I am quite frankly and honestly of the opinion that the "new deal" starting from 1925 was not honoured in quite the correct distribution and moreover you are the one to suffer particular injustice. (If I may say so although I ought not to say so) de Broglie was the one to accommodate a mistake in sign the other way round. To divide a prize between him and me, and another one between you and Heisenberg, would have been more equitable.[78]

In 1938 Schrödinger had to leave Graz in a great hurry and had left his priceless notes on the first ideas of wave mechanics of 1925–26 locked in a filing cabinet together with his Nobel and Planck gold medals. On this subject Anny said:

> Yes, I think most of his things have gotten lost in Graz. It is amazing that we could have saved this amount. He had a steel cabinet, a filing cabinet, where he kept some things separate which were very valuable to him. And when we had to leave Graz we didn't know what to do with all these precious things — the Nobel medal and Planck medal and a few others as well. So he just put them in on the end of this cabinet and when I saw this cabinet again, the Russians had crushed in the drawers but they saw it was only paper so they didn't mind. They left it as it was. I really found the golden medals. That is really amazing.[50]

In recent years, some Nobel Prize gold medals awarded to the greatest scientists have been auctioned for several million pounds. There is no doubt that Schrödinger's Nobel medal is very valuable.

In 1956 Schrödinger was made a full Professor of the Institute for Theoretical Physics at the University of Vienna, which was not the case in 1936 when he was made an Honorary Professor. He gave a full course on General Relativity for the academic year of 1957–58 and held tutorials for students at his new home in Pasteur-Gasse, not far from the Institute.

He then retired from teaching. In a short autobiography that was published after his return to Austria, in a new print of his book *What is Life?* he said: "Austria had treated me generously in every respect, and thus my academic career ended happily at the same Physics Institute where it had begun."[49]

Born and Schrödinger continued their detailed and friendly correspondence after they had retired. While they were both at English-speaking institutions they corresponded in English but now, after returning to Germany and Austria, they reverted to writing in German. Only Einstein had rivalled the volume of Born's correspondence with Schrödinger. In the very last letters between them they were still, after all these years, discussing priority for the Nobel Prize. On 24 October 1960 Schrödinger wrote to Born:

> I have proposed Marietta Blau twice to Stockholm. The second time I asked them to at least let me know why it is out of the question, despite her undoubted priority. The other day I noticed that for Fritz London's theory of the homopolar bond, someone else received the Chemistry Award. And why was R.W. Wood in disgrace — just because his experiments were years ahead of Danish theory?[76(t)]

Here he was referring to the Austrian physicist Marietta Blau who had invented a photographic method to measure nuclear particles with high energy. Born responded on 6 November 1960:

> I find your criticism of the Nobel people largely justified, even though I know from a long talk with Oskar Klein how terribly difficult it is to choose the right name from the many suggested. As far as the London-Heitler theory is concerned, Linus Pauling received the Chemistry Award, at the same time as me. Now it is true that Pauling did something important and he is also an impressive personality, a brave person whom I rate highly. Nevertheless, this year I have proposed to give the Physics Prize to Heitler (unfortunately, London is no longer alive) with a detailed presentation of the facts and with criticism of the Pauling matter.[76(t)]

It is interesting that, as late as 1960, Schrödinger and Born were still discussing who should have won the Nobel Prize for the application of wave mechanics to the description of the chemical bond. By this time, Schrödinger's invention of the orbital was being taught to chemistry students throughout the world and had become the standard method for explaining the bonding in molecules.

Schrödinger was an unconventional scientist who had unusually broad intellectual interests. Just one year before he died, in an introduction to his *My View of the World*, he wrote from Alpbach with some reflection:

> In 1918 when I was thirty one I had good reason to expect a chair of theoretical physics at Czernowitz (in succession to Geitler). I was prepared to do a good job lecturing on theoretical physics with, as my supreme model, the magnificent lectures given by my beloved teacher Fritz Hasenöhrl, who had been killed in the War; but for the rest to devote myself to philosophy, being deeply imbued at the time with the writings of Spinoza, Schopenhauer, Mach, Richard Semon and Richard Avenarius. My guardian angel intervened: Czernowitz soon no longer belonged to Austria. So nothing came of it. I had to stick to theoretical physics and, to my astonishment, something occasionally emerged from it.[79]

Grave of Erwin Schrödinger in Alpbach, Austria.

Death

Schrödinger died in Vienna on 4 January 1961 of "general arteriosclerosis". He was buried in Alpbach, the place he loved the most, and his time-dependent Schrödinger equation was inscribed on his grave. As he was not a follower of the Catholic religion there was some hesitation about burying him in the graveyard but when the priest heard he was a founding member of the Pontifical Academy of Sciences he agreed to the burial. Time and time again, membership of that Academy came to help Schrödinger. Normally, graves in Austria are only kept for a limited period but that rule was broken for one of the very greatest Austrians. Anny returned to Dublin, obtained some earth from the garden at their house in Clontarf and added that to the grave.

Following the death of her husband, Anny received a large numbers of condolence letters from notable people including many mentioned in this book who were still alive. This included Max and Hedi Born, Paul Dirac, Werner Heisenberg, Walter Heitler, Pascual Jordan, Bruno Kreisky, Magda von Laue, Georges Lemaître, Lise Meitner, Karl Popper, Emeke and Peter Pringsheim, Karl Przibram, Tess Simpson, Walter Thirring and Ellen Weyl. Anny also formally adopted Schrödinger's daughter Ruth Braunizer (née March) after his death and this enabled Ruth to become responsible for the estate of her father after the death of Anny in 1965. Arthur March died in 1957 while Hilde lived to the age of 87 and died in 1987.

There were many obituaries written on Schrödinger. Perhaps the comment from Dirac was most insightful:

> Of all the physicists I met, I felt that Schrödinger was the one that I felt to be most similar to myself. I found myself getting into agreement with Schrödinger more readily than anyone else. I believe the reason for this is that Schrödinger and I both had a very strong appreciation of mathematical beauty, and this appreciation dominated all our work. It was an act of faith with us that any equations which describe fundamental laws of nature must have great mathematical beauty in them. It was like a religion to us.[80]

Several prizes and lectureships have been given Schrödinger's name. In addition to the Schrödinger Prize of the Austrian Academy of Sciences, which Schrödinger would have been pleased to learn was awarded to Marietta Blau in 1962, there is the Erwin Schrödinger Award of the German Helmholtz Research Association. This Award recognises outstanding scientific achievements and technological innovations at the interface between various disciplines in medicine, the natural sciences and engineering. The main award of the World Association of Theoretical and Computational Chemists is the Schrödinger Medal.

Trinity College Dublin, in association with the Austrian Embassy and the National Bank of Austria, organises an annual Schrödinger Lecture to commemorate his "What is Life?" lectures first given in Trinity College in 1943. The Erwin Schrödinger International Institute for Mathematics and Physics was founded in Vienna in 1992 and has an annual lecture in his name which is directed towards a general audience of mathematicians and physicists. The 2011 Schrödinger Lecture on Group Theory was given by Professor Martin Bridson who holds the Whitehead Professorship of Mathematics at Magdalen College, Oxford and which is named after Schrödinger's close friend Henry Whitehead. Imperial College London also has an annual Schrödinger Lecture. When Ruth Braunizer was alive she regularly attended these lectures. In addition, there is a company in the USA called Schrödinger which develops and distributes software for

The Austrian 1000 Schilling bank note.

drug discovery and materials design which is firmly based on solutions to Schrödinger's equation.

In life Schrödinger received many honours and that continued in death. His face was placed on the Austrian 1000 Schilling bank note in 1983. This featured his wave function ψ on the front and the University of Vienna on the back. On the centenary of his birth in 1987 his face was also put on the 5 Schilling stamp. In 1970 a large crater on the dark side of the Moon was named Schrödinger by the International Astronomical Union.

The Austrian 5 Schilling stamp.

CHAPTER SIX

Schrödinger's Legacy

Chemistry

Schrödinger's great work was his equation that he first published in 1926 when working at the University of Zurich. This breakthrough won him the Nobel Prize in 1933. His move to Max Planck's Chair at Berlin was very prestigious but he did not produce any highly significant research papers while in Berlin. This was perhaps partly due to his spending much of his time lecturing on his revolutionary development of wave mechanics during this period.

Schrödinger only spent three years at Oxford from 1933–36 (and again briefly in 1938) but during his first period there he wrote his papers on entanglement. In recent years, this idea has been highly influential with physicists working on quantum information and quantum computing. His "Schrödinger Cat" paper, also written in Oxford, has done much to popularise science. The idea has appeared in several television programmes and has even been used as a term in the title of a subsequent Nobel Lecture in Physics.[1] His *What is Life?* little book published in Dublin inspired several brilliant physical scientists to change their field to molecular biology even though some of the concepts and terminologies in his book were wrong or misleading.

However, it is in the whole field of chemistry, and the wider area of molecular science, where Schrödinger's greatest work is having the most direct impact. This is because the Schrödinger equation, or variants of it, provides the underlying theory for calculating accurately all the observable properties of atoms, molecules and condensed phases.[2] This is the field of quantum chemistry. The progress in being able to perform quantum chemistry calculations was initially slow after 1926 due to the numerical difficulties of computing the required wave functions and dealing with the repulsion between electrons. However, by 2021 the breadth and reach of Schrödinger's wave mechanics is truly enormous and the use of electronic computers has been central in accelerating this progress.[3]

The first solution by Schrödinger of his equation gave, essentially, perfect results for the energy levels and properties of the electron in the hydrogen atom and also for atomic ions containing one electron.[4] A key aspect was to see if this success extended to atoms with two electrons such as helium. If the theory then worked well the whole field of atoms and molecules with many electrons would be open for applications of the theory. Schrödinger, however, was not interested in numerical computations and this crucial progress was achieved by others.

The helium atom is the simplest system with two electrons. It has the additional difficulty, not seen in the hydrogen atom, of two electrons with the same charge which repel each other. For this reason there is no analytical mathematical solution to Schrödinger's equation for the helium atom and numerical computations are needed. Soon after the publication of the Schrödinger equation, the helium atom was tackled successfully by different researchers who reported calculations for the ionisation potential for helium of increasing accuracy.[5] This culminated in an almost perfect agreement for the ionisation energy of helium between the wave mechanics calculations and experiment by Hylleraas in 1929.[6] The work of Heitler and London in Zurich on the hydrogen molecule, which also contains two electrons, demonstrated that wave mechanics could be applied successfully to the chemical bond.[7] These studies thus set the foundations for the field of quantum chemistry.

In the following years progress in making wave mechanics useful for chemical applications was based more on conceptual ideas as opposed to accurate computations due to the difficulty of accounting for the electron correlation in the wave functions. However, useful progress was made, particularly by Pauling, Slater, Mulliken, Lennard-Jones, Hund and others, on showing how the overlaps of Schrödinger's orbitals could give qualitative explanations of molecular shapes and bonds.

With the arrival and availability of electronic computers after the Second World War the theories of quantum chemistry were successively improved until, in the 21st century, they have become a standard approach used by chemists and molecular scientists the world over for calculating essentially all the observable quantities that can be measured on molecules.[3] This includes the structural, electrical, magnetic and optical properties of molecules, the thermodynamics and strengths of chemical bonds, spectra, and the rates of chemical reactions. At the heart of all these calculations is the Schrödinger equation.

This book has emphasised Nobel Prizes as being a central feature in inspiring research and rewarding scientific progress. As Schrödinger first heard he had won the Nobel Prize when he was called to the office of the President of Magdalen College, these awards have been highlighted in this book on *Schrödinger in Oxford*.

Furthermore, the significant progress made over time in a particular scientific field can be calibrated, to a certain extent, from the Nobel Prizes that have been won and these are described in detail in the Nobel Prize Archives.[8] The Nobel Prize for Chemistry has been awarded on a regular basis for research linked to quantum chemistry. The first Nobel Prize awarded to a quantum chemist was to Linus Pauling in 1954 "for his research into the nature of the chemical bond and its application to the elucidation of the structure of complex substances". Pauling had gone to Zurich in 1926 to learn from Schrödinger about the new wave mechanics and there he met Heitler and London. Pauling found that their valence bond approach, in which a wave function is expressed in an approximate form using atomic orbitals placed on different atoms, was particularly powerful in explaining the structure of organic molecules. By the 1950s at the California Institute of Technology, Pauling had also started to use X-ray crystallography for understanding the structure of biological molecules and this research was mentioned in his Nobel citation.

Then the Nobel Prize for Chemistry was awarded in 1966 to Robert Mulliken "for his fundamental work concerning chemical bonds and the electronic structure of molecules by the molecular orbital method". From the late 1920s, Mulliken and others, such as John Lennard-Jones and Friedrich Hund, had been applying molecular orbital theory to understand the structure and spectra of small molecules. In this method, the wave functions are expanded as linear combinations of atomic orbitals which has mathematical advantages in setting up equations that can be solved in a self-consistent way.

As described in Mulliken's Nobel Lecture,[8] the molecular orbital approach was formulated into a rigorous computational procedure by Mulliken's colleague at the University of Chicago, Clemens Roothaan. His self-consistent field equations were well suited for solving using electronic computers. They were also based on previous formulations by the mathematical physicists Hartree and Fock. In the discussion following his 1927 Solvay Lecture, Schrödinger mentions the promise of the work of Hartree in applying his wave mechanics to systems with many electrons.[9] Clemens Roothaan was a Dutch physicist who was imprisoned by the Nazis in the Netherlands. After he was freed at the end of the Second World War he emigrated to the USA.

Mention should also be made of the Chemistry Nobel Laureate for 1971. Gerhard Herzberg won the Prize "for his contributions to the knowledge of electronic structure and geometry of molecules, particularly free radicals". A German by birth, he had worked with Max Born in Göttingen and Lennard-Jones in Cambridge and, like so many scientists, was forced to leave Germany in the 1930s. He then founded a major laboratory at Ottawa in Canada. He was not a quantum chemist but a spectroscopist who determined the structures and properties of molecules from observing how they absorbed electromagnetic radiation. He used the predictions of the quantum chemists very effectively in explaining his experimental results, especially for free radicals which are molecules that have unpaired electrons.

Another experimentalist who made significant use of quantum chemistry was William Lipscomb. He was a student of Pauling and won the Nobel Prize in 1976 "for his studies on the structure of boranes illuminating problems of chemical bonding". Boron compounds have some similarities to organic carbon compounds and the unexpected structures of several new boranes, compounds of boron with hydrogen, were predicted by quantum chemistry methods and then observed by Lipscomb's team at Harvard University.

The Chemistry Nobel Prize for 1981 was awarded to the quantum chemists Kenichi Fukui, from Kyoto in Japan, and Roald Hoffmann, from Cornell University in the USA, "for their theories, developed independently, concerning the course of chemical reactions". They explained how molecular orbitals can change during certain types of chemical reactions and hence the chemical products of the reaction can be predicted. Understanding the symmetry of the molecular orbitals proved to be very powerful in this work. Hoffmann was born in Poland and as a young boy was hidden by his mother from the German forces during the Second World War. He is yet another example of a refugee from war-torn Europe who succeeded in science in the USA.

Experimentalists who made significant use of the results of quantum chemistry calculations on chemical reactions were Dudley Herschbach (Harvard), Yuan T. Lee (Berkeley and Taiwan) and John Polanyi (Toronto) who won the 1986 Prize "for their contributions concerning the dynamics of chemical elementary processes". They developed molecular beam and spectroscopic methods to study the details of molecular collisions and chemical reactions, and the results were explained by using potential energy surfaces. These surfaces are defined as the electronic energy of a molecular system expressed in terms of the inter-atomic distances. The first such potential energy surface was obtained in approximate form for the simplest of all reactions between a hydrogen atom and a hydrogen molecule. This was achieved in the early 1930s through the work of Fritz London, Henry Eyring and Michael Polanyi, the father of John Polanyi.[10] At that time, London was an assistant to Schrödinger in Berlin and Michael Polanyi was a close colleague.

Chemical reactions were also the subject of the Nobel Prize in 1992 which was awarded to Rudolph Marcus at the California Institute of Technology "for his contributions to the theory of electron transfer reactions in chemical systems". Marcus used wave mechanics to explain how electrons jump from one molecule to another. This process has many applications and is especially important in biology.

The Nobel Prize for Chemistry in 1998 was the one that most prominently declared the arrival of a matured form of quantum chemistry. This was awarded to the British scientist John Pople "for his development of computational methods in quantum chemistry" and to Walter Kohn "for his development of density-functional theory". Building on earlier work of a colleague, Frank Boys, in Cambridge, UK, Pople and his group developed a general computer program that could be widely used for performing quantum chemistry calculations on molecules. This enabled experimental groups to back up their observations with calculations they could perform themselves and took Schrödinger's wave mechanics right through the world of molecular science.

Like Schrödinger, Walter Kohn was brought up in Vienna. He came from a Jewish family and managed to escape from Austria on one of the last Kindertransports to England in August 1939 at the age of 16. Both his parents were subsequently murdered in Auschwitz. He attended a local school in England but, like several others who moved to Britain from Germany or Austria, he was interned in 1940 and sent to Canada. He then learned mathematics and physics from lectures given by academics who were also interned. After studying at the University of Toronto, he undertook his PhD research with Julian Schwinger at Harvard. He won major awards for applying wave mechanics to various problems in solid state physics.

Walter Kohn's Density Functional Theory (DFT) was a highly significant breakthrough in quantum chemistry as it enabled mathematical and computational simplifications to be applied to treat interactions between electrons. Instead of having a wave function that depended on all the Cartesian coordinates of every electron in a molecule, Kohn, and his co-worker Hohenberg, showed that the electronic energy could be expressed in terms of the total electron density which depended just on three coordinates (x, y, z).[11]

Kohn's theory has taken Schrödinger's wave mechanics to new, more complex, scientific areas. The first two DFT papers published by Walter Kohn have by 2021 received over a remarkable 80,000 citations in peer-reviewed publications from the scientific community.[11,12] It has become apparent that DFT can be useful for the understanding of almost every scientific area involving atoms, molecules or condensed phases. It has even been used in geology to predict the temperature at the centre of the earth[13] and now has broad applications in engineering such as in evaluating the strength of metals and alloys from first principles.[14] In addition, there have been many applications of DFT to biological molecules.[15]

Nobel Prize winner Walter Kohn, refugee from Austria.

Walter Kohn came to visit the writer of this book in 2006 to discuss Schrödinger's period in Oxford. We had dinner in the Great Hall at Magdalen College where Schrödinger dined in 1933 just before he heard, in the office of the President of the College, that he had won the Nobel Prize. In that office, some 73 years later, Kohn told me in moving detail the story of Schrödinger's escape from Austria in 1938 and his own escape one year later just one month before the start of the Second World War. In 2016 Walter Kohn died at the age of 93 and the last paper he published was in 2014.

In 1999 the Nobel Prize for Chemistry was awarded to Ahmed Zewail. Like Pauling and Marcus, he was from the California Institute of Technology. Zewail's Nobel citation was "for his studies of the transition states of chemical reactions using femtosecond spectroscopy". He developed laser methods on a time scale of 10^{-15} seconds to study the details of chemical reactions. To explain his experimental results his research group and others solved numerically the time-dependent Schrödinger equation using a method known as wave packets. Zewail, an Egyptian by birth, also visited Magdalen College in 2007 to give the opening talk in a conference on femtochemistry, the field he created.

Another experimentalist who worked closely with theoreticians is Gerhard Ertl who won the Nobel Prize in 2007 for "for his studies of chemical processes on solid surfaces". Ertl did his work at the Fritz-Haber Institut of the Max Planck Society in Berlin which is quite close to where Schrödinger was based from 1927–33. Ertl's collaborators in Berlin made much use of Walter Kohn's density functional theory to explain how molecules dissociate and react on solid surfaces, a very important process in catalysis.[16]

Then in 2013 the Nobel Laureates were Martin Karplus, Michael Levitt and Arieh Warshel "for the development of multiscale models for complex chemical systems". Like Schrödinger and Kohn, Karplus was brought up in Vienna and had to leave Austria after the Anschluss. He emigrated to the United States where he pioneered how to model molecular processes as a function of time and did the first such molecular dynamics simulations of proteins.

In a centenary celebration of Schrödinger's birth in 1987, Karplus emphasised that the promise proposed in Schrödinger's book *What is Life?*, that biology might be explained with the laws of chemistry and physics, was finally being achieved through molecular simulations.[17] Levitt and Warshel developed an efficient theory that combined Schrödinger's wave mechanics with classical molecular mechanics and allows simulations of complex problems in materials science and biochemistry to be carried out. Their methods are now being used by pharmaceutical companies worldwide to help in the search for new drugs where, for example, small molecule candidates bind to proteins and inhibit their function.[18]

Perhaps some of the more recent developments in the applications of Schrödinger's wave mechanics to chemical problems, including highly accurate methods for treating electron correlation, the quantum Monte Carlo method (which maps the Schrödinger equation onto a diffusion equation which can be solved with a random walk algorithm), the efficient inclusion of relativistic effects which are needed for heavy atoms, the use of artificial intelligence and machine learning procedures in quantum chemistry, and the quantum dynamics of chemical reactions will receive Nobel Prizes in due course.

Physics

The above discussion emphasises the highly significant influence of Schrödinger's wave mechanics on the research leading to numerous Nobel Prizes in Chemistry. The same is true also for many Nobel Prizes in Physics. The distinguished philosopher of physics Max Jammer wrote:

> Schrödinger's brilliant paper was undoubtedly one of the most influential contributions ever made in the history of science. It deepened our understanding of atomic phenomena, served as a convenient foundation for the mathematical solution of problems in atomic physics, solid state physics and, to some extent, nuclear physics, and finally opened new avenues for thought. In fact, the subsequent development of non-relativistic quantum theory was to no small extent merely an elaboration and application of Schrödinger's work.[19]

Numerous Nobel Prizes for Physics have been awarded which involved the direct use of Schrödinger's wave mechanics and there are some in which it was one of several different theoretical methods which were employed. There have been very few Nobel Prizes in Physics since 1933 where the research leading to the award was not influenced in some way by Schrödinger.

Max Born's Nobel Prize of 1954 "for his fundamental research in quantum mechanics, especially for his statistical interpretation of the wave function" came directly from his interpretation of Schrödinger's theory. The Prize awarded to Wolfgang Pauli in 1945 "for the discovery of the Exclusion Principle, also called the Pauli Principle" was for work that was essential to the extension of wave mechanics to atoms and molecules containing more than one electron.

Other examples include the Nobel Prize won by Brian Josephson in 1973 in which he applied wave mechanics to predict a new electron tunnelling effect in superconductors, and the Prize awarded to Philip Anderson, Nevill Mott and John Van Vleck in 1977 "for their fundamental theoretical investigations of the electronic structure of magnetic and disordered systems" in which direct use of wave mechanics was made.

Without Schrödinger's wave mechanics, it is very likely that key devices such as transistors and semiconductors would not have been developed. Understanding the electronic structure of the materials in these devices, and the consequent energy levels and band gaps, is essential for the needs of modern electronics, and it is wave mechanics that provides this understanding and quantitative predictions. Without these advances in microelectronics there would be no computers, smartphones and other essential electronic devices of the modern quantum age.

This field has received several Nobel Prizes in Physics including the 1956 award to William Shockley, John Bardeen and Walter Brattain "for their researches on semiconductors and their discovery of the transistor effect". Subsequent Nobel Prizes also acknowledged further vital developments such as that awarded in 2000 to Jack Kilby for the integrated circuit and to Zhores Alferov and Herbert Kroemer "for developing semiconductor heterostructures used in high-speed- and opto-electronics". Another example is the 2007 Prize awarded to Albert Fert and Peter Grünberg for the discovery of giant magnetoresistance. Their novel materials, which are the basis for modern computer data storage, are constructed from thin layers of different metals. An understanding of the electronic structure of these materials, which determines their resistance, arises once again from Schrödinger's wave mechanics. Quantum devices was also the subject of the 2014 Prize awarded to the Japanese physicists Isamu Akasaki, Hiroshi Amano and Shuji Nakamura for their work on the light-emitting diode (LED). An LED is a semiconductor light source which emits light when a current is passed through it. It has revolutionised lighting in many different contexts.

A prime example of a quantum device is the laser. The frequency of laser light is determined by the difference in energy of quantum mechanical levels, and the intensity of the laser radiation is calculated from the wave functions of those levels. The Nobel Prize in Physics for 1964 was awarded to Charles Townes, Nicolay Basov and Aleksandr Prokhorov for the invention of the laser. Lasers are now used in many modern devices and also to discover new physics. This latter aspect has been acknowledged several times by the Nobel Committees. The 1997 Prize was awarded to Steven Chu, Claude Cohen-Tannoudji and William Phillips for their development of methods to cool and trap atoms with laser light, while the 2005 Prize was awarded to John Hall and Theodor Hänsch for their "contributions to the development of laser-based precision spectroscopy, including the optical frequency comb technique" and to Roy Glauber for "his contribution to the quantum theory for optical coherence".

In 2018, the Prize was won by Gérard Mourou and Donna Strickland for their method of generating high-intensity, ultra-short optical laser pulses, and by Arthur Ashkin for the invention of optical tweezers with applications to the study of biological systems. An optical tweezer is an instrument that uses a highly focussed laser beam to move individual microscopic objects and has even been applied to cells and bacteria. With his interest in biology, this is something Schrödinger would have approved of.

In the centres of physics research throughout the world today there is a major effort to invent and develop new devices with novel properties all deriving from quantum mechanics. Examples are the fields of quantum electronics, quantum materials, quantum cryptography, quantum teleportation, quantum information processing, and quantum computing. There are sure to be new Nobel prizes in some of these areas in the future.

Fundamental effects in magnetism derive from the understanding of Schrödinger's wave mechanics. Nuclear Magnetic Resonance (NMR) has developed into a prime method for determining the structures of complex molecules. It involves applying magnetic fields to perturb the quantum states deriving from spins of the atomic nuclei. Isidor Rabi, who studied with Bohr, Pauli and Heisenberg, was awarded the Nobel Prize for Physics in 1944 for the first NMR experiments on atoms. Then, in 1952, the Prize went to Edward Purcell and Felix Bloch for research leading to the extension of NMR to more complex systems. Bloch was a student in Zurich in 1926 when he witnessed the first announcement by Schrödinger of his equation and he also used wave mechanics to develop the fundamental theory for the periodic electronic structure of solids.

There have also been Chemistry Nobels for NMR including that to Richard Ernst in 1991 and Kurt Wüthrich in 2002 who extended the method to the determination of the structures of biomolecules. The technique of Magnetic Resonance Imaging (MRI) derives from NMR and has become a widely used diagnostic tool in medicine. This was acknowledged not by the Nobel Committees for Physics or Chemistry but by the Committee for Physiology or Medicine who gave the award for 2003 to Paul Lauterbur and Sir Peter Mansfield.

Other novel magnetic effects have also won Nobels. Robert Laughlin proposed a new wave function which won him the 1998 Physics Prize for explaining the Fractional Quantum Hall effect, observed by Horst Störmer and Daniel Tsui, in which the conductance of electrons is quantized by a magnetic field. The Integer Quantum Hall effect had been observed by Klaus von Klitzing who was awarded the Prize in 1985.

Several of the Physics Nobel awards for other experimental work, including the strange properties of systems such as liquid helium at very low temperatures, require wave mechanics to aid with the explanation of the results. Indeed the 1978 Prize was awarded to Paul Dirac's great friend Pyotr Kapitsa for his work in this field. Schrödinger had lectured to the Kapitsa Club some 46 years before in Cambridge.

Novel quantum mechanical effects such as superconductivity and superfluidity have also received Nobels. The theory for these effects was pioneered by Schrödinger's assistant Fritz London when he was working in Oxford in the 1930s. Sadly, the Nobel list does not include London himself as he died at the young age of 54 before he could be awarded the Prize for this work or for his quantum theory of the chemical bond.

The Prize for 2003 was won by Alexei Abrikosov, Vitaly Ginzburg and Sir Anthony Leggett for contributions to the theory of superfluids, while Georg Bednorz and Alexander Müller won the 1987 award for their discovery of superconductivity in ceramic materials. Superconductivity has been observed at increasingly higher temperatures, and the detailed quantum mechanical explanation is still under much debate.

David Wineland's Nobel Prize lecture of 2012 was entitled "Superposition, Entanglement, and Raising Schrödinger's Cat" in which he discusses the quantum phenomenon of entanglement, a term first used by Schrödinger in his 1935 paper published when he was in Oxford.[1] The lecture also demonstrated that the statement made in a paper by Schrödinger,[20] "We never experiment with just one electron or atom or small molecule", has been shown to be incorrect with the most modern instrumentation that Wineland and others have developed. Wineland won the Prize with Serge Haroche for "ground-breaking experimental methods that enable measuring and manipulation of individual quantum systems."

The development of new materials whose properties require Schrödinger's wave mechanics for their understanding has also been acknowledged by several Nobel prizes. Andre Geim and Konstantin Novoselov won the Physics Prize in 2010 for discovering graphene, a novel two-dimensional carbon material. Graphene has many unusual electronic properties which require not just the Schrödinger equation for an explanation but also the Dirac equation. Research leading to a Chemistry Nobel Prize helped to initiate this field. The 1996 Laureates Sir Harry Kroto, Richard Smalley and Robert Curl were the first to observe the C_{60} molecule, which has the highly symmetric shape of a football. The special stability of this "fullerene" carbon molecule had been previously predicted from molecular orbital wave mechanical theory. Following on from this work, larger carbon structures called nanotubes were made and this initiated the huge fields of nanoscience and nanotechnology which involve understanding the structures and properties of materials on the length scale of 10^{-9} metres.

In 2016 the Nobel Laureates in Physics were David Thouless, Duncan Haldane and Michael Kosterlitz who developed new topological concepts in condensed matter physics. Their work gave a fundamental understanding of novel "topological materials", which have a deformable electronic structure different to that found in ordinary insulators and metals. Schrödinger's wave mechanics formed the underlying theory of their work.

Wave mechanics has also been an important part of the theories used in particle physics, a field which has received many Nobel Prizes. Even Richard Feynman's path integrals, widely used as a theoretical tool in the prediction of elementary particles, can be mapped onto the Schrödinger equation.

The Nobel Prize in Physics in 2020 was awarded to the Oxford mathematician Roger Penrose "for the discovery that black hole formation is a robust prediction of the general theory of relativity". Penrose has often written on Schrödinger's work. He stated that Schrödinger's *What is Life?* "must surely rank among the most influential of scientific writings in the twentieth century".[21] Penrose also has written warmly about Schrödinger's book *Space-Time Structure* which introduced him as a student to relativity.[21,22] On quantum mechanics Penrose said:

> The theory has, indeed, two powerful bodies of fact in its favour, and only one thing against it. First, in its favour are all the marvellous agreements that the theory has had with every experimental result to date. Second, and to me almost as important, it is a theory of astonishing and profound mathematical beauty. The one thing that can be said against it is that it makes absolutely no sense.[23]

There are similarities between these two mathematical physicists, Schrödinger and Penrose, who both worked in Oxford, have been active in the communication of science and branched out into wider topics after publishing fundamental theoretical work that was to win them the Nobel Prize. Penrose was only the second person to be awarded the Nobel Prize for Physics while in an academic appointment at Oxford. Schrödinger was the first.

Nowadays, science communication to the general public has become a major expectation from leading scientists but it had far less emphasis in the 20th century. Schrödinger was an exception and pioneered this activity in Ireland. He achieved a very popular status and his talks and publications still have influence there today. He wrote and spoke with great clarity. Walter Thirring said:

> In his scientific papers his style was always superb. He told me how to do this. When I was writing a paper I showed it to him. He said, no, that's not good, you have to do it quite differently, you have to go to your room, lock the door and read it loudly, one sentence after the other. After each sentence you think a little bit about whether this is really the best way you can do it or whether by reshuffling the sentence you can make a little bit clearer what you want to say.[24]

Daughter

Schrödinger's daughter Ruth, who was brought up in Schrödinger's house in Dublin together with her mother Hilde and her godmother Anny, became the executor of his estate after Anny's death in 1965. Ruth was a proud supporter and defender of Erwin Schrödinger. She arranged for Anny's name to be added to Schrödinger's grave in Alpbach, the place in Austria he loved the most. Over the years, Ruth attended many symposia and lectures given in the name of her father. She also contributed to publications and was not afraid to correct public statements about him which she considered to be incorrect.

The archives of Schrödinger's papers are mainly kept in two places. The majority were retrieved after the Second World War from Graz and, following an interview of Anny by the scientific historian Thomas Kuhn, they were kept by the University of Vienna. They were initially put on microfilm but more recently have been made available by the University on the internet in open access and high resolution.[25] This is an excellent resource for scholars of quantum mechanics and of Schrödinger. A smaller amount of archive material was left at the house in Alpbach where Schrödinger spent much time in the last years of his life. His daughter Ruth enabled scholars, including the author of this book, to view these documents and they have recently been put on loan by the Braunizer family at the Brenner Archive Research Institute at the University of

Innsbruck together with the papers of his close colleague and husband of Hilde, Arthur March.

In 1956 Ruth married Arnulf Braunizer, an Austrian engineer. He told the writer of this book that as a ten-year-old boy he had witnessed Hitler's visit to Austria at the time of the Anschluss in 1938. He also said that he was conscripted into the German forces and was involved in an anti-aircraft artillery unit in the last days of the Second World War. Arnulf was highly critical of the Nazis. He was a staunch supporter of Ruth and the name of Erwin Schrödinger. They had four children and subsequently lived in Alpbach. They often visited Ireland, where Ruth had several friends from her childhood, and they also kept a house at Painswick in the Cotswolds in England.

This book has intentionally kept away from the details of Schrödinger's extra-marital affairs. His relationship with Hilde March, which was carried out with the full consent of his colleague Arthur March, is widely known. It is probable that Schrödinger would have found his own unusual personal life to be challenged in the modern academic world. This aspect is often commented on by students when the writer of this book gives lectures on the career of Schrödinger. He is not the only famous theoretical physicist to have a complicated personal life. His great friend Albert Einstein also had several relationships.[26]

The biography of Schrödinger by Walter Moore, *Schrödinger: Life and Thought*,[27] made many assumptions on Schrödinger's personal life. This resulted in several reviews of Moore's book which highlighted, in sensational terms, Schrödinger's love life and not his great scientific achievements. His daughter Ruth Braunizer had the view that some of the details in Moore's book, especially on Schrödinger's personal relationships, were incorrectly interpreted from Schrödinger's handwritten diaries and poems. Following the publication of an essay and a fictitious novel featuring Schrödinger by Neil Belton,[28] Ruth Braunizer wrote in forthright terms on 11 June 2005 to the *Irish Times*, which had previously published so many articles on her father:

> I had not heard of Walter Moore until he — an elderly gentleman accompanied by his wife, Pat — knocked at my door in the mid-1980s

and introduced himself as my father's biographer. As an ardent admirer of my father he considered himself fully competent to take on the task; that he had never met my father did not seem to bother him at all...

I had my doubts, and they were confirmed when it turned out that he had trouble deciphering the vast amount of material he wanted to see. It would have been essential for him to study the German language and the Gothic script and the Gabelsberg shorthand before undertaking what he had set out to do...

No one who did not live in the Nazi regime could imagine what life was like then, or understand the situation that prevailed in Germany and Austria under Hitler. When Moore realised he was short of time he amplified the sparse information he could gather by filling the gaps with fiction. And by doing this he concocted a new character, one that did not resemble the one he had set out to describe...

It was one of the darkest periods in German history. Many of my father's colleagues were being dismissed and replaced by others. The criteria were acceptance of the regime's ideas and proof of pure Aryanism.

My father could have fulfilled both. But he was not prepared to accept the new regime and co-operate with it... My father left Berlin, where he had spent many happy years; and because he did not foresee — like many Austrians, among them Jews — what was going to happen in the near future, he returned to Austria. When Austria ceased to exist he was immediately dismissed, for obvious reasons. This time he had to choose between internment and escaping. He chose the latter.[29]

Arnulf Braunizer had also looked through Schrödinger's private diaries and many poems in detail. He confirmed to the author of this book that the diaries were handwritten, often in shorthand, very hard to read and consequently open to wide interpretation.[30] In Chapter 1 of this book, the evidence is discussed on whether there really was an unknown lady who inspired Schrödinger's invention of his equation at Arosa around Christmas time of 1925, as suggested by Moore.[27]

On the topic of his relationship with women, Schrödinger, in a brief memoir, had a rather evasive tone:

I must refrain from drawing a complete picture of my life, as I am not good at telling stories; besides, I would have to leave out a very

> substantial part of this portrait, *i.e.*, that dealing with my relationships with women. First of all it would no doubt kindle gossip, secondly it is hardly interesting enough for others, and last but not least I don't believe anyone can or may be truthful enough in those matters.[31]

Schrödinger mentions gossip in this memoir but it has to be said he brought this on himself. Even in his 60s he was having open liaisons at scientific meetings. From a conference in Geneva, Max Born wrote on 5 September 1952 to his wife Hedi:

> In this pension there is Schrödinger — but not with Anny. There is another female with him, black haired, small, quite pretty. They say "du" to one another and are very intimate. What a fellow! We all have our meals together at the same table.[32]

It is reported in Moore's book that Schrödinger fathered, in addition to Ruth, at least two other children who were both born in Ireland.[27] The mother of one was Sheila May, an actress, and the other mother has been given the pseudonym Kate Nolan.[33] Her daughter Linda was eventually taken by her mother to southern Africa. Linda's son is the theoretical physicist Terry Rudolph who is now, most appropriately, Professor of Quantum Physics at Imperial College London and has written several scientific papers related to Schrödinger's ideas on entanglement.

The writer of the current book met Walter Moore just once. This was at the University of Sydney, Australia in 1994 when I gave a seminar on my research using Schrödinger's equation to calculate the dynamics and rates of chemical reactions. In the question period after the seminar a senior gentleman asked me: "Were you using the time-independent or the time-dependent Schrödinger equation?" I replied that it was the time-independent equation. After the seminar he introduced himself to me as Walter Moore, a Professor of Physical Chemistry in Sydney. Before writing the book on Schrödinger, Moore had written one of the standard texts on physical chemistry and was well known in that field.[34]

On 27 April 2008 a symposium was organised by Magdalen College, Oxford, on the occasion of its 550th anniversary, to celebrate the seven members of the College who, by that time, had won the Nobel Prize for science. There were talks on Howard Florey, Erwin Schrödinger, Charles Sherrington, John Eccles, Peter Medawar and Robert Robinson, the last four of which had all known Schrödinger at Magdalen.

There was also a lecture by Sir Tony Leggett. After Schrödinger, he was the second member of Magdalen to win the Nobel Prize for Physics and had been a Fellow by Examination at the College from 1963–67. Leggett's lecture was "Does the everyday world really obey quantum mechanics?" and he discussed some of the ideas raised by the famous "Schrödinger Cat" paper written when Schrödinger was a Fellow of Magdalen College in 1935. Tony Leggett is now the Chief Scientist at the Institute for Condensed Matter in the University of Illinois at Urbana-Champaign.

Also present at the symposium was Ruth Braunizer, returning to Magdalen where she had been christened in the College Chapel some 74 years before in 1934. In addition, Gustav Born, the son of Max Born, attended the symposium. He had been a research student of Howard Florey at Magdalen after the war and had then lived in the home of Francis Simon who had recommended Magdalen to his father as a good

Left to right: Arnulf Braunizer, Ruth Braunizer, David Clary and Gustav Born, at Magdalen College, Oxford on 27 April 2008.

college for a graduate student studying medicine.[35] Despite the fact that Ruth Braunizer and Gustav Born were both children of famous refugee fathers who were close friends and had lived in Oxford and Cambridge, respectively, from 1933–36 they had no recollections or records of ever meeting before. The moment of their very first meeting is recorded in the photograph shown on the previous page which was taken at the 2008 Symposium at Magdalen College.

Ruth and Gustav had many lively discussions while staying in the President's Lodgings during the period of the symposium at Magdalen College. Gustav said he remembered well when he was 12 years old and had to share a bedroom with Anny Schrödinger in the summer of 1933 in the Tyrol when many of the quantum scientists got together at the time of the very difficult political situation that had developed back in

Ruth Braunizer in 2008 in the office of the President of Magdalen College, where her father heard he won the Nobel Prize on November 9, 1933.

Germany. Gustav said he had never forgotten that Anny kept him awake all night due to her loud snoring.

There is another photograph shown on the previous page of Ruth Braunizer in the office of the President of Magdalen College, the place where her father heard he won the Nobel Prize in 1933. The photograph also displays a fine sketch of Schrödinger which was drawn by the artist Peter Edwards who had painted previously the portrait of the Irish poet Seamus Heaney, another Magdalen Fellow who won the Nobel Prize, in this case for literature.

Ruth Braunizer died at the age of 84 in the summer of 2018 and her husband Arnulf died a few weeks later at the age of 90. Gustav Born also died in 2018 at the fine age of 96. He was doing work connected with the Council for At-Risk Academics right up until he died. This is the modern version of the Academic Assistance Council which helped so many scientific refugees, including his father and Schrödinger, in the 1930s.

Opening of the Schrödinger Building at the Oxford Science Park in 2018. Left to right: Piers Scrimshaw-Wright (Science Park CEO), Sam Gyimah (Minister for Universities and Science), David Clary, and Rory Maw (Bursar of Magdalen College).

Epilogue

Magdalen College owns and manages the Oxford Science Park where there are many companies developing innovations mostly deriving from research done at the University of Oxford. In 2018 a major new building was built there and named after Schrödinger. This was the first time that Schrödinger had been commemorated in Oxford. The ground for the building was cut by the UK Minister for Science and Universities Jo Johnson, the brother of the subsequent Prime Minister Boris Johnson.

The Schrödinger building was opened by the newly appointed UK Science Minister Sam Gyimah (see the photograph on the previous page). In the speeches for the opening of the building he and the writer of this book, who was then President of Magdalen College, emphasised that many new innovations in science and technology would not have been possible without the work of Schrödinger. One of the new companies that moved into the Schrödinger building was Vaccitech, the company that subsequently developed the Oxford-AstraZeneca vaccine which saved many lives worldwide in the Covid-19 pandemic.

A year after the opening of the Schrödinger Building two more buildings in the Oxford Science Park were named after Charles Sherrington and Henry Whitehead, both Fellows of Magdalen College who feature in this book and were highly regarded by Schrödinger. In the same month Sir Peter Ratcliffe, a current Fellow of Magdalen, was awarded the Nobel Prize for Physiology or Medicine for his research on hypoxia. He is the tenth member of this College to win the Nobel Prize.

Erwin Schrödinger held academic appointments at Jena, Stuttgart, Breslau, Zurich, Berlin, Oxford, Graz, Ghent, Dublin and Vienna. It was only when he was at the University of Zurich in 1926 when he published his truly great scientific work. The 1930s was an extraordinary time of disruption in Europe for Schrödinger and for so many of his colleagues. His years at Oxford are a significant part of his story and this book *Schrödinger in Oxford* has been written with the intention to put his whole career in context.

Unfortunately, Schrödinger did not settle well in his time in Oxford. He did not have a permanent academic position and he was not prepared to wait until a suitable statutory professorship became available. Schrödinger's first term in Oxford in 1933 was dominated by the award of his Nobel Prize and his subsequent visit to Stockholm to receive the Prize. In the next few months he went to the USA where he fully expected to be offered a position alongside his great friend Albert Einstein at the Institute for Advanced Study at Princeton. He then made two extended visits to Spain where he was very warmly received. After that he was negotiating with the authorities in Austria to return to his home country in 1936. Schrödinger himself admitted subsequently that this was his most foolish mistake.[31] Many of his contempories considered his statement which seemed to support the Führer, and was published just after the Anschluss in 1938, was also very badly judged. It was clear, however, that he was under extreme pressure at that time from the Nazi authorities.

President Gordon of Magdalen College tried heroically to help Schrödinger establish himself in Oxford but he was not successful. However, he did publish his papers on entanglement and Schrödinger's Cat during his Oxford period. The world of science continues to be influenced by these works. It must be said that Schrödinger's time in Oxford was an important part of the complicated journey of this extraordinary man, without doubt one of the greatest scientists of the 20th century. In addition, few individuals are having a greater influence on 21st century science and technology through their scientific work. Schrödinger's wave mechanics is the foundation of every area of chemistry, a large amount of physics and his influence on biology has been huge.

Perhaps Erwin Schrödinger is best summed up by Max Born who said: "His private life seemed strange to bourgeois people like ourselves. But all this did not matter. He was a most loveable person, independent, amusing, temperamental, kind and generous, and he had a most perfect and efficient brain."[36]

References

Chapter 1: Schrödinger's Breakthrough

1. A. Hermann, *Erwin Schrödinger: Die Wellenmechanik* (Battenberg, Stuttgart, 1963).
2. E. Schrödinger, Quantisierung als Eigenwertproblem (Erste Mitteilung), *Ann. Der Phys.*, **79**, 361 (1926).
3. N. Bohr, On the Constitution of Atoms and Molecules, *Phil. Mag.*, **26**, 1 (1913).
4. M. Born, *My Life: Recollections of a Nobel Laureate* (Taylor and Francis, London, 1978).
5. L. de Broglie, A Tentative Theory of Light Quanta, *Phil. Mag.*, **47**, 446 (1924).
6. F. Bloch, Heisenberg and the Early Days of Quantum Mechanics, *Physics Today*, **29**, 23 (1976).
7. E. Schrödinger, Quantisierung als Eigenwertproblem (Zweite Mitteilung), *Ann. Der Phys.*, **79**, 489 (1926).
8. E. Schrödinger, Über das Verhältnis der Heisenberg-Born-Jordanschen Quantenmechanik in der meinen, *Ann. Der Phys.*, **79**, 734 (1926).
9. E. Schrödinger, Quantisierung als Eigenwertproblem (Dritte Mitteilung). Störungstheorie, mit Anwendung auf den Starkeffekt der Balmerlinien, *Ann. Der Phys.*, **80**, 437 (1926).
10. E. Schrödinger, Quantisierung als Eigenwertproblem (Vierte Mitteilung), *Ann. Der Phys.*, **81**, 109 (1926).
11. D.C. Clary, 100 years of Atomic Theory, *Science*, **341**, 244 (2013).
12. G.E. Uhlenbeck and S. Goudsmit, Spinning Electrons and the Structure of Spectra, *Nature*, **117**, 264 (1926).

13. G.W. Kellner, Die Ionisierungsspannung des Heliums nach der Schrödingerschen Theorie, *Z. Phys.*, **44**, 91 (1927).
14. W. Heitler and F. London, Wechselwirkung Neutraler Atome und Homöopolare Bindung Nach der Quantenmechanik, *Z. Phys.*, **44**, 455 (1927).
15. L. Pauling, Schrödinger's Contributions to Chemistry and Biology, in *Schrödinger Centenary Celebrations of a Polymath*, C.W. Kilmister, Ed. (Cambridge University Press, Cambridge, 1987).
16. P.A.M. Dirac, The Quantum Theory of the Electron, *Proc. Roy. Soc. A*, **117**, 610 (1928).
17. P.A.M. Dirac, Quantum Mechanics of Many-Electron Systems, *Proc. Roy. Soc. A*, **123**, 714 (1929).
18. Schrödinger Archiv, Österreichische Zentralbibliothek für Physik, Wien. (t) refers to a translation from German to English.
19. H. Thirring, Interview by T.S. Kuhn, American Institute of Physics. Niels Bohr Library & Archives, College Park, USA (1963).
20. E. Schrödinger, Höhenverteilung der durchdringenden atmosphärischen Strahlung, *S. Ber. Akad. Wiss. Wien*, **121**, 2391 (1912).
21. E. Schrödinger, Zur Theorie der Fall-und Steigversuche an Teilchen mit Brownscher Bewegung, *Phys. Z.*, **16**, 289 (1915).
22. E. Schrödinger, Zur Akustik der Atmosphäre, *Phys. Z.*, **18**, 445 (1917).
23. E. Schrödinger, Die Ergebnisse der Neueren Forschung über Atom und Molekularwärmen, *Naturwiss.*, **5**, 537 (1917).
24. E. Schrödinger, Die Energiekomponenten des Gravitationsfeldes, *Phys. Z.*, **19**, 4 (1918).
25. E. Schrödinger, Ein Lösungssystem der allgemeinen kovarianten Gravitationsgleichungen, *Phys. Z.*, **19**, 20 (1918).
26. E. Schrödinger, Theorie der Pigmente von grosster Leuchtkraft, *Ann. der Phys.*, **62**, 603 (1920).
27. E. Schrödinger, Versuch zur modellmässigen Deutung des Terms der scharfen Nebenserien, *Z. Phys.*, **4**, 347 (1921).
28. E. Schrödinger, Über die spezifische Wärme fester Körper bei hoher Temperatur und über die Quantelung von Schwingungen endlicher Amplitude, *Z. Phys.*, **11**, 170 (1922).
29. E. Schrödinger, Über eine bemerkenswerte Eigenschaft der Quantenbahnen eines einzelnen Elektrons, *Z. Phys.*, **12**, 13 (1922).
30. A. Schrödinger, Interview by T.S. Kuhn, American Institute of Physics. Niels Bohr Library & Archives, College Park, USA (1963).

31. M. Atiyah, Hermann Weyl 1885–1955, *Biogr. Mem. Nat. Acad. Sci.*, **82** (2002).
32. S.N. Bose, Plancks Gesetz und Lichtquantenhypothese, *Z. Phys.*, **26**, 178 (1924).
33. A. Einstein, Quantentheorie des Einatomigen Idealen Gases, *Königliche Preussische Akademie der Wissenschaften*, 261 (1924).
34. E. Schrödinger, Die Energiestufen des Idealen Einatomigen Gas Modells, *Sitzungsberichte der Preussischen Akademie der Wissenschaften: Physikalisch-mathematische Klasse*, 23 (1926).
35. E. Schrödinger, Zur Einsteinschen Gastheorie, *Phys. Z.*, **27**, 95 (1926).
36. M.H. Anderson, J.R. Ensher, M.R. Matthews, C.E. Wieman and E.A. Cornell, Observation of Bose-Einstein Condensation in a Dilute Atomic Vapor, *Science*, **269**, 198 (1995).
37. W.J. Moore, *Schrödinger: Life and Thought* (Cambridge University Press, Cambridge, 1989).
38. E.P. Fischer, We Are All Aspects of One Single Being: An Introduction to Erwin Schrödinger, *Soc. Res.*, **51**, 809 (1984).
39. E.P. Fischer and C. Lipson, *Thinking About Science: Max Delbrück and the Origins of Molecular Biology* (Norton, New York, 1988).
40. A. Pais, *Inward Bound: Of Matter and Forces in the Physical World*, p. 252 (Clarendon Press, Oxford, 1986).
41. K. von Meyenn, Ed., Letter in *Eine Entdeckung von ganz ausserordentlicher Tragweite: Schrödingers Briefwechsel zur Wellenmechanik und zum Katzenparadoxon*, Band 1 and 2 (Springer, Berlin/Heidelberg, 2011). (t) refers to a translation from the German to the English.
42. W. Rotach, *Reihenentwicklungen einer willkürlichen Funktion nach Hermite'schen und Laguerre'schen Polynomen*, Doctoral Thesis, ETH Zurich (1925).
43. W. Heisenberg, Über Quantentheoretische Umdeutung Kinematischer und Mechanischer Beziehungen, *Z. Phys.*, **33**, 879 (1925).
44. O. Klein, Quantentheorie und Fünfdimensionale Relativitätstheorie, *Z. Phys.*, **37**, 895 (1926).
45. W. Gordon, Der Comptoneffekt nach der Schrödingerschen Theorie, *Z. Phys.*, **40**, 117 (1926).
46. E.A. Hylleraas, Neue Berechnung der Energie des Heliums im Grundzustande, Sowie des Tiefsten Terms von Ortho-Helium, *Z. Phys.*, **54**, 347 (1929).
47. W. Pauli, Über das Wasserstoffspektrum vom Standpunkt der neuen Quantenmechanik, *Z. Phys.*, **36**, 336 (1926).

48. C.P. Enz, W. Pauli's Scientific Work, in *The Physicist's Conception of Nature*, J. Mehra, Ed. (Reidel, Dordrecht, 1973).
49. R. Peierls, Wolfgang Ernst Pauli: 1900–1958, *Biogr. Mems. Fell. R. Soc.*, **5**, 186 (1960).
50. C.P. Enz, *No Time to be Brief: A Scientific Biography of Wolfgang Pauli* (Oxford University Press, Oxford, 2002).
51. W. Heisenberg, Interview by T.S. Kuhn and J.L. Heilbron, Session I, American Institute of Physics. Niels Bohr Library & Archives, College Park, USA (1962).
52. W. Heisenberg, The Development of Quantum Mechanics, in *Nobel Lectures Physics 1922–1941* (Elsevier, Amsterdam, 1965).
53. M. Born, Zur Quantenmechanik der Stossvorgänge, *Z. Phys.*, **37**, 863 (1926).
54. D.C. Clary, Quantum Dynamics of Chemical Reactions, *Science*, **321**, 789 (2008).
55. P. Halpern, *Einstein's Dice and Schrödinger's Cat* (Basic Books, New York, 2015).
56. N.T. Greenspan, *The End of the Certain World: The Life and Science of Max Born* (Basic Books, New York, 2005).
57. W. Heisenberg, Ueber den Anschaulichen Inhalt der Quantentheoretischen Kinematik und Mechanik, *Z. Phys.*, **43**, 172 (1927).
58. E. Schrödinger, *Collected Papers on Wave Mechanics together with his Four Lectures on Wave Mechanics* (AMS Chelsea Publishing, Providence, 1928).
59. E. Schrödinger, An Undulatory Theory of the Mechanics of Atoms and Molecules, *Phys. Rev.*, **28**, 1049 (1926).
60. K. von Meyenn, Die Rezeption der Wellenmechanik und Schrödingers Reise nach Amerika im Winter 1926/27, *Gesnerus*, **39**, 261 (1982).
61. G. Bacciagaluppi and A. Valentini, *Quantum Theory at the Crossroads: Reconsidering the 1927 Solvay Conference* (Cambridge University Press, Cambridge, 2009).
62. G. Farmelo, *The Strangest Man: The Hidden Life of Paul Dirac, Quantum Genius* (Faber and Faber, London, 2009).
63. K. Gavroglu, *Fritz London: A Scientific Biography* (Cambridge University Press, Cambridge, 1995).
64. N.F. Mott, Walter Heinrich Heitler: 1904–1981, *Biogr. Mems. Fell. R. Soc.*, **28**, 141 (1982).
65. W. Heitler, Erwin Schrödinger: 1887–1961, *Biogr. Mems. Fell. R. Soc.*, **7**, 221 (1961).

66. L. Pauling, *The Nature of the Chemical Bond and the Structure of Molecules and Crystals: An Introduction to Modern Structural Chemistry* (Cornell University Press, Ithaca, 1939).
67. F. London and M. Polanyi, Über die atomtheoretische Deutung der Adsorptionskräfte, *Naturwissen.*, **18**, 1099 (1930).
68. H. Eyring and M. Polanyi, Über einfache Gasreaktionen, *Z. Phys. Chem.*, **12**, 279 (1931).
69. F. London, Zur Theorie und Systematik der Molekularkräfte, *Z. Phys.*, **63**, 245 (1930).
70. Kapitsa Club Minutes, Cockcroft papers, Churchill Archive Centre, Churchill College, Cambridge.
71. E. Schrödinger, Diracsches Elektron im Schwerefeld I., *Sitzungsberichte der Preussischen Akademie der Wissenschaften: Physikalisch-mathematische Klasse*, 105 (1932).
72. P.A.M. Dirac, Prof. Erwin Schrödinger, For.Mem.R.S., *Nature*, **189**, 355 (1961).
73. J. Navarro, Electron Diffraction chez Thomson: Early Responses to Quantum Physics in Britain, *Brit. J. Hist. Sci.*, **43**, 245 (2010).
74. F.B. Pidduck, Laguerre's Polynomials in Quantum Mechanics, *J. Lond. Math. Soc.*, **4**, 163 (1929).
75. C. Eckart, Interview by J.L. Heilbron, American Institute of Physics. Niels Bohr Library & Archives, College Park, USA (1962).
76. E. Schrödinger, Über die Umkehrung der Naturgesetze, *Sitzungsberichte der Preussischen Akademie der Wissenschaften: Physikalisch-mathematische Klasse*, 144 (1931).
77. K. Liu *et. al.*, Characterization of a Cage Form of the Water Hexamer, *Science*, **381**, 501 (1996).
78. M. Walker, The Rise and Fall of an "Aryan" Physicist, in *Nazi Science*, p. 5 (Springer, Boston, 1995).
79. P. Ball, *Serving the Reich: The Struggle for the Soul of Physics Under Hitler* (University of Chicago Press, Chicago, 2014).
80. B.R. Brown, *Planck: Driven by Vision, Broken by War* (Oxford University Press, Oxford, 2015).
81. A. Robinson, *Einstein on the Run: How Britain Saved the World's Greatest Scientist* (Yale University Press, New Haven, 2019).
82. J.L. Heilbron, *The Dilemmas of an Upright Man: Max Planck and the Fortunes of German Science* (Harvard University Press, Massachusetts, 2000).

83. J. Borkin, *The Crime and Punishment of I.G. Farben* (Barnes and Noble, New York, 1978).
84. P.K. Hoch, The Reception of Central European Refugee Physicists of the 1930s: U.S.S.R., U.K., U.S.A., *Ann. Sci.*, **40**, 217 (1983).

Chapter 2: To Oxford and the Nobel Prize

1. L.W.B. Brockliss, *The University of Oxford: A History* (Oxford University Press, Oxford, 2016).
2. L.W.B. Brockliss, Ed., *Magdalen College Oxford: A History* (Magdalen College, Oxford, 2008).
3. Lord Cherwell papers, D224–5, Nuffield College Archive, Oxford. (t) refers to a translation from German to English.
4. Ch. 1, Ref. 56.
5. Ch. 1, Ref. 4.
6. G. Kerber, A. Dick and W. Kerber, *Erwin Schrödinger 1887–1961, Documents, Materials and Pictures* (Austrian Central Library for Physics, Vienna, 2015).
7. Ch. 1, Ref. 37.
8. Ch. 1, Ref. 41.
9. M.J. Klein, *Paul Ehrenfest: The Making of a Theoretical Physicist* (Elsevier, Oxford, 1985).
10. Ch. 1, Ref. 62.
11. Archives, Magdalen College, Oxford.
12. T. Colacicco, in *Education of Italian Elites, 19th–20th Century Case Studies*, A. Gaudio, Ed. (Aracne, Rome, 2018).
13. C. Malaparte, *Kaputt*, C. Foligno, Trans. (NYRB, New York City, 1948).
14. Archive of the Society for the Protection of Science and Learning, Special Collections, Oxford University Library Services.
15. Historical Newspapers, Gale Primary Sources, Oxford University Library Services.
16. Ch. 1, Ref. 1.
17. Nobel Prize Archives, The Nobel Foundation, Stockholm, Sweden.
18. W. Heisenberg, Mehrkörperproblem und Resonanz in der Quantenmechanik, *Z. Phys.*, **38**, 411 (1926).
19. K.F. Bonhoeffer and P. Harteck, Experimente über Para- und Orthowasserstoff, *Naturwiss.*, **17**, 182 (1929).

20. W.E. Garner and J.E. Lennard-Jones, Molecular Spectra and Molecular Structure, *Trans. Faraday Soc.*, **25**, 611 (1929).
21. C.D. Anderson, The Positive Electron, *Phys. Rev.*, **43**, 491 (1933).
22. Ch. 1, Ref. 16.
23. Ch. 1, Ref. 43.
24. Staatsbibliothek Preussischer Kulturbesitz, Berlin, Nachlass Born 704, BI 90–94.
25. Ruth Braunizer, private communication, Schrödinger Archive, Alpbach.
26. Paul A.M. Dirac Papers, Florida State University Libraries, Special Collections and Archives.
27. Science News, *Science-Supplement*, **78**, 6 (1933).
28. Nobel Prizes for Quantum Theory Investigations, *Nature*, **132**, 775 (1933).
29. Ch. 1, Ref. 18.
30. J. Hendry, Ed., *Cambridge Physics in the Thirties* (Adam Hilger, Bristol, 1984).
31. A. Hodges, *Alan Turing: The Enigma*, p. 415 (Hutchinson, London, 1983).
32. Ch. 1, Ref. 78.
33. D.C. Cassidy, *Beyond Uncertainty: Heisenberg, Quantum Physics, and the Bomb* (Bellevue Literary Press, New York, 2010).
34. H. Kragh, *Simply Dirac* (Simply Charly, New York, 2016).
35. E. Schrödinger, *What is Life? With Mind and Matter and Autobiographical Sketches* (Cambridge University Press, Cambridge, 2012).

Chapter 3: Life and Work in Oxford

1. Ch. 2, Ref. 25.
2. A. Bennett, *Writing Home* (Faber and Faber, London, 2014).
3. Ch. 1, Ref. 4.
4. Ch. 2, Ref. 6.
5. P. Ayres, *Shaping Ecology: The Life of Arthur Tansley* (Wiley-Blackwell, New Jersey, 2012).
6. Arthur Tansley Papers, University Library, University of Cambridge.
7. S. Zuckerman, *From Apes to Warlords: the Autobiography of Solly Zuckerman (1904–1946)* (Hamish Hamilton, London, 1978).
8. K. Popper, *Unended Quest: An Intellectual Autobiography* (William Collins, Glasgow, 1974).
9. Ch. 1, Ref. 30.
10. W.J.M. Mackenzie, *Explorations in Government. Collected Papers: 1951–68*, pp. xxii–xxiii (Macmillan Press, London, 1975).

11. Ch. 2, Ref. 15.
12. P.M.S. Blackett Papers, Royal Society Archives, London.
13. Chapel Archives, Magdalen College, Oxford.
14. Ch. 1, Ref. 37.
15. A. March, Mathematische Theorie der Regelung nach der Korngestalt bei Affiner Deformation, *Z. Kristallogr.*, **81**, 285 (1932).
16. A. March, On the Adsorption Theory of the Electrokinetic Potential, *Trans. Faraday Soc.*, **31**, 1468 (1935).
17. Ch. 1, Ref. 81.
18. Ch. 2, Ref. 15.
19. R. Fox, Einstein in Oxford, *Notes Rec. R. Soc. Lond.*, **72**, 293 (2018).
20. A. Robinson, Einstein in Oxford, *Physics World*, June 2019. Quoted from *Thinking as a Hobby*, by W. Golding.
21. W. McRea, Eamon de Valera, Erwin Schrödinger and the Dublin Institute, in *Schrödinger: Centenary Celebrations of a Polymath*, C.W. Kilmister, Ed. (Cambridge University Press, Cambridge, 1987).
22. Ch. 1, Ref. 18.
23. L.M. Lederman and D. Teresi, *The God Particle: If the Universe is the Answer, What is the Question?*, Chapter 5 (Dell, New York City, 1993).
24. Ch. 2, Ref. 3.
25. Director's Office: Faculty files, Shelby White and Leon Levy Archives Center, Institute for Advanced Study, Princeton NJ, USA. (t) refers to a translation from German to English.
26. Ch. 1, Ref. 62.
27. E.P. Wigner, *The Recollections of Eugene P. Wigner, as told to Andrew Szanton* (Plenum, 1992).
28. Ch. 2, Ref. 11. Magdalen College. Summary of Events 1933–34.
29. J.I.D. Diaz, A propósito del potencial de paredes infinitas: el joven Gamow, Schrödinger en España y algunos comentarios matemáticos, *Rev. R. Acad. Cienc. Exact. Fis. Nat. (Esp)*, **109**, 33 (2016).
30. E. Perez, Erwin Schrödinger in the Second Spanish Republic 1934–35, *Biographies in the History of Physics*, C. Forstner and M. Walker, Eds., (Springer, Berlin/Heidelberg, 2020).
31. Fellows' Admissions Book, Magdalen College, Oxford.
32. Ch. 2, Ref. 11.
33. Ch. 2, Ref. 3. 11 Jan 1955.

34. P.K. Hoch and E.J. Yoxen, Schrödinger at Oxford: A Hypothetical National Cultural Synthesis which Failed, *Ann. Sci.*, **44**, 593 (1987).
35. R. Fox and G. Gooday, Eds., *Physics in Oxford 1839–1939* (Oxford University Press, Oxford, 2005).
36. Ch. 1, Ref. 41, pp. 524–5.
37. B. Bleaney, The Physical Sciences in Oxford, 1918–1939 and Earlier, *Notes Rec. R. Soc. Lond.*, **48**, 247 (1994).
38. Ch. 1, Ref. 30.
39. E. Schrödinger, Discussion of Probability Relations between Separated Systems, *Math. Proc. Cam. Phil. Soc.*, **31**, 555 (1935).
40. E. Schrödinger, Probability Relations between Separated Systems, *Math. Proc. Cam. Phil. Soc.*, **32**, 446 (1936).
41. E. Schrödinger, Die Gegenwärtige Situation in der Quantenmechanik, *Naturwissen.*, **23**, 807, 823 & 844 (1935). (t) refers to a translation from German to English.
42. A. Einstein, B. Podolsky and N. Rosen, Can Quantum Mechanical Description of Physical Reality be Considered Complete?, *Phys. Rev.*, **47**, 777 (1935).
43. M. Born, *The Born-Einstein Letters 1916–1955* (Macmillan Press, New York, 1971).
44. Ch. 1, Ref. 41.
45. Dr. Arnold Berliner and *Die Naturwissenschaften*, *Nature*, **136**, 506 (1935).
46. F. London and H. London, The Electromagnetic Equations of the Supraconductor, *Proc. Roy. Soc. A*, **149**, 71 (1935).
47. Ch. 1, Ref. 63.
48. Ch. 1, Ref. 56.
49. Sidgwick Papers Archive, Lincoln College, Oxford.
50. W.H.E. Schwarz *et al.*, Hans G.A. Hellmann (1903–1938) A Pioneer of Quantum Chemistry, Bunsen-Magazin (1 and 2) 10-21 and 60-70 (1999).
51. Ch. 2, Ref. 2.
52. Ch. 2, Ref. 14.
53. J.H. Sanders, Nicholas Kurti: 1908–1998, *Biogr. Mems. Fell. R. Soc.*, **46**, 299 (2000).
54. J. Morrell, *Science at Oxford 1914–1939: Transforming an Arts University* (Clarendon Press, Oxford, 1997).
55. K. Mendelssohn, *The World of Walther Nernst: the Rise and Fall of German Science, 1864–1941* (University of Pittsburgh Press, Pittsburgh, 1973).

Chapter 4: Return to Austria and Escape Back to Oxford

1. Ch. 1, Ref. 30.
2. Ch. 1, Ref. 18.
3. Pontificia Academia Scientiarum, Yearbook 2004, Vatican City.
4. E. Schrödinger, *Science and the Human Temperament* (George Allen and Unwin, London, 1935).
5. Pontificia Academia Scientiarum, Acta Annus I Volumen I, Ex Aedibus Academicis in Civitae Vaticana (1937).
6. Ch. 2, Ref. 6.
7. Ch. 3, Ref. 54.
8. R.H. Beyler, M. Eckert and D. Hoffmann, The Planck Medal, in *The German Physical Society in the Third Reich: Physicists between Autonomy and Accommodation*, D. Hoffmann and M. Walker, Eds., A.M. Hentschel, Trans. (Cambridge University Press, Cambridge, 2007).
9. Minutes of DPG Board, 10 March 1937. See Ref. 8.
10. Ch. 2, Ref. 3.
11. Bicentenary of the Birth of Galvani: Celebration at Bologna, *Nature*, **140**, 836 (1937).
12. Ch. 3, Ref. 12.
13. P.A.M. Dirac, The Cosmological Constants, *Nature*, **139**, 323 (1937).
14. H. Dingle, Modern Aristotelianism, *Nature*, **139**, 784 (1937).
15. P. Neville, *Hitler and Appeasement: The British Attempt to Prevent the Second World War* (Hambledon, London, 2005).
16. Ch. 2, Ref. 15.
17. W. Höflechner, *History of the Karl-Franzens University Graz: From the Beginning until 2005* (Leykam, Graz, 2006).
18. Ch. 2, Ref. 14.
19. G. Holfter, Ludwig Hopf (1884–1939), in *Voices from Exile: Essays in Memory of Hamish Ritchie*, I. Wallace, Ed. (Koninklijke Brill Nv, Leiden, 2015).
20. L.G. Schwoerer, Lord Halifax's Visit To Germany: November 1937, *Historian*, **32**, 353 (1970).
21. G. Holfter and H. Dickel, *An Irish Sanctuary: German Speaking Refugees in Ireland 1933–45* (De Gruyter, Berlin, 2017).
22. Ch. 3, Ref. 10.

23. Grazer Tagespost, 30 March 1938. (t) refers to a translation from German to English.
24. Prof. E. Schrödinger and the University of Graz, *Nature*, **141**, 929 (1938).
25. Ch. 1, Ref. 41.
26. Ch. 3, Ref. 34.
27. Ch. 1, Ref. 72.
28. E. Jones, *The Life and Work of Sigmund Freud* (Pelican Books, London, 1964).
29. L. Bass, Schrödinger: A Philosopher in Planck's Chair, *Brit. J. Phil. Sci.*, **43**, 111 (1992).
30. M. Walker, National Socialism and German Physics, *J. Cont. Hist.*, **24**, 63 (1989).
31. D. Hoffmann, Between Autonomy and Accommodation: The German Physical Society During the Third Reich, *Phys. Perspect.*, **7**, 293 (2005).
32. J. Vogel-Prandtl, *Ludwig Prandtl: A Personal Biography Drawn from Memories and Correspondence* (Universitätsverlag Göttingen, Göttingen, 2014).
33. D. Hoffmann, H. Rössler and G. Reuther, "Lachkabinett" und "grosses Fest" der Physiker. Walter Grotrians "Physikalischer Einakter" zu Max Plancks 80. Geburtstag, *Ber. Wissenschaftsgesch.*, **33**, 30 (2010).
34. Ch. 1, Ref. 82.
35. D. Hoffmann and M. Walker, The German Physical Society under National Socialism, *Physics Today*, **57**, 52 (2004).
36. Ch. 1, Ref. 56.
37. R.L. Sime, *Lise Meitner: A Life in Physics* (University of California Press, Berkeley, 1996).
38. A. Kramish, *The Griffin: The Greatest Untold Espionage Story of World War II* (Houghton, Boston, 1986).
39. K. Kniefacz and H. Posch, Universität Wien, https://geschichte.univie.ac.at/en/articles/expulsion-teachers-and-students-1938.
40. J. Feichtinger, H. Matis, S. Sienell and H. Uhl, *The Academy of Sciences in Vienna 1938 to 1945* (Austrian Academy of Sciences Press, Vienna, 2014).
41. G. Heiss, The University of Vienna from Nazification to De-Nazification, *Dig. Dis.*, **17**, 267 (1999).
42. Ch. 1, Ref. 37.
43. Ch. 2, Ref. 11.
44. Ch. 1, Ref. 30.
45. K.D. McRae, *Nuclear Dawn: F.E. Simon and the Race for Atomic Weapons in World War II* (Oxford University Press, Oxford, 2014).

46. Private communication to the author from Lucy Baxandall, granddaughter of Francis Simon.
47. Papers of Sir Francis (Franz) Eugen Simon, Royal Society Archive, London. (t) refers to translation from German to English.
48. Dublin Institute for Advanced Studies, https://www.dias.ie/2010/09/15/stp-theoreticalhistory1935-1940/.
49. The Papers of Professor Paul Dirac, Churchill Archive Centre, Churchill College, Cambridge.
50. S. Lee, *Sir Rudolf Peierls: Selected Private and Scientific Correspondence. Vols. 1 and 2* (World Scientific, Singapore, 2007). (t) refers to translation from German to English.
51. Dr. E. Schrödinger, *Nature*, **142**, 1155 (1938).
52. I. Prigogine, The Meaning of Entropy, in *Evolutionary Epistemology*, W. Callebaut and R. Pinxten, Eds. (Springer, Dordrecht, 1987).
53. E. Schrödinger, The Proper Vibrations of the Expanding Universe, *Physica*, **6**, 899 (1939).
54. E. Schrödinger, Nature of the Nebular Red Shift, *Nature*, **144**, 593 (1939).
55. Franklin D. Roosevelt Presidential Library and Museum, New York.
56. S. Sigurdsson, Physics, Life and Contingency: Born, Schrödinger and Weyl in Exile, in *Forced Migration and Scientific Change: Émigré German-Speaking Scientists and Scholars after 1933*, M.G. Ash and A. Söllner, Eds. (Cambridge University Press, Cambridge, 1996).
57. Ch. 2, Ref. 35.
58. The Papers of Lise Meitner, Churchill Archive Centre, Churchill College, Cambridge.

Chapter 5: To Dublin and Final Days in Vienna

1. Irish Newspaper Archives, Dublin.
2. Ch. 2, Ref. 14.
3. The Jewish Telegraph Agency, New York, 20 May, 1940.
4. Royal Irish Academy: New Professor of Theoretical Physics, *Nature*, **145**, 544 (1940).
5. Ch. 1, Ref. 50.
6. Ch. 1, Ref. 18.
7. W. Thirring, in *Erwin Schrödinger – 50 Years After*, W.L. Reiter and J. Yngvason, Eds., pp. 1–7 (European Mathematical Society, Zurich, 2013).

8. Ch. 1, Ref. 37.
9. E. Schrödinger, A Method of Determining Quantum Mechanical Eigenfunctions and Eigenvalues, *Proc. R. Irish Acad. Section A: Math. Phys. Sci.*, **46**, 9 (1940).
10. Ch. 4, Ref. 49.
11. N.F. Mott and H.S.W. Massey, *The Theory of Atomic Collisions* (Clarendon Press, Oxford, 1933).
12. W. Heitler, *Quantum Theory of Radiation* (Clarendon Press, Oxford, 1936).
13. Ch. 1, Ref. 41, p. 580.
14. N.F. Mott, Walter Heinrich Heitler: 1904–1981, *Biogr. Mems. Fell. R. Soc.*, **28**, 142 (1982).
15. The Papers of Professor Max Born, Churchill Archives Centre, Churchill College, Cambridge. (t) refers to a translation from German to English.
16. Ch. 1, Ref. 4.
17. Ch. 2, Ref. 11.
18. R.W. Clark, *Tizard* (MIT Press, Massachusetts, 1965).
19. Sir Henry Tizard, K.C.B., F.R.S., *Nature*, **150**, 148 (1942).
20. O. Hahn and F. Strassmann, Über den Nachweis und das Verhalten der bei der Bestrahlung des Urans mittels Neutronen entstehenden Erdalkalimetalle, *Naturwissen.*, **27**, 11 (1939).
21. N. Arms, *A Prophet in Two Continents: The Life of F.E. Simon* (Pergamon Press, Oxford, 1966).
22. Ch. 4, Ref. 50.
23. A. Speer, *Inside the Third Reich* (Weidenfeld & Nicolson, London, 1970).
24. A. Speer, Testimony at the Nuremberg Trial Proceedings, Volume 16, 1946.
25. Ch. 4, Ref. 37.
26. Ch. 4, Ref. 38.
27. C. DeWitt, J. Edelstein and B. Tekin, The Three Physicists, *Physics Today*, **74**, 42 (2021).
28. C.S. Lewis and W. Hooper, Ed., *God in the Dock: Essays on Theology and Ethics* (William B. Eerdmans, Michigan, 1972).
29. F.G. Donnan Papers, University College London.
30. Ch. 1, Ref. 56.
31. E. Schrödinger, *What is Life? The Physical Aspect of the Living Cell* (Cambridge University Press, Cambridge, 1944).
32. E.J. Yoxen, Where does Schrödinger's "What is Life?" belong in the History of Molecular Biology, *Hist. Sci.*, **17**, 17 (1979).

33. F.H.C. Crick, Recent Research in Molecular Biology, *Brit. Med. Bull.*, **21**, 183 (1965).
34. J.D. Watson, *The Double Helix* (Atheneum, New York City, 1968).
35. Archives of the Dublin Institute for Advanced Studies, Ireland.
36. M. Wilkins, *The Third Man of the Double Helix: The Autobiography of Maurice Wilkins* (Oxford University Press, Oxford, 2003).
37. N. Williams, Irene Manton, Erwin Schrödinger and the Puzzle of Chromosome Structure, *J. Hist. Biol.*, **49**, 425 (2016).
38. N.W. Timofeev-Ressovsky, K.G. Zimmer and M. Delbrück, Über die Natur der Genmutation und der Genstruktur, Nach. von der Gesell. der Wissen. zu *Göttingen: Mat.-Phys. Klasse, Fachgruppe VI, Biologie Bd. 1*, **13**, 189 (1935).
39. Ch. 1, Ref. 39.
40. J.B.S. Haldane, A Physicist Looks at Genetics, *Nature*, **155**, 375 (1945).
41. The Haldane papers. Wellcome Collection, London.
42. Ch. 1, Ref. 15.
43. M.F. Perutz, Erwin Schrödinger's What is Life? and Molecular Biology, in *Schrödinger: Centenary Celebrations of a Polymath*, C.W. Kilmister, Ed. (Cambridge University Press, Cambridge, 1987).
44. G. Jones, Catholicism, Nationalism and Science, *Irish Rev.*, **20**, 47 (1997).
45. Ch. 3, Ref. 12.
46. P.G.O. Freund, Review of Schrödinger: Life and Thought, *Physics Today*, **44**, 92 (1991).
47. M. Frayn, *Copenhagen* (Anchor Books, New York, 2000).
48. D.C. Cassidy, *Uncertainty: the Life and Science of Werner Heisenberg* (W. H. Freeman, New York, 1992).
49. Ch. 2, Ref. 35.
50. Ch. 1, Ref. 30.
51. M. Eckert, *Arnold Sommerfeld: Science, Life and Turbulent Times 1868 – 1951*, T. Artin, Trans. (Springer, Berlin/Heidelberg, 2013).
52. K. Hentschel, Distrust, Bitterness and Sentimentality: On the Mentality of German Physicists in the Immediate Post-War Period, in *The German Physical Society in the Third Reich: Physicists between Autonomy and Accommodation*, D. Hoffmann and M. Walker, Eds., A.M. Hentschel, Trans. (Cambridge University Press, Cambridge, 2012).
53. R. Braunizer, Memories of Dublin – Excerpts from Erwin Schrödinger's Diaries, in *German Speaking Exiles in Ireland 1933–1945*, G.M.B. Holfter, Ed. (Brill, Leiden, 2006).

54. Ch. 1, Ref. 55.
55. Ch. 2, Ref. 6.
56. C. Sherrington, *Man on his Nature* (Cambridge University Press, Cambridge, 1940).
57. J.C. Eccles, Ed., *Brain and Conscious Experience* (Springer-Verlag, New York, 1966).
58. M. Born, *Natural Philosophy of Cause and Chance, being the Waynflete Lectures delivered in the College of St. Mary Magdalen, Oxford, in Hilary Term 1948* (Clarendon Press, Oxford, 1949).
59. N.T. Greenspan, *The Atomic Spy: The Dark Lives of Klaus Fuchs* (Viking, New York City, 2020).
60. F. Close, *Trinity: The Treachery and Pursuit of the Most Dangerous Spy in History* (Penguin, London, 2019).
61. Ch. 3, Ref. 43.
62. R. Flower, Gustav Victor Rudolph Born: 1921–2018, *Biogr. Mems. Fell. R. Soc.*, **68**, 23 (2020).
63. Ch. 4, Ref. 47.
64. Ch. 2, Ref. 26.
65. Fellowship Certificates, Royal Society Archive, London.
66. Ch. 4, Ref. 58.
67. Letter from Max von Laue to Erwin Schrödinger, 2 August 1933, Archive of the Max Planck Society, Werner Heisenberg Estate, Kalliope Network, Berlin.
68. https://www.pro-physik.de/nachrichten/verantwortung-und-gedenken.
69. M. Born, Max Karl Ludwig Ernst Planck: 1858–1947, *Biogr. Mems. Fell. R. Soc.*, **6**, 161 (1948).
70. Ch. 4, Ref. 45.
71. M. Walker, *German National Socialism and the Quest for Nuclear Power, 1939–49* (Cambridge University Press, Cambridge, 1989).
72. U. Deichmann, The Expulsion of German-Jewish Chemists and Biochemists and their Correspondence with Colleagues in Germany after 1945: The Impossibility of Normalization? in *Science in the Third Reich*, M. Szöllösi-Janze, Ed. (Berg, Oxford, 2001).
73. Peter Medawar papers, Wellcome Collection, London.
74. L. Infeld, Visit to Dublin, *Scientific American*, **181**, 11 (1949).
75. A. Einstein, *Letters on Wave Mechanics*, K. Przibram, Ed. (Philosophical Library, New York, 1967).
76. Ch. 1, Ref. 41.

77. Ch. 4, Ref. 40.
78. Ch. 2, Ref. 24.
79. E. Schrödinger, *My View of the World* (Cambridge University Press, Cambridge, 1963).
80. H. Kragh, *Dirac: A Scientific Biography* (Cambridge University Press, Cambridge, 1990).

Chapter 6: Schrödinger's Legacy

1. D.J. Wineland, Superposition, Entanglement, and Raising Schrödinger's Cat, Nobel Lecture, Nobel Prize Foundation, 2012.
2. C.J. Craimer, *Essentials of Computational Chemistry* (Wiley, New Jersey, 2004).
3. F. Neese *et al.*, Chemistry and Quantum Mechanics in 2019: Give us Insight and Numbers, *J. Am. Chem. Soc.*, **141**, 2814 (2019).
4. Ch. 1, Ref. 2.
5. Ch. 1, Ref. 13.
6. Ch. 1, Ref. 46.
7. Ch. 1, Ref. 14.
8. Ch. 2, Ref. 17.
9. Ch. 1, Ref. 61.
10. Ch. 1, Refs. 67 and 68.
11. P. Hohenberg and W. Kohn, Inhomogeneous Electron Gas, *Phys. Rev.*, **136**, B864 (1964).
12. W. Kohn and L.J. Sham, Self-Consistent Equations Including Exchange and Correlation Effects, *Phys. Rev.*, **140**, A1133 (1965).
13. J.P. Brodholt and L. Vocadlo, Applications of Density Functional Theory in the Geosciences, *MRS Bulletin*, **31**, 675 (2006).
14. A. Paul and T. Birol, Applications of DFT + DMFT in Materials Science, *Ann. Rev. Mat. Res.*, **49**, 31 (2019).
15. D.J. Cole and N.D.M. Hine, Applications of Large-Scale Density Functional Theory in Biology, *J. Phys. Condens. Matt.*, **28**, 393001 (2016).
16. M. Bonn *et al.*, Phonon- Versus Electron-Mediated Desorption and Oxidation of CO on Ru(0001), *Science*, **285**, 1042 (1999).
17. M. Karplus, Molecular Dynamics: from $H+H_2$ to Biomolecules, in *Schrödinger: Centenary Celebrations of a Polymath*, C.W. Kilmister, Ed. (Cambridge University Press, Cambridge, 1987).

18. A. Lodola and M. De Vivo, The Increasing Role of QM/MM in Drug Discovery, *Adv. Protein Chem. Struct. Biol.*, **87**, 337 (2012).
19. M. Jammer, *The Conceptual Development of Quantum Mechanics*, 2nd edition (McGraw-Hill, New York, 1989).
20. E. Schrödinger, Are there Quantum Jumps?, *Brit. J. Phil. Sci.*, **3**, 233 (1952).
21. R. Penrose, Foreword, Ch. 2, Ref. 35.
22. E. Schrödinger, *Space-Time Structure* (Cambridge University Press, Cambridge, 1950).
23. R. Penrose, Gravity and State Vector Reduction, in *Quantum Concepts in Space and Time*, R. Penrose and C.J. Isham, Eds. (Oxford University Press, Oxford, 1986).
24. Ch. 5, Ref. 7.
25. Ch. 1, Ref. 18.
26. R. Highfield and P. Carter, *The Private Lives of Albert Einstein* (Faber and Faber, London, 1993).
27. Ch. 1, Ref. 37.
28. N. Belton, *A Game with Sharpened Knives* (W&N, London, 2005).
29. Ch. 5, Ref. 1.
30. A. Braunizer, private communication to D.C. Clary at Magdalen College, Oxford, 2008.
31. Ch. 2, Ref. 35.
32. Ch. 5, Ref. 15.
33. J. Gribbin, *Erwin Schrödinger and the Quantum Revolution* (Transworld, London, 2012).
34. W.J. Moore, *Physical Chemistry* (Longman, London, 1998).
35. F. Simon, letter to M. Born, 9 Jan 1946, Ch. 4, Ref. 47.
36. Ch. 1, Ref. 4, p. 270.

Bibliography

N. Arms, *A Prophet in Two Continents: The Life of F.E. Simon* (Pergamon Press, Oxford, 1966).

P. Ayres, *Shaping Ecology: The Life of Arthur Tansley* (Wiley-Blackwell, New Jersey, 2012).

G. Bacciagaluppi and A. Valentini, *Quantum Theory at the Crossroads: Reconsidering the 1927 Solvay Conference* (Cambridge University Press, Cambridge, 2009).

P. Ball, *Serving the Reich: The Struggle for the Soul of Physics Under Hitler* (University of Chicago Press, Chicago, 2014).

N. Belton, *A Game with Sharpened Knives* (W&N, London, 2005).

A. Bennett, *Writing Home* (Faber and Faber, London, 2014).

J. Borkin, *The Crime and Punishment of I.G. Farben* (Barnes and Noble, New York, 1978).

M. Born, *My Life: Recollections of a Nobel Laureate* (Taylor and Francis, London, 1978).

M. Born, *The Born-Einstein Letters 1916–1955* (Macmillan Press, New York, 1971).

L.W.B. Brockliss, Ed., *Magdalen College Oxford: A History* (Magdalen College, Oxford, 2008).

L.W.B. Brockliss, *The University of Oxford: A History* (Oxford University Press, Oxford, 2016).

B.R. Brown, *Planck: Driven by Vision, Broken by War* (Oxford University Press, Oxford, 2015).

D.C. Cassidy, *Uncertainty: the Life and Science of Werner Heisenberg* (W. H. Freeman, New York, 1992).

D.C. Cassidy, *Beyond Uncertainty: Heisenberg, Quantum Physics, and the Bomb* (Bellevue Literary Press, New York, 2010).

R.W. Clark, *Tizard* (MIT Press, Massachusetts, 1965).
F. Close, *Trinity: The Treachery and Pursuit of the Most Dangerous Spy in History* (Penguin, London, 2019).
C.J. Craimer, *Essentials of Computational Chemistry* (Wiley, New Jersey, 2004).
J.C. Eccles, Ed., *Brain and Conscious Experience* (Springer-Verlag, New York, 1966).
J.C. Eccles and W.C. Gibson, *Sherrington: His Life and Thought* (Springer, Berlin/Heidelberg, 1979).
M. Eckert, *Arnold Sommerfeld: Science, Life and Turbulent Times 1868–1951*, T. Artin, Trans. (Springer, Berlin/Heidelberg, 2013).
A. Einstein, *Letters on Wave Mechanics*, K. Przibram, Ed. (Philosophical Library, New York, 1967).
C.P. Enz, *No Time to be Brief: A Scientific Biography of Wolfgang Pauli* (Oxford University Press, Oxford, 2002).
G. Farmelo, *The Strangest Man: The Hidden Life of Paul Dirac, Quantum Genius* (Faber and Faber, London, 2009).
J. Feichtinger, H. Matis, S. Sienell and H. Uhl, *The Academy of Sciences in Vienna 1938 to 1945* (Austrian Academy of Sciences Press, Vienna, 2014).
G. Ferry, *Max Perutz and the Secret of Life* (Chatto and Windus, London, 2007).
E.P. Fischer and C. Lipson, *Thinking About Science: Max Delbrück and the Origins of Molecular Biology* (Norton, New York, 1988).
R. Fox and G. Gooday, Eds., *Physics in Oxford 1839–1939* (Oxford University Press, Oxford, 2005).
M. Frayn, *Copenhagen* (Anchor Books, New York, 2000).
K. Gavroglu, *Fritz London: A Scientific Biography* (Cambridge University Press, Cambridge, 1995).
J. Götschl, Ed., *Erwin Schrödinger's World View: The Dynamics of Knowledge and Reality* (Springer, Berlin/Heidelberg, 1992).
N.T. Greenspan, *The Atomic Spy: The Dark Lives of Klaus Fuchs* (Viking, New York City, 2020).
N.T. Greenspan, *The End of the Certain World: The Life and Science of Max Born* (Basic Books, New York, 2005).
J. Gribbin, *Erwin Schrödinger and the Quantum Revolution* (Transworld, London, 2012).
P. Halpern, *Einstein's Dice and Schrödinger's Cat* (Basic Books, New York, 2015).
J.L. Heilbron, *The Dilemmas of an Upright Man: Max Planck and the Fortunes of German Science* (Harvard University Press, Massachusetts, 2000).

J. Hendry, Ed., *Cambridge Physics in the Thirties* (Adam Hilger, Bristol, 1984).
A. Hermann, *Erwin Schrödinger: Die Wellenmechanik* (Battenberg, Stuttgart, 1963).
R. Highfield and P. Carter, *The Private Lives of Albert Einstein* (Faber and Faber, London, 1993).
A. Hodges, *Alan Turing: The Enigma* (Hutchinson, London, 1983).
D. Hoffmann, *Erwin Schrödinger* (Teubner, Berlin, 1984).
D. Hoffmann and M. Walker, Eds., *The German Physical Society in the Third Reich: Physicists between Autonomy and Accommodation*, A.M. Hentschel, Trans. (Cambridge University Press, Cambridge, 2007).
W. Höflechner, *History of the Karl-Franzens University Graz: From the Beginning until 2005* (Leykam, Graz, 2006).
G. Holfter and H. Dickel, *An Irish Sanctuary: German Speaking Refugees in Ireland 1933–45* (De Gruyter, Berlin, 2017).
M. Jammer, *The Conceptual Development of Quantum Mechanics*, 2nd edition (McGraw-Hill, New York, 1989).
E. Jones, *The Life and Work of Sigmund Freud* (Pelican Books, London, 1964).
G. Kerber, A. Dick and W. Kerber, *Erwin Schrödinger 1887–1961: Documents, Materials and Pictures* (Austrian Central Library for Physics, Vienna, 2015).
C.W. Kilmister, Ed., *Schrödinger: Centenary Celebrations of a Polymath* (Cambridge University Press, Cambridge, 1987).
M.J. Klein, *Paul Ehrenfest: The Making of a Theoretical Physicist* (Elsevier, Oxford, 1985).
H. Kragh, *Dirac: A Scientific Biography* (Cambridge University Press, Cambridge, 1990).
H. Kragh, *Simply Dirac* (Simply Charly, New York, 2016).
A. Kramish, *The Griffin: The Greatest Untold Espionage Story of World War II* (Houghton, Boston, 1986).
S. Lee, *Sir Rudolf Peierls: Selected Private and Scientific Correspondence. Vol. 1 and 2* (World Scientific, Singapore, 2007).
W.J.M. Mackenzie, *Explorations in Government. Collected Papers: 1951–68*, (Macmillan Press, London, 1975).
K.D. McRae, *Nuclear Dawn: F. E. Simon and the Race for Atomic Weapons in World War II* (Oxford University Press, Oxford, 2014).
J. Medawar and D. Pyke, *Hitler's Gift: The True Story of the Scientists Expelled by the Nazi Regime* (Arcade, New York, 2000).
K. Mendelssohn, *The World of Walther Nernst: the Rise and Fall of German Science, 1864–1941* (University of Pittsburgh Press, Pittsburgh, 1973).

K. von Meyenn, Ed., *Eine Entdeckung von ganz ausserordentlicher Tragweite. Schrödingers Briefwechsel zur Wellenmechanik und zum Katzenparadoxon*, Band 1 and 2 (Springer, Berlin/Heidelberg, 2011).

W.J. Moore, *Schrödinger: Life and Thought* (Cambridge University Press, Cambridge, 1989).

W.J. Moore, *Physical Chemistry* (Longman, London, 1998).

J. Morrell, *Science at Oxford 1914–1939: Transforming an Arts University* (Clarendon Press, Oxford, 1997).

N.F. Mott and H.S.W. Massey, *The Theory of Atomic Collisions* (Clarendon Press, Oxford, 1933).

P. Neville, *Hitler and Appeasement: The British Attempt to Prevent the Second World War* (Hambledon, London, 2005).

A. Pais, *Inward Bound: Of Matter and Forces in the Physical World* (Clarendon Press, Oxford, 1986).

A. Pais, *Subtle is the Lord: The Science and the Life of Albert Einstein* (Oxford University Press, Oxford, 2005).

K. Popper, *Unended Quest: An Intellectual Autobiography* (William Collins, Glasgow, 1974).

A. Robinson, *Einstein on the Run: How Britain Saved the World's Greatest Scientist* (Yale University Press, New Haven, 2019).

E. Schrödinger, *My View of the World* (Cambridge University Press, Cambridge, 1963).

E. Schrödinger, *Science and the Human Temperament* (George Allen and Unwin, London, 1935).

E. Schrödinger, *Space-Time Structure* (Cambridge University Press, Cambridge, 1950).

E. Schrödinger, *What is Life? The Physical Aspect of the Living Cell* (Cambridge University Press, Cambridge, 1944).

W.T. Scott, *Erwin Schrödinger: An Introduction to his Writings* (University of Massachusetts Press, Massachusetts, 1967).

C. Sherrington, *Man on his Nature* (Cambridge University Press, Cambridge, 1940).

R.L. Sime, *Lise Meitner: A Life in Physics* (University of California Press, Berkeley, 1996).

A. Speer, *Inside the Third Reich* (Weidenfeld & Nicolson, London, 1970).

M. Szöllösi-Janze, Ed., *Science in the Third Reich* (Berg, Oxford, 2001).

J. Vogel-Prandtl, *Ludwig Prandtl: A Personal Biography Drawn from Memories and Correspondence* (Universitätsverlag Göttingen, Göttingen, 2014).

M. Walker, *German National Socialism and the Quest for Nuclear Power, 1939–49* (Cambridge University Press, Cambridge, 1989).

J.D. Watson, *The Double Helix* (Atheneum, New York City, 1968).

V. Weisskopf, *The Joy of Insight: Passions of a Physicist* (Basic Books, New York, 1991).

E.P. Wigner, *The Recollections of Eugene P. Wigner, as told to Andrew Szanton*, (Plenum, 1992).

M. Wilkins, *The Third Man of the Double Helix: The Autobiography of Maurice Wilkins* (Oxford University Press, Oxford, 2003).

S. Zuckerman, *From Apes to Warlords: the Autobiography of Solly Zuckerman (1904–1946)* (Hamish Hamilton, London, 1978).

Permissions

The author is grateful for the following permissions to publish photographs:

Photographs of Erwin Schrödinger, by kind permission of the Braunizer family.

Photograph of Dirac, Heisenberg and Schrödinger at Stockholm Railway Station, December 1933, by kind permission of Niels Bohr Library & Archives, American Institute of Physics (Sarah Weirich).

Photograph of Schrödinger and Lindemann, by kind permission of Mrs Kathrin Baxandall (daughter of Sir Francis Simon).

Photograph of Walter Kohn, by kind permission of Walter Kohn Papers, UArch FacP 34. Department of Special Collections, Davidson Library, University of California, Santa Barbara, USA (Danelle Moon, Director, Special Research Collections).

Photograph of Hans Thirring, by kind permission of the Universität Wien (Ulrike Denk, Stellvertretende Leiterin Universitätsarchiv).

Photograph of Ruth Braunizer, and also with Arnulf Braunizer and Gustav Born, by kind permission of Heather Clary.

The author is grateful for the following permissions to publish photographs, texts of letters, emails or reports:

Letters and photographs at the Schrödinger Archive, Austrian Central Library for Physics, Vienna (Christof Capellaro). Reprinted with kind permission of the Braunizer family.

Letters from Erwin Schrödinger, by kind permission of the Braunizer family.

Letters from Francis Simon, by kind permission of Mrs Kathrin Baxandall (daughter of Sir Francis Simon).

Letters from Albert Einstein, Hermann Weyl and Abraham Flexner about negotiations with Schrödinger, by kind permission from the Shelby White and Leon Levy Archives Center of the Institute for Advanced Study, Princeton, NJ, USA (Erica Mosner, Archival Specialist).

Letters from Lord Lindemann, by kind permission of the English-Speaking Union (Tim Morris, Director of Finance).

Letter from Nevil Sidgwick to Schrödinger, by kind permission of Lincoln College, Oxford (Lindsay McCormack, Archivist).

Letters associated with the Academic Assistance Council and the Society for Protection of Science and Learning, by kind permission of CARA (the Council for At-Risk Academics, Stephen Wordsworth, Executive Director).

Letters from Max Born, by kind permission of Sebastian Born (grandson of Max Born).

Letter and texts from W.J.M. Mackenzie, by kind permission of James Mackenzie (son of W.J.M. Mackenzie).

Letters from Sir Rudolf Peierls, by kind permission of Jo Hookway (daughter of Sir Rudolf Peierls), and also assistance from Professor Sabine Lee.

Email message from Professor Ernst Peter Fischer to the author, by kind permission of Professor Fischer.

Reproductions of reports in *Nature*, the *Annalen der Physik* and *Naturwissenschaften*, with permission through the Copyrights Clearance Center RightsLink Service.

Reports in the *Irish Press*. Licence acquired from the Irish Newspaper Archives, Dublin.

Index

Printed in the United States
by Baker & Taylor Publisher Services